The Building Conservation Directory 2001

A guide to specialist suppliers, consultants and craftsmen in traditional building conservation, refurbishment and design

CATHEDRAL
COMMUNICATIONS LIMITED

THE BUILDING CONSERVATION DIRECTORY 2001
The Ninth Edition of the Directory

Published Spring 2001

ISBN 1 900915 17 0

PUBLISHED BY
Cathedral Communications Limited
High Street, Tisbury,
Wiltshire, England SP3 6HA
Tel 01747 871717
Fax 01747 871718
E-mail bcd@cathcomm.demon.co.uk
Website www.buildingconservation.com

MANAGING DIRECTOR
Gordon Sorensen

EXECUTIVE EDITOR
Jonathan Taylor

MARKETING DIRECTOR
Elizabeth Coyle-Camp

ADVERTISING DEPARTMENT
Anthony Male
Nicholas Rainsford
Dan Wilson

PRODUCTION & ADMINISTRATION
Jenny Brown
Maggie Flower
Jane Martin
Hannah Moffat

TYPESETTING & DESIGN
Xendo, London

PRINTING
Optichrome, Woking

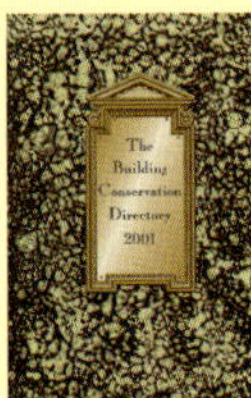

SNAKE SKIN COVER The faux snake skin effect used to decorate the cover of this edition of the Directory is an oil glaze paint effect which was popular in the Regency and early Victorian periods. It was created by Nola Marshall (see page 201).

The many companies and specialist groups advertising in The Building Conservation Directory have been invited to participate on the basis of their established involvement in the field of building conservation and the suitability of some of their products and services for historic buildings. Some of the participants also supply products and services to other areas of the building market which have no application in the building conservation field. The inclusion of any company or individual in this publication should not necessarily be regarded as either a recommendation or an endorsement by the publishers. Although every effort has been made to ensure that information in this book is correct at the time of printing, responsibility for errors or omissions cannot be accepted by the publishers or any of the contributors.

The paper used for *The Building Conservation Directory* is Harrier 2000 Silk from wood fibre originating from sustainable forests.

THE BUILDING CONSERVATION DIRECTORY

The Building Conservation Directory provides specifiers of works to historic buildings, their contents and surroundings a starting point in the search for appropriate products and services, and expert advice. Approximately 1,000 different companies and organisations are represented in this edition.

The many technical editorial articles are written by leaders in this field and cover a wide range of practical issues. The articles are not intended to be comprehensive but rather to raise awareness and stimulate dialogue amongst those involved with old buildings. Other reference information provided points the way on current legislation, continuing education and sources for further information and advice.

How to find what you need

1. The table of contents opposite together with the products and services index and the advertisers index at the back should point you to the product, service or supplier you are looking for. Product Selector tables, in yellow, listing suppliers and their products and services head up each main section.
2. Follow the index or Product Selector page reference to the appropriate section or company and start the specification process. It may be helpful to contact more than one supplier. And please remember to tell each that you found them in *The Building Conservation Directory.*
3. If you still can't find what or who you need, don't despair. You can visit our regularly updated website at **www.buildingconservation.com** or ring us on **01747 871717** and we'll try to put you in touch with a supplier who can help.

All suppliers in the Directory pay a fee to be included and although Cathedral Communications does not formally 'approve' or 'recommend' them we do screen out inappropriate suppliers and products to maintain the established integrity of the Directory. This ensures that it remains a useful and credible forum in which appropriate suppliers can promote their businesses. Directory users should seek more detailed information and advice from suppliers before undertaking any sensitive project.

We are always looking for ideas to improve the Directory so please write to let us know if you have any suggestions for improvements to its content or presentation which will help you in your work with old buildings.

And don't forget, our website at **www.buildingconservation.com** is the primary Internet gateway to the building conservation and restoration industry. Please see the inside back cover of this edition for more information, or why not click in have a look for yourself?

ACKNOWLEDGEMENTS

We are most grateful to all those who have contributed to this, the ninth edition of *The Building Conservation Directory*, without whose help its publication would not have been possible.

In particular we would like to thank Liz Forgan, Chairman, Heritage Lottery Fund for contributing this year's foreword, all our advertisers for their continuing support, and all those who have contributed editorial information and illustrations or who have helped with production including:

Nigel Armstrong, R W Armstrong & Sons Limited
Tim Buxbaum
Liz Carlisle, Chiltern Open Air Museum
Judy Cligman, Hertiage Lottery Fund
Kevin Davies
Lydia Davies, Heritage Lottery Fund
Tim Davies, PricewaterhouseCoopers
Jane Davies, Jane Davies Conservation
Robert Demaus, Demaus Building Diagnostocs Ltd
Peter Ellis, Rose of Jericho
Julia Fairchild, Stonehealth Limited
Alexandra Fairclough, Institute of Historic Building Conservation
David Farrell, Rowan Technologies Ltd
Iain Forbes, Forbes Leslie Network
Elizabeth Garrod, Building Research Establishment
David Gibbon, Gibbon Lawson McKee Ltd
Francis J Golden, Highmead VAT Consultancy
Linda Hall
Michael Heaton, Archaeological Site Investigations
Julian Holder, Scottish Centre for Conservation Studies,
Edinburgh College of Art
Frank Kelsall, The Architectural History Practice Limited
John Linden, DoE Northern Ireland
Graham Lott, British Geological Survey
Iain McCaig, North Shropshire District Council
Nola Marshall
Vanessa Mato, Text 100
Matthew Mullee, La Playa
Robin Nugent, Broadway Malyan Cultural Heritage
Lisa Oestreicher, Architectural Paint Analysis
Richard Oxley, Oxley Conservation
Gail Pollock, Environment and Heritage Service, Northern Ireland
Ian Pritchett, IJP Building Conservation Ltd
Trevor Proudfoot, Cliveden Conservation Workshop Ltd
Graham Reed, Historic Scotland
Bryan Rowbotham
David Rowe, Cadw
Ian Sims, STATS Consultancy
Ben Sinclair, Norgrove Studios
Jagjit Singh, Environmental Building Solutions Ltd
David Smith, Mineralogy Department of the
Natural History Museum
Roy Switsur, Cambridge Dating Unit, Anglia Polytechnic University
David Taylor, Historic Scotland
David Walton, Sindall Ltd
David Wrightson, Acanthus Lawrence & Wrightson Architects

CATHEDRAL
COMMUNICATIONS LIMITED

CONTENTS

‘When we build,

let us think that we build ***forever’***

John Ruskin 1819–1900

FOREWORD

RUSKIN lamented the decline in craftsmanship caused by the Industrial Revolution and dreamt of the thorough reform of the arts and crafts, the replacement of cheap mass production by conscientious and meaningful handiwork. The craftsmanship which he advocated remains as important today as 150 years ago, if we are to hand on our heritage in good heart to future generations.

No one knows better than conservationists the continuing need for work created with care and pride. Your contribution to our built heritage adds to the spiritual and intellectual welfare of our nation as a whole. Fortunately, since the advent of the lottery, the Heritage Lottery Fund (HLF) has been able to provide new funding to help achieve this vision; helping to imbue the nation with a sense of place and a sense of self-worth.

Although in its comparative infancy, the Heritage Lottery Fund has made an important impact by supporting projects in communities across the whole of the UK. Following years of neglect, we believe that Britain's heritage is enjoying a true renaissance. New uses and new life are being put back into areas and buildings that have lost their traditional economic base. Lottery grants are helping to restore not just the fabric of buildings but the fabric of local communities.

The government rightly requires the Heritage Lottery Fund to look for wider public benefits than simply the safeguarding of Britain's rich heritage for its own sake. To secure these objectives, HLF evolved from being a merely reactive body, processing applications for grants, to an organisation which acts more strategically to address areas of need that are not being met by existing funders. As well as conservation, our mission is to finance public access to as much of our heritage as possible, and award grants which actively engage more people in the enjoyment and support of their heritage. The future of the heritage is in the hands of people as well as in bricks and mortar.

Since 1999 we have refocused three successful programmes to reach areas of deprivation, the *Townscape Heritage Initiative*, the *Joint Scheme for Places of Worship* in England and the *Urban Parks Programme.* Over £1.5 billion in grant has been awarded, to more than 4,000 projects, many of which are in the most deprived local authority areas. At an average grant rate of 46 per cent, this means HLF has levered at least another £1.7 billion into the economy through partnership funding, directly creating over 2,300 jobs per year.

The Building Conservation Directory plays a vital role disseminating information between the various branches of the conservation world. It plays a key part in the new renaissance in heritage – so under-funded in the past and now part of a really vibrant future.

Tabernacle Community Centre, Haverfordwest
This fine Victorian school which is listed Grade II* is being converted to provide much needed community facilities with the help of the Heritage Lottery Fund
Photograph © James Morgan, Heritage Lottery Fund

Liz Forgan
Chairman, Heritage Lottery Fund

THE CONCEPT OF CHARACTER IN HISTORIC BUILDINGS

JULIAN HOLDER

Character, like so many of the central concepts we use on a daily basis in conservation, is a somewhat nebulous concept. It is also one we rarely stop to think about in abstract. Not only is it hard to define but it shares with related concepts such as integrity and honesty, a family resemblance by employing what Ruskin termed 'the pathetic fallacy.' That is to say we apply concepts properly belonging to human beings to inanimate objects. Can a building really be 'compromised,' its 'integrity' questioned, its 'character' altered? It all rather conjures up the image of a shy Edwardian bather embarrassed to be caught half-way through changing into a swim-suit in a bathing engine on the South Coast.

When using these concepts we ask those reading our letters of objection, our proofs of evidence, and our conservation plans, to take them on trust and engage in a debate partially defined, controlled, and organised around such anthropomorphic concepts. To accept the concepts ensures that all the participants are already treating buildings as people, as living breathing beings, whose fate we care about, and not simply as bricks and lime mortar.

At its best this is a linguistic slight of hand, based on custom and practice going back to Ruskin, Morris and other members of the Arts and Crafts Movement. It is a direct, and frequently effective appeal to the emotions of those who make decisions in planning committees up and down the land. At its worst it is a transparently bullying misappropriation which fails to impress the hard headed and leaves conservation looking distinctly amateur.

However, if we take the human analogy at face value perhaps it is not so inappropriate. We are perturbed when a person's character changes out of all recognition and we no longer know who they are. Character, at least for human beings, is meant to be fixed and stable, something we 'settle into,' and any alteration of this, unless, as in literature, the redemption of a bad character, is seen as unfortunate. So it is for buildings. We believe that we know them, their age, their history, their appearance. Should this change as a result of new research, possibly leading to a new appearance then we feel let down, sometimes confused, and even angry. Sometimes, as we learn more we value more – a building is upgraded, a friend more valued. Yet it works both ways. How often do we say of a person that 'such-and-such' an event, usually a new partner or new interest following a death, has 'been the making of them'. Such may be the defence of some high profile restorations such as that of Stirling Great Hall, or the recovery of the original interior paint scheme at Bolsover Castle. However, like all concepts, 'character' is historically constructed and has its own history.

In his recent work *Words and Buildings*, Adrian Forty has performed a valuable service in clarifying the historical development of many central concepts in architecture. In so doing he chronicles the development of the closed language of contemporary architecture which has alienated the architectural profession from the public it serves. Many of these, such as 'character', are as applicable to conservation as they are to architecture. Interestingly, the use of such concepts in conservation has not alienated us from the public in the way that their use in the development of modernist architecture has. Why should this be? How has the architectural avant-garde managed to use the same concepts to exclude the public that conservation seems to have deployed to include them?

Forty argues that 'character', with a background in literary debate, entered the architectural vocabulary in the 18th century, being found first in the writings of French architect Germain Boffrand, such as *Livre d'Architecture* (1745). For Boffrand the concept of 'character' was clearly related to function, or genre, as when he writes that "Different buildings, by their arrangement, by their construction, by the way they are decorated, should tell the spectator their purpose." This definition, which sounds almost Modernist, seems at odds with how we use the concept today when few would worry about a building such as Chatelherault where a kennel and stable block is given something of the form of a Palladian villa.

Yet what Boffrand does point to is the assumption, implicit in the concept of 'character', that there is a truthful, or honest part of a building, a character based on function which should be expressed. Whilst it may be possible to use Boffrand's concept of 'character' to deal with buildings up until the mid 18th century, thereafter it is an enterprise fraught with danger. Why? Because under the influence of the Industrial Revolution, the expansion of the population, the development of the State, and the increasing concentration of capital in the hands of a few, the stability of the few core building types – church and manor house, cathedral and palace – became challenged by the factory and the need for mass housing.

As a result, what buildings meant and how they could be read in functional terms underwent a profound change from which they have never recovered. Meaning was a problem that John Ruskin felt acutely when he wrote of his despair at the fate of the Gothic Revival being used for Victorian gin-palaces. Character then, for Boffrand, meant learning the established forms and decoration for a set number of building types so that their function and status could be learnt, deployed as necessary, and then be readily identified. It was a practice closely allied with the Classical language of architecture where particular orders, and their meaning, could be used to emphasise the character and function of a building – Doric used to characterise a powerful building dedicated to a God, Corinthian a delicate building dedicated to a young Goddess. Shortly after Boffrand's initial attempt to pin down the meaning of character, J-F Blondel's *Cours d'Architecture* of 1766 developed Boffrand's concept by defining 64 building genres, with 38 different characters. Necessarily, such a mechanistic approach was also doomed to failure from the start, despite acknowledging a wide variety of characters.

With a wider view of what constitutes architecture, we now appreciate that character can be changed so easily its varieties are almost infinite such that Boffrand's attempt to clarify 'character' seems inadequate, and indeed unnecessary. To read such subtle alterations in character calls for a high level of visual acuity which, if we are not careful, can label us as remote, other-worldly cranks who are obsessed by minor details. Yet all too often it is the details which contribute the essential elements to our reading of character. A new door on a small terraced house as at Wirksworth, a new window inserted in a previously blank wall on a house in Chelsea, a change in roof height on a railway station in Edinburgh, or a new development allowed within the curtilage of a church in Herefordshire can all conspire to change the character of a building or site that is, or should be, protected and passed on to future generations as an authentic record of the past.

Hard enough to control on listed buildings, in conservation areas only the Draconian step of Article 4 directions, followed up with enforcement action, can hold back the slow tide of ill-informed alterations. Conservation areas are probably one of the most popular aspects of conservation for the majority of the public. Here examples like the Roe Green Estate in Kingsbury, north London, shine out like a beacon in the darkness to demonstrate what can be done to revive the character of an early 20th century housing estate and allow us to value one of the most neglected and despised building types – council housing.

'Character' is what we are trying to save – and it is inbuilt, not applied. More perhaps than the great Burra Charter shibboleth we are all meant to bow down before at present – 'cultural significance' – character is crucial, and it is crucial to cultural significance. However the complexity of the concept starts when we begin to consider which character we are seeking to conserve on any particular building, or area. Is it its original character – in which case are we right to demolish later additions or alterations? Or is it all the complex accretions of a building over time? And what of the repairs necessary to arrest decay and maintain the character of the original? Do we disguise some of these to maintain the character of the building we care for? Or are we content to let 'time and tide' take its inexorable toll on the building and weather new elements back to the old?

Beyond the historical conception of character as related to meaning, in terms of the 'plain-language' school of philosophy, character also seems to be implicitly related to age. If you don't believe this try a simple

The conversion of the former Bankside Power Station, London, into the Tate Modern could not have been achieved without considerable alteration to its character. Yet it represents a building type excluded from traditional notions of architecture until after the Industrial Revolution.

Lyme Park, Cheshire, its classical façade by G Leoni was completed at a time when the definition of 'Architecture' was restricted to high status buildings, and character was understood as function

test: to what extent can a new building be said to have 'character' – beyond that of 'fatal newness' (as Ruskin put it)? So, character gives a privilege to the older building in the same way that guidelines for listing do. Not only does old usually mean scarce, but it also means baring the signs of that age on its fabric.

If this is so, it is something which the dominant philosophy of conservative repair, and especially SPAB principles, conspires with. Minimal intervention argues that as much of the original fabric of a building as possible is saved. These will be the elements which carry the marks of age, not merely the marks of the tooling, but of the weathering, decay, and consequent repair of the fabric. It raises the interesting issue of the restoration of Modern Movement buildings. In *The Architects' Journal* several years ago I asked the rhetorical question "Would you pay good money to visit the rusting remains of a Modern Movement building?" Probably not. Yet the day may be coming closer when this happens. For a younger generation of conservation students, a building such as Gillespie Kidd and Coia's Cardross Seminary can be viewed as a great ruin, a thought not too abhorent to its chief designer. If an increased emphasis on the manipulation of space is one of the defining characteristics of Modernism can this be appreciated if Cardross were left as a ruin, or Brynmawr, or Bankside Power Station? Certainly it is the sheer volume of the great Turbine Hall of Bankside, now celebrated as the Tate Modern, which seems to have impressed most visitors and much of the buildings essential character was retained in the recent conversion until they stuck an internally lit hat on top and turned the symbol of its former function into an artist's flagpole.

But it is not only Modern Movement buildings that raise this question, it is also inevitably subject to local policies. In London it has been thought acceptable to gently soot-wash brick repairs to the parapets of Georgian terraces in order to blend the repair in with the original work. The justification can only be to retain the sense of the original character, and of that character being based on age and the accumulation of soot – a view which Riegl would have perhaps approved of. However, in Edinburgh indented stone repairs are deliberately left in their new state to contrast with the older stone. What is at issue here seems to be a desire for an instant heritage which cannot wait for 'time and tide' to mellow repairs. Recent research in support of stone cleaning has established that, whilst people (and for once we are not talking about the conservation fraternity talking to itself) like buildings to be cleaned, they do not like them too clean – some patina of age is necessary it seems for old buildings to become old, revered and develop character. Once again this seems unlikely to be the case with buildings of the Modern Movement whose materials age in a way still largely unappreciated by the public. The shock of the new confuses as in the restoration of Modern Movement buildings, such as the recently restored Sonneveld House in Rotterdam of 1933, pose all too clearly that perennial question 'What time is this place?'

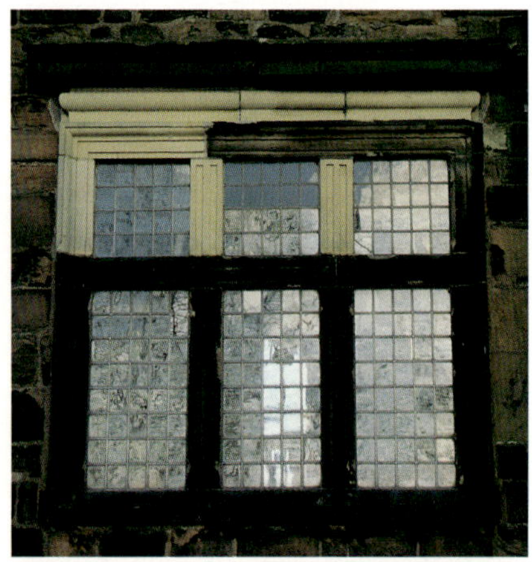

***Stone repair to a window in Carlisle.** Can we wait for the age-related character of the new repair to happen naturally or should the new stone be soot-washed to tone in with the old?*

Clearly the concept of character, as historically constructed in 18th and 19th century architectural discourse, has little point of contact with us today. Forms have long lost their meaning and their attachment to a narrow and high-culture set of building types. The deliberate attempt to eliminate character from modern architecture in the 20th century has partially resulted in the concept being taken over by the conservation movement and used to defend our diminishing stock of old buildings against the wrecker's ball. But with the challenge of the conservation of 20th century buildings of non-traditional appearance and materials we will need to think about it far more carefully in the future. Ironically, having wrested the concept of character from contemporary architecture during the last century, we are perhaps now in the situation where we need to borrow some other concepts, still current in contemporary discourse, in order to defend Modernist buildings. If Modernist buildings of the 20th century no longer have character in the way that those of traditional design and materials have what concepts do we use to defend them?

RECOMMENDED READING

Adrian Forty, *Words and Buildings; a vocabulary of modern architecture* (London, 2001)

K O Garrigan, *Ruskin on Architecture; his thought and influence* (London, 1973)

Sir John Summerson, *The Classical Language of Architecture* (London, 1980)

John Onians, *Bearers of Meaning; the classical orders in antiquity, the middle ages, and the Renaissance* (Cambridge, 1988)

Peter J Larkham, *Conservation and Management in U.K. suburbs*, in, R Harris and PJ Larkham (eds), *Changing suburbs; foundation, form, and function* (London, 1999)

Alois Riegl, *The Modern Cult of Monuments; its character and its origins*, reprinted in Oppositions, no 25, 1982, special issue on Monument/memory and the mortality of architecture

Robin Webster, *Stone Cleaning; and the nature, soiling and decay of stone* (London, 1992)

JULIAN HOLDER is the Director of the Scottish Centre for Conservation Studies at Edinburgh College of Art. He was previously Casework Officer of the Twentieth Century Society, worked on the re-survey of listed buildings for Cadw, and was a consultant to English Heritage on 20th century building types.

CONSERVATION ARCHITECTS

Barely a drawing board in sight at the busy modern practice of Acanthus Lawrence & Wrightson Architects

Photograph © Clive Hicks, Acanthus Lawrence & Wrightson Architects

PROFESSIONAL SERVICES

Firm	Services	Page
SCOTLAND		**PAGE**
AOC Archaeology Group	dn	19
A R P Lorimer & Associates	ar hs qs su	24
Addyman & Kay Ltd	ae hi ms po	15
Benjamin Tindall Architects	ar cm id ur	26
Building Design Partnership	ar pc qs st su ur	27
David Narro Associates	st	40
Elliott & Company Structural Engineers	st	40
Gibbon, Lawson, McKee Limited	ar cm su	29
Gray, Marshall & Associates	ar	30
Law & Dunbar-Nasmith	ar hi	32
Lincoln & Campbell Associates Ltd	ar	32
Murdoch Green Kensalls	cm qs su	44
Nicholas Groves-Raines Architects Ltd	ar	33
Peter Stephen & Partners	st	41
Simpson & Brown Architects	ar hi id ms su	35
YORKSHIRE		
Allen Tod Architects	ar	24
Byrom Clark Roberts	ar st su	27
Building Design Partnership	ar pc qs st su ur	27
CoDA Conservation	ar st	28
David Lewis Associates	ar su	28
Martin Stancliffe Architects	ar cm su	33
Sheffield Dendrochronology Laboratory	dn	19
Structural Perspectives	ae hi ms po	16
W R Dunn & Co	ar cm hs po	48
NORTH OF ENGLAND		
Blackett-Ord Consulting Engineers	ar hi st su	40
Elaine Rigby Architects	ar	29
Johnston & Wright	ar	31
Robin Kent Architecture & Conservation	ar hi su	34
Spence & Dower Architects	ar	35
NORTH-WEST		
Anthony Blacklay & Associates	ar	25
Brock Carmichael Associates	ar pc	26
Building Design Partnership	ar pc qs st su ur	27
Byrom Clark Roberts	ar st su	27
Donald Insall Associates Ltd	ar cm hi id ms pc po ur	29
Field Archaeology Centre	ae hi ms ph	15
Graham Holland Associates	ar	30
Haigh Architects	ar	30
Hodkinson Mallinson Partnership	su	46
Lloyd Evans Prichard	ar cm po su	32
Murdoch Green Kensalls	cm qs su	44
Purcell Miller Tritton	ar hi id pc su	34
Robinsons Preservation Limited	nd tt	177
Ryder & Dutton	he hi nd su	48
Fred Tandy	st	41
Wm Langshaw & Sons	cm	68
WALES		
Abbey Masonry and Restoration	ma	104
David Harvey Architects	ar su	28
Davies Sutton Architecture Limited	ar he po su	28
Dean & Cheason – Architects	ar	28
Ede Surveyors & Architects	cm he ms po su	47
Graham Holland Associates	ar	30
Griff Davies Architectural Design & Conservation	ar hi pc po su	47
Heritage & Archaeology Research Practice	dn	19
Highmead VAT Consultancy	vt	38
James Brotherhood & Associates	ar cm id ms pc	30
PricewaterhouseCoopers	vt	38
WEST MIDLANDS		
Archive	ms	50
Arroll & Snell	ae ar he la ms po su	25
The Brown-Matthews Partnership	ar pc	26
Demaus Building Diagnostics Ltd	nd st su tt	50
Donald Insall Associates Ltd	ar cm hi id ms pc po ur	29
Hawkes Edwards & Cave	ar	30
John Hunt Associates, Interior Architecture	cm id pc	200
Jubb & Jubb Ltd	su	46
Peter Yiangou Associates	ar su	34
Pidduck & Whittaker	ar	34
Ridout Associates	nd su	177
Stainburn Taylor Architects	ar	35
T J Crump Oakwrights Limited	cm	74
Wakemans	pc qs su	48
Ward & Dale Smith Chartered Building Surveyors	su	48
EAST MIDLANDS		
Anderson and Glenn	ar ho	39
Anthony Short and Partners	ar id ms su	25
Tim Benton Architect	ar	36
Drew-Edwards Keene	ar	29
Field Archaeology Centre, The University of Manchester	ae hi ms ph	15
Hirst Conservation	ae	45
Historic Buildings Conservation	po	39
Latham Architects	ar ur	32
Network Archaeology	ae	15
University of Nottingham, Tree-ring Dating Laboratory	dn	19
SOUTH-EAST – NORTH OF THE THAMES		
Acanthus Clews Architects Ltd	ar cm po su	24
Andrew Martin Associates	pc ur	38
Archer Partnership	ar cm id pc	25
Austin Trueman Associates	st	40
Bedford Timber Preservation Company	nd	177
Boniface Associates	ar mi ms nd pc po su	45
Cambridge Dating Unit	dn	19
Cliveden Conservation Workshop Ltd	cm ma	108
David Pitts Chartered Architects	ar	28
Derek Rogers Associates	ar ms su	28
Donald Insall Associates Ltd	ar cm hi id ms pc po ur	29
Feilden + Mawson	ar hi id pc su	29
G B Geotechnics Ltd	nd st su	50
The Hertfordshire Roofing & Renovation Company	su	59
I J P Building Conservation	ms nd	61
La Playa	in	183
McCurdy & Co Ltd	st	73
Mansfield Thomas & Partners	ar hi pc qs su	32
The Morton Partnership Ltd	ar st su	41
Munters Property Damage Restoration Ltd	nd	177
Network Archaeology Ltd	ae	15
Oscar Faber Consulting Engineers	cm he hs nd st	188
Oxford Archaeological Unit	ae hi ms po xu	15
Oxford Dendrochronology Laboratory	dn	19
Oxley Conservation	su	47
Peter Codling Architects	ar su	33
Plowman Craven & Associates	ms su	50
Press & Starkey	qs	45
Purcell Miller Tritton	ar hi id pc su	34
Ridge and Partners	su	47
Sheppard & Co	su	47
STATS Consultancy	he ma	45
Thornburrow Holdsworth	ar	36
The Victor Farrar Partnership	ar he po su	36
Michael Walton	ar	36
The Whitworth Co-Partnership	ar su	36
LONDON		
A & Q Construction Management	ar cm qs su	24
Abercrombie Planning and Conservation	pc su	38
AMBO	ar id po	24
Architectural Archaeology	ae	15
Austin Trueman Associates	st	40
Avanti Architects Ltd	ar	25
Baily Garner	ar cm qs su	45
Blampied and Partners Ltd	ar cm	26
Broadway Malyan Cultural Heritage	ae ar he in po	26
Peregrine Bryant	ar	26
The Budgen Partnership	st	40
Building Design Partnership	ar pc qs st su ur	27
C G M S	ar pc	38
Cameron Taylor Bedford	cm hs st su	40
Carden & Godfrey Architects	ar id po	27
Charles Knowles Design	ar	27
David Ashton Hill Architects	ar po	28
David Brown & Partners	ar cm id ms	28
David Gibson Architects	ar	28
Donald Insall Associates Ltd	ar cm hi id ms pc po ur	29
Drivers Jonas	cm ms pc su	46
Ellis & Moore	st	40
Feilden + Mawson	ar hi id pc su	29
Gibberd Conservation	ar id	29
Giles Quarme & Associates	ar hi ms pc su	31
Gilmore Hankey Kirke Ltd	ar	30
The Halpern Partnership	ar	30
HOK International Limited	ar	30
Institute of Archaeology, University College, London	dn	19
James Dunnett Architects	ar	29
Julian Harrap Architects	ar hi id su	31
KAW Design	id	200
Lincoln & Campbell Associates Ltd	ar	32
MRDA	ar he ms po	32
Mansfield Thomas & Partners	ar hi pc qs su	33
The Morton Partnership	ar st su	41
Murdoch Green Kensalls	cm qs su	44
Pollard Thomas & Edwards Architects	ar	33
Purcell Miller Tritton	ar hi id pc su	34
Reliable Effective Research	hi	14
Retrouvius Architectural Salvage	id	160
Richard Griffiths Architects	cm ar id pc po	34
Roger Mears Architects	ar	35
Rothermel Thomas Chartered Architects and Planners	ar pc po	38
T P B Planning	ar id pc po ur vt	38
Taywood Engineering	cm ma su	39
Thomas Ford & Partners	ar su	35
TFT Cultural Heritage	ar cm he hs su	48
Watkinson & Cosgrave	su	48
SOUTH-EAST – SOUTH OF THE THAMES		
Adrian Cox Associates	st sv	40
Alan Dickinson Chartered Building Surveyor	ae hi ms su	46
AMBO	ar id po	24
The Architectural History Practice Limited	hi po	14
Bailey Partnership	cm su	45
Broadway Malyan Cultural Heritage	ae ar he in po	26
The Chapman Bathurst Partnership Ltd	su	188
Cherished Land	xu	178
Chris Romain Architecture	ar	27
Christopher Rayner Architects	ar	27
Clague	ar su	27
Cluttons	su	46
The Conservation Studio	pc po	27
Donald Insall Associates Ltd	ar cm hi id ms pc po ur	29
Edward Sargent Conservation Architect	ar	35
Gifford and Partners	ae cm st sv	41
H G P Greentree Allchurch Evans Ltd	ar pc	30
Hutton+Rostron Environmental Investigations Limited	cm he nd su	50
I W Payne & Company Ltd	st	61
Ian Russell Engineering	nd st su	41
Ingram Consultancy Ltd	he ma su	39
John D Clarke & Partners	ar	31
Julian R A Livingstone Chartered Architect RIBA	ar	32
King Sumners Partnership	cm hs qs	44
Kingsland Surveyors Limited	ms	50
T C R MacMillan-Scott	ar	32
Mather & Smith Ltd / M J Allen Group	cm	149
Murdoch Green Kensalls	cm qs su	44
P W P Architects	ar cm ms pc su	33
Purcell Miller Tritton	ar hi id pc su	34
Resurgam	ar he ma nd po	39
Rickards Conservation	su	47
Roger Joyce Associates	ar	34
Shambrooks	qs	45
Stuart Page Architects	ar id	35
Anthony Swaine Architect	ar he po	35
The VAT Consultancy	vt	38
John Wardle	st	41
Weald & Downland Open Air Museum	hi	14
SOUTH-WEST		
Acanthus Associated Architectural Practices	ar po ur	24
Archaeological Site Investigations	ae hi ms po xu	15
Architecton	ar	25
Bare Leaning & Bare	hs qs	44
Barlow Schofield Partnership	ar	26
Bill Harvey Associates	st	40
Bosence & Co	cm su	45
Carrek Limited	cm	60
Chedburn Design & Conservation	ar	27
Cluttons	su	46
Davis Blackburn Chartered Building Surveyors	su	46
Donald Insall Associates Ltd	ar cm hi id ms pc po ur	29
Nicholas Durnan	he	114
Ede Surveyors & Architects	cm he ms po su	47
The Hartley Conservation Partnership	su	46
The Heritage Practice	ar	30
Hill Beild Associates	ae hi ms su	46
John Stark & Crickmay Partnership	ar	31
Jonathan Rhind. Architects	ar id	31
King Sturge Heritage	su	48
MRDA	ar he ms po	32
Mann Williams	st	41
Michael Drury Architects	ar	33
Mildred, Howells & Co	qs	44
National Monuments Record	hi ms	14
Pathfinders	ae ms	15
Michael Pearce - Conservation Consultant	ar pc	33
Robert Seymour Conservation	ar	34
Rose of Jericho	ma	170
St Blaise Ltd	cm ma	65
Watson Bertram & Fell	ar su	36
Wessex Archaeology	ae ms	16
Wilsons	lg	38
NORTHERN IRELAND		
Belfast Tree-ring Laboratory	dn	19
Building Design Partnership	ar pc qs st su ur	27

KEY

- ae archaeologists
- ar architects
- cm contract management
- dn dendrochronology
- he building conservation technologists
- hi historical researchers
- ho horticultural consultants
- hs health and safety consultants
- id interior designers and consultants
- in insurance
- la landscape architects
- lg legal services
- ma materials analysis
- mi miscellaneous services
- ms measured surveys
- nd non-destructive investigations
- pc planning consultants
- po conservation plans and policy consultants
- ps planning solicitors and barristers
- qs quantity surveyors
- st structural engineers
- su surveyors
- tt structural timber testing
- ur urban designers
- vt VAT consultants
- xu exhumation of human remains

Regional designation is according to office location. Many firms operate nationally.

DATING OLD BUILDINGS

FRANK KELSALL

Two buildings in Fleet Street, the former Union Bank by George Aitchison Sr of 1856 and the former Legal and General Insurance Office of 1885 by Sir Robert Edis. These are dated by illustrations in the ***Illustrated London News*** *and the* ***Builder*** *and by documentary research in the bank and insurance companies' archives.*

The office of the Ancient Monuments Society is in a delightful small Edwardian building (listed Grade II), in an alley just south of St Paul's Cathedral. The date, function and architect are not difficult to establish for the building has its former use, as 'St Ann's Vestry Hall', inscribed on the frieze and the more eagle-eyed can spot a stone at low level marked 'Banister Fletcher & Sons, Architects, 1905'. Dating a building by inscription is a long tradition, though few name the architect in such brief form as that on the Town Hall at Blandford Forum which reads 'Bastard, Architect, 1734'. The trouble with inscriptions, useful though they are, is that you cannot be sure that they are right (many have been added by later owners) or that they date more than a particular feature or phase of development. The datestone has to be treated with the same critical eye as the rest of the building.

Historic buildings need historians. That might seem axiomatic, but surprisingly few of the half million or so listed buildings have ever been thoroughly investigated. The rise of a specialist role of architectural historian has gone hand-in-hand with the growth of the conservation movement over the last half-century. What do architectural historians do? How can they contribute both to an understanding of architecture of all periods and to the selection of what we should seek to conserve? Architectural historians find out about buildings; who built them and when; what they were for; how they have been altered and take the form they do now; what people and events have been associated with them. They assemble evidence and interpret it. Dating is an essential first step.

Those who listed historic buildings for many years worked to the acronym DAMPFISHES, later BDAMPFISHES. 'D' for date, 'A' for architect, 'M' for materials and so on. The primacy given to date was preceded in later years only by a 'B' for a brief description of what sort of building the list description covered. As the listing criteria set out in Government guidance PPG15 show, the older a building is, the more likely it is to be listed. Date is inevitably crucial to any understanding of a listed building.

The same street frontage as shown in Tallis's ***London Street Views*** *of 1847, showing what was here before: earlier timber-framed buildings and Thomas Hopper's insurance office, built in 1838 and demolished 1885.*

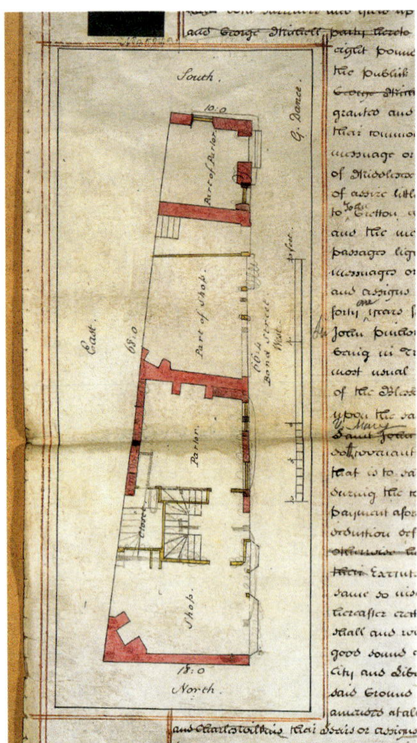

A plan from the Corporation of London's archives, attached to one of their leases, with a plan as in 1796 and pencilled changes made before the next lease in 1810. The plan is signed by George Dance, not as architect of the building but as surveyor to the City Corporation.

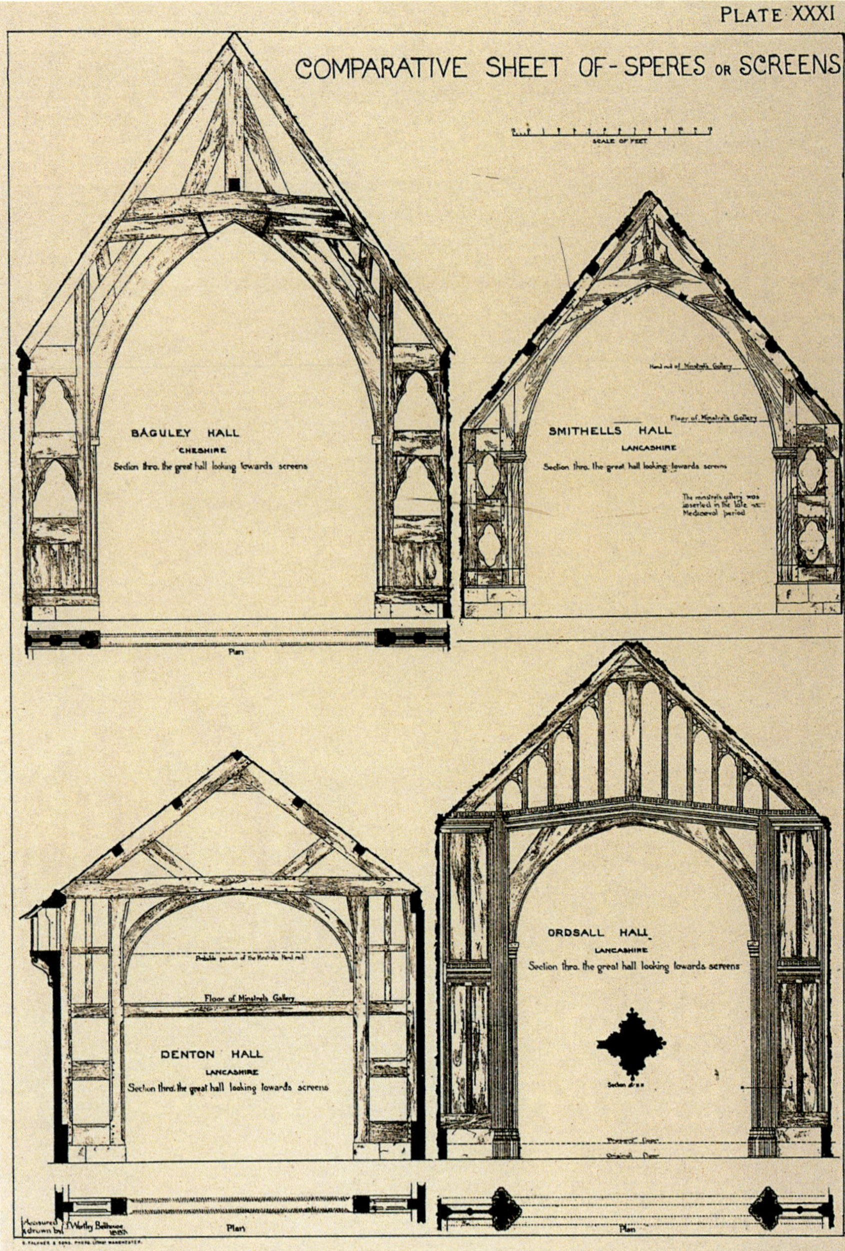

Survey drawing for Henry Taylor's **Old Halls in Lancashire and Cheshire**, 1884, showing comparative sections through four spere trusses.

Generally speaking, the older the building, the more likely it is that dating will have to be by comparison with other known and dated examples. The traditional typologies developed by architectural historians, especially for timber framed vernacular buildings, are now being given greater precision by dendrochronology. The mouldings on stonework on the other hand remain a matter for comparative analysis alone. Such details provide a rich source of information. In London for example, a plain 18th century terrace house can usually be dated to within five or ten years simply by looking at it with an informed eye. Furthermore, there are rich documentary sources from which, in much of London north of the river at least, it is possible to date a house even more accurately.

In many cases an approximate date can be given after a first inspection. There are clues in storey heights and the relation of these heights to ground level, in window spacing, in roof form and pitch, in plan and the position of chimney stacks as well as in the architectural detail. Although published some 20 years ago, J T Smith's guide *On the Dating of Houses from External Evidence* is still a most helpful guide. His more recent study, *English Houses 1200–1800: The Hertfordshire Evidence (1992)*, has an excellent discussion of the difficulties which many buildings present; from the antiquarian copyist who makes his building look older than it is, to the enthusiastic revealer of the timber frame who removes most of the physical evidence of his building's history by stripping it back to the original.

Historians always like to confirm a date suggested by the physical evidence against any available documentary sources. At one time this was regarded as almost unnecessary, but the revolution brought about by Howard Colvin's *A Biographical Dictionary of English Architects 1660–1840* has changed the nature of post medieval architectural history. From the first edition, published in 1954 to the thoroughly indexed third edition, published in 1995, we have been able to locate dates and architects quickly and to find the references to back them up. There are later biographical dictionaries but (except for a notable local attempt in Suffolk) none provide such comprehensive lists of works.

Colvin, of course, includes only buildings known to have been designed by architects. Major buildings are usually easy to date. *The Builder*, now *Building*, established in 1843, is one of many architectural periodicals that deal with buildings of a later date than those covered by Colvin.

The great advantage of documentary research is that it gives more than a date: it provides information about the building process; how the design evolved; how the building has changed since first built and for architectural history on a wider front, how the building was used and by whom. All this complements the information derived from the building itself.

Before launching straight into primary research it is sensible to see what is already known and what might be available. The statutory list is often itself a help. It should give an analysis of how a building has developed as well as a description. More recent lists often include a bibliographical note, useful in identifying articles in *Country Life* or local journals and sometimes references to the *Builder* or other primary sources. Few lists are as detailed as that for Barrow-in-Furness where many entries give dates and attributions from the local building act plans. Then there is Pevsner, of course, the inimitable series of county by county guides to the buildings of Britain. But the absence of a reference does not mean that none exists. More research now can be done via the Internet, with useful websites at the British Library, Historical Manuscripts Commission and the National Monuments Record. The NMR with some three million photographs and 50,000 measured surveys should always be tapped. But in many cases it is the local library and record office which is the principal source; here, in addition to topographical works and the *Victoria County History* there will be the journals of local antiquarian societies and other printed works.

When trying to establish a date from primary sources it is often easiest to work backwards. Map evidence can be crucial and the Ordnance Survey is always the best place to start, comparing the various editions. Some industrial areas have had very large scale maps made of them which give the plans of churches and public buildings as well. Earlier maps vary in quality and usefulness; it is always important to remember that a building shown on a site does not necessarily mean that it is the building that is there now. Working backwards through street directories and ratebooks can also tell you when a building first appeared and gaps in the series, or significant changes in value, can be a clue to alterations and rebuilding.

Owners may have title deeds and these should be examined. Some areas have had land registration since the 18th century (of these, Middlesex and West Yorkshire are the best known), but the registers are not easy to handle. For most of that part of London which used to be in Middlesex, original building leases should be registered, a source profitably tapped by the *Survey of London,* that Rolls-Royce of architectural surveys. Where land was owned by one of the great estates – the Crown, aristocratic landlords, corporate bodies – then with luck the estate records will survive, now often in a local archive office, or, for the Crown Estate, at the Public Record Office. Locating these records is often something with which the National Register of Archives can help.

What architectural historians like to find, of course, are drawings and illustrations. There are many topographical records and photographs. These will not necessarily help to date a building, though they may establish limits before or after which changes have taken place. As antiquarian interest in old buildings developed, this changed the nature of drawings from that of 'seats', as in many county histories, to that of architectural record, as in the huge collection of some 12,000 drawings produced by the Buckler family and now in the British Library. Further into the 19th century more archaeological drawings were produced. All such drawings need to be assessed carefully. A recent study of JS Crowther's drawings of Cheshire churches has shown that he doctored them to fit his preferences for what the churches should have looked like and it is known that TH Shepherd, usually a very reliable topographical draughtsman, removed existing accretions from a drawing of 1851 to 'restore' the terrace to its Palladian symmetry of 1738.

Most drawings were not produced for artistic, antiquarian or archaeological purposes but for practical reasons, at the time of building or subsequent alteration. These are often a better clue to the date of a building than the topographical illustration. They can range from beautiful perspectives, designed to attract the patron and critic, to the technical detail of working drawings. The greatest collection of design drawings is that of the Royal Institute of British Architects in the British Architectural Library, now being merged with the archive of the Victoria and Albert Museum. Otherwise designs can often survive with clients, in family or corporate collections.

Not all designs were carried out, in full or in part, but the unbuilt is often as fascinating as the built, as shown in *Unbuilt Oxford* or *London As It Might Have Been*. For Victorian and later buildings designs were often published in architectural magazines such as the *Builder* (for which there is a splendid published illustrations index for 1843–1883) and the British Architectural Library's 'grey books' are an important guide to published illustrations of 20th century buildings. When searching for illustrations and contemporary descriptions, however, the relevant trade literature can be as important as the architectural. A brewery might appear in the *Transactions of the Institute of Brewing* or a hotel in the *Caterer and Hotel Proprietors Gazette*.

The process of dating, like all other research into architectural history, is interactive between the building and the documents. Each helps interpret the other. It is the architectural historian's job to work at this interface and join other professionals in formulating proposals for what should happen to historic buildings. Recent Heritage Lottery Fund guidance on conservation plans has emphasized the importance of research and understanding in drawing up plans, and this is paralleled by the advice in PPG15 that applicants for listed building consent should show that they understand their buildings. How this is done is perhaps less important than that it should be done and done well.

FRANK KELSALL worked for the Greater London Council as an architectural historian from the 1960s to 1986. He then joined English Heritage as an inspector of historic buildings. Since early retirement in 1998 he has acted as casework adviser to the Ancient Monuments Society and, with Dr James Anderson, has founded the Architectural History Practice, a specialist research service for both public and private clients.

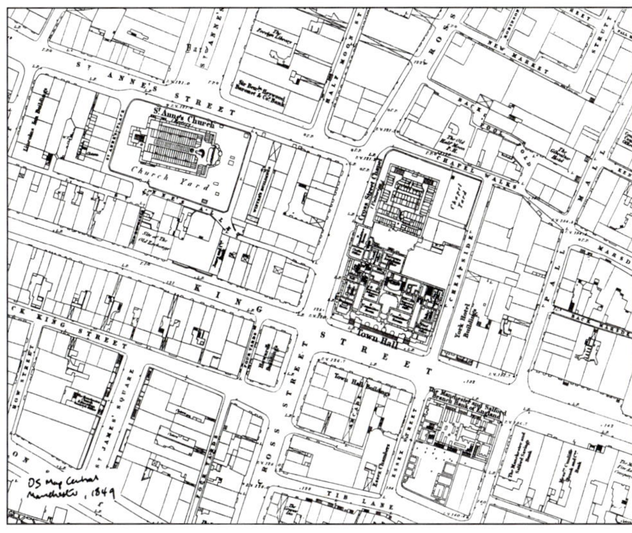

Black and white Ordnance Survey map of Central Manchester
The early Ordnance Survey maps, especially those of industrial towns at 1:1056, are full of detail. This map of central Manchester in 1849 shows the ground plans of St Anne's Church and the branch Bank of England (both still there) and the Cross Street Chapel and the old Town Hall (now both demolished).

Colour drawing from Corbett Estate sales brochure
Many people live in speculatively built houses but few speculative builders produced brochures. This is one of the grander houses on the Eltham Park Estate built by Archibald Cameron Corbett, first Lord Rowallan, and advertised in his sales brochure of 1913.

ARCHITECTURAL HISTORIANS

▶ THE ARCHITECTURAL HISTORY PRACTICE LIMITED

Phillimore Cottage, Thorncombe Street, nr Bramley, Surrey GU5 0LU
Tel/Fax 01483 208633 E-mail jla@architecturalhistory.co.uk
Website www.architecturalhistory.co.uk
Contacts James Anderson or Frank Kelsall

ARCHITECTURAL HISTORIANS: The Architectural History Practice offers independent advice on the special interest of historic buildings and the importance of such buildings in the planning system. The practice's advice is research-based and it specialises in a wide range of archival and documentary investigations in association with other professionals concerned with the care and management of the heritage. Clients include public sector bodies, private owners, architects, planners and developers. Recent projects have included Heritage Lottery Fund applications, advice on buildings subject to listed building and planning controls and recording work in partnership with architectural and archaeological consultants.

HISTORICAL RESEARCHERS

▶ RELIABLE EFFECTIVE RESEARCH

17 Homestead Road, Fulham, London SW6 7DB
Fax 020 7736 3741 (Agency)

SPECIALIST IN HISTORICAL RESEARCH: Robert E Rodrigues through Reliable Effective Research (RER) provides a personal and efficient research service for those wishing to establish a better understanding of their property in its historical context. RER helps owners understand the history and form of their building using data derived from a variety of sources. Desktop research is relied on extensively to reduce the need for disruptive site visits. Please write or fax RER to discuss how you can get to know your property better for historical reasons, or for enlightening background information for conservation project planning.

▶ STRUCTURAL PERSPECTIVES

48 Holdsworth Road, Holmfield, Halifax HX2 9SZ
Tel/Fax 01422 240789

RECORDING AND ANALYSIS OF HISTORIC BUILDINGS: *See also: display entry and profile entry in Archaeologists section, page 16.*

▶ WEALD AND DOWNLAND OPEN AIR MUSEUM

Singleton, Chichester, West Sussex PO18 0EU
Tel 01243 811363 Fax 01243 811475
E-mail wealddown@mistral.co.uk Website www.wealddown.co.uk

CONSERVATION TRAINING, RESEARCH AND SUPPLIES: Research library designed for use by professionals. *See also: profile entry in Courses & Training section, page 221.*

When contacting companies listed here, please let them know that you found them through *The Building Conservation Directory*

ARCHAEOLOGISTS

▶ ADDYMAN & KAY LTD

Studio 206, Abbey Mill Business Centre, Paisley PA1 1TJ
Tel/Fax 0141 561 0268
E-mail addyman_kay1@hotmail.com
Contact Thomas Addyman MA (Conservation of Historic Buildings), Archaeologist

BUILDINGS ANALYSIS, ARCHAEOLOGY AND CONSERVATION: Addyman & Kay specialise in assessment, analysis and recording of historic buildings and archaeological sites, having expertise in strategic planning issues, correct use of traditional materials, conservation prioritisation and assessment of significance. The office has long term experience of working with conservation professionals, architects and contractors, together with in-house CAD computer imagery, EDM survey of buildings and topography. Recent projects include a complete drawn survey/programme of analysis of Queensberry House, Scottish Parliament site, Edinburgh, advising the project architects. Conservation plan for St Giles' Cathedral, Edinburgh; a long term programme of recording/monitoring conservation works at Rosslyn Chapel, Midlothian; and a stone by stone CAD generated survey of Riding House Range, Bolsover Castle, Derbyshire.

▶ ARCHAEOLOGICAL SITE INVESTIGATIONS

Furlong House, 61 East Street, Warminster, Wiltshire, BA12 9BZ
Tel/Fax 01985 847791
E-mail mike@archaeology.demon.co.uk
Website www.archaeology.demon.co.uk

ARCHAEOLOGIST: Archaeological consultant specialising in the assessment, evaluation and survey of historic buildings and the management and design of archaeological projects. Affiliate of the Institute of Historic Building Conservation.

▶ ARCHITECTURAL ARCHAEOLOGY

20 Coleman Road, London SE5 7TG
Tel 020 7703 7519
E-mail mwsamuel@aol.com / klhamlyn@aol.com
Contact Mark Samuel PhD

ARCHAEOLOGICAL SERVICES: Architectural Archaeology provides a range of services necessary for statutory PPG15 work including assessment reports, accurate recording of architectural features and supervision of recording work. Architectural Archaeology offers particular expertise in the rapid analysis of stone buildings, including dating, recognition and graphic reconstruction of architectural features employing conventional and CAD methods. In-house writing and editing services provided. Clients include: Dean & Chapter of St Paul's Cathedral, William Whitfield & Partners (Paternoster Square), Museum of London, Guildford Borough Council.

▶ FIELD ARCHAEOLOGY CENTRE

The University of Manchester, Oxford Road, Manchester M13 9PL
Tel 0161 275 2314 Fax 0161 275 2315 E-mail umfac@man.ac.uk
Website www.art.man.ac.uk/fieldarcheologycentre
Robina McNeil BA MIFA Greater Manchester Archaeological Unit
John Walker BA FSA The University of Manchester Archaeological Unit

ARCHAEOLOGY, BUILDINGS, CONSERVATION SPECIALISTS: The centre houses two independent units with expertise in planning archaeology, environmental assessments, field archaeology, building archaeology, analysis, research, interpretation and management of the archaeological resource and specialising in urban regeneration, building conservation, sustainable development, buildings and monuments at risk strategies and, conservation plans. Both units have access to resources and staff across the University. In addition to strategic planning, consultancy, contract work and research the units offer an advisory and information service on their cultural heritage and are committed to a strong publication programme.

▶ GIFFORD AND PARTNERS

Carlton House, Ringwood Road, Woodlands, Southampton SO40 7HT
Tel 023 8081 7500 Fax 023 8081 7600

ARCHAEOLOGISTS: *See also: profile entry in Structural Engineers section, page 41.*

▶ HIRST CONSERVATION

Laughton, Sleaford, Lincolnshire NG34 0HE
Tel 01529 497449 Fax 01529 497518

ARCHAEOLOGY AND CONSERVATION: *See also: display entry on the inside front cover and profile entry in Building Contractors section, page 62.*

▶ OXFORD ARCHAEOLOGICAL UNIT

Janus House, Osney Mead, Oxford OX2 0ES
Tel 01865 263800 Fax 01865 793496
Website www.oau-oxford.com
Contact Julian Munby or Bob Williams

ARCHAEOLOGISTS: Oxford Archaeological Unit undertakes historic buildings investigations and surveys, documentary studies and archaeology. Listed building and scheduled monument consents, environmental assessments and planning. They are consultants to Historic Royal Palaces and other clients include the National Trust, English Heritage, Victoria and Albert Museum, Union Railways, local authorities, Oxford Colleges and many churches. For further information please contact Julian Munby or Bob Williams.

▶ PATHFINDERS

25 The Hollow, Lower Woodford, Salisbury, Wiltshire SP4 6NJ
Tel 01722 782881 Mobile 07818 098234
E-mail pathfinders@compaqnet.co.uk
Contact Bill Moffat

ARCHAEOLOGISTS: Pathfinders is an association of sole traders each with more than ten years experience in all forms of site work across the south of England and abroad. Since its formation in 2000 Pathfinders has successfully completed works for Terrain Archaeology, ASI and Bath Archaeological Trust. Pathfinders is able to undertake; historic building recording and survey, CAD assignments, desk-based assessments, watching briefs, evaluations, survey, excavations, finds analysis and report preparation.

▶ STRUCTURAL PERSPECTIVES

48 Holdsworth Road, Holmfield, Halifax HX2 9SZ
Tel/Fax 01422 240789

INDUSTRIAL ARCHAEOLOGY: Structural Perspectives provides a specialist service to the owners of historic industrial and engineering structures and sites. The company helps meet the requirements which the statutory authorities place upon owners of listed and scheduled properties and provides essential information for refurbishment and reuse. They also offer assistance with planning applications, listed building consents and grant funding applications. Services include: comprehensive recording and analysis of sites and buildings; civil engineering assessments of historic structures, and historic structural materials including cast, wrought iron and masonry. Clients have included; Historic Scotland, Phoenix Trust, GVA Grimley, British Gas (Properties), Rackwood Mining, J Banks Ltd and Railtrack plc. *See also: display entry on this page.*

▶ WESSEX ARCHAEOLOGY

Portway House, Old Sarum Park, Salisbury, Wiltshire SP4 6EB
Tel 01722 326867 Fax 01722 337562 E-mail postmaster@wessexarch.co.uk
Website www.wessexarch.co.uk
Contact Sue Davies or John Dillon

ARCHAEOLOGISTS: Wessex Archaeology provides a full range of archaeological services countrywide, including consultancy, heritage management, desk assessment, fieldwork, analysis and publication. High quality digital recording and modelling techniques are frequently used for strategic management and analytical projects, including building surveys. Recent building and conservation management projects include a conservation plan and management strategy for the Castle of Gibraltar, and management surveys of the Royal Dockyards of Plymouth and Portsmouth, surveys of medieval church spires and 16th century mansions, and of industrial, military and vernacular buildings in central southern England.

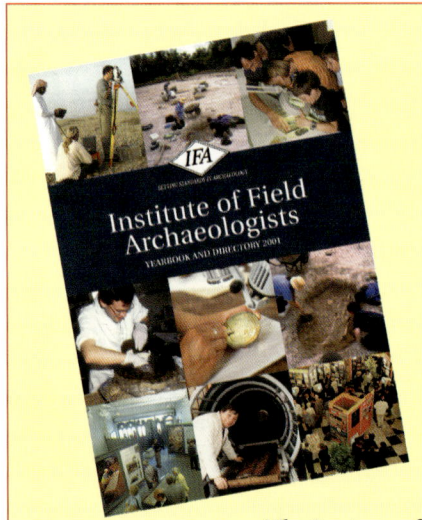

DATING TECHNOLOGY

ROY SWITSUR

Upon encountering a new site, the archaeologist immediately requires information about its age in order to set it in context with other sites. In research into our heritage the conservationist or architect may be able to date the general period of a building he is working with from either the situation, materials of construction, type of timber joints or other stylistic features. Almost certainly the century or portion of a century when it was built may be assigned with some certainty. However, as more and more work is done and increasing numbers of structures with complex constructional phases are encountered, the general features may not be sufficient to give the accuracy in dating that is currently required. If research into other sources of information also fails to throw light on the building's history, resort may be made to the various scientific methods of dating. This article outlines three of the most important methods currently used for dating buildings or, in a complex situation, the order of construction within the building. These are; dendrochronology (or 'tree-ring' dating), radiocarbon dating and thermoluminescence dating.

Each method has a distinct role in the investigation of historic buildings. None is infallible and before embarking on an extensive dating survey, due thought must be given to what might be achieved and which methods might be the more successful. If necessary, seek advice.

Whilst earlier types of wooden joints may be copied in later buildings and earlier styles may be reintroduced in later periods to confound the conservationist or historian, any re-use of older materials should become obvious by the use of the chronometrical methods described here. The incorporation of ancient bog oak into a building, no matter how intricately carved or jointed would immediately become obvious to the chronologist, as would timber renovations.

DENDROCHRONOLOGY

Dendrochronology is the oldest method, having been introduced over a century ago by an American astronomer, Professor A E Douglass. He wanted to know whether the number of sunspots affected weather on Earth. If this were so the width of the annual growth rings would show changes in synchronism with the sunspot numbers. He established a laboratory in the university of Arizona, at Tucson, to study tree-rings. Unfortunately, after many years of analysis he was not able to confirm the correlation he sought. Nevertheless. the laboratory was able to demonstrate many interesting properties of ring widths and their relationship with various aspects of climate and other natural phenomena, and of course, their use in the accurate dating of timber. His laboratory is still one of the leading centres in world dendrochronology. It was not until 1939 that the science was taken seriously in Europe, mainly through the efforts of Professor Huber in Germany, and not until after World War II that such studies became established in the UK. The main centres in Britain researching this field are located at universities in Belfast, Cambridge, East Anglia, London, Nottingham and Sheffield with several freelance practitioners *(see page 19)*. Whilst most dendrochronological research is still concerned with climatic change, where the precise dating provided by the growth rings is of vital importance, all units in this country are proficient in performing dendrochronological surveys of buildings.

Oak is the species of prime interest and it is possible to date wood back to over 7,000 years with a precision, in appropriate circumstances, of a single year. This is most impressive and makes dendrochronology the main dating method for structures containing oak timber. The method relies upon the response of trees to the weather during the growing season, which runs from March to October. In a 'good' growing season the trees within a large climatically homogeneous region all respond by putting on a wide growth ring within the cambium which separates the sapwood from the bark. In a 'poor' growing season the trees all respond so that only a very narrow growth ring is formed. In more typical growing seasons a ring of intermediate width is produced. (It should be noted that there is no direct linear relationship between ring width and, say, sunshine, or other weather components) Thus a 'good' or 'poor' growing season is defined with reference to the amount of growth produced. For example, the year 1976 had a gloriously hot, long summer with most rainfall arriving in autumn but the trees did not appreciate it and all oaks produced a distinctively narrow ring. Again the summer of 1915 was cold and wet, quite different from 1976, yet the trees also produced a distinctly narrow growth ring. So it will be seen that seasons that are hot and dry as well as those that are cold and wet will produce a narrow ring so that such a ring is not diagnostic of the weather. Year by year the trees throughout the region produce a similar pattern of wide and narrow rings in response to the weather changes. It is this pattern that allows the accurate dating. The pattern of ring widths on a specimen taken from a building is matched, using a computer with a 'master chronology' often several centuries long for the particular area. This regional chronology will have been painstakingly built up from many thousands of measurements and by cross-matching many overlapping patterns of timbers. The youngest patterns are obtained from living trees, where the felling date of the final ring is known. Progressively older patterns are obtained from trees in recent buildings, older buildings, archaeological sites and ancient bog oaks. Because of local, non-climatic causes of change of growth width, the chronologies around the country vary somewhat and the best dating match is always obtained from a local regional master chronology.

The dendro-date is thus the year in which the final ring of the specimen grew (the year in which the tree was felled, but not necessarily the year in which the building was constructed). In order to obtain an accurate match and hence a date, it is important to have at least 80 rings on the specimen that is to be dated. With fewer rings the pattern might have repeat matches at different points in the time scale and so give rise to multiple possible dates. This has implications for some vernacular structures in which rapidly grown, wide-ringed oaks 30 to 40 years old, were used. In such instances it might be possible to date the wall plate which often contain far more rings. In practice it is found that 100 or 120 growth rings are most likely to provide a unique match. However, because of the local ecological, non-climatic effects on the tree ring, it is not possible to guarantee that any particular specimen will give a date. In order to have greater certainty it is important to obtain several samples, in the form of cores drilled from the timber, and to construct a 'site chronology' for the building. The number of cores required will depend upon the complexity of the structure, but some ten cores per building phase is preferred. These are normally taken by the dendrochronologist in co-operation with the historian and the position of the cores is carefully marked on the building plan for future analysis of the results. The core leaves a small hole in the timber of about 15mm in diameter which may be plugged with a timber dowel.

Although this method is capable of dating to the individual year, in practice several factors conspire to reduce the precision in dating the construction, sometimes drastically, and it is important to be aware of the limitations. Whilst in the middle ages it was the practice to use the timber 'green' usually within a year of the felling date, in more recent times the timber is usually allowed to dry out, sometimes for decades before use. Furthermore, carpenters, aware of the effects of insect attacks, would deliberately remove the sapwood and even some heartwood. The number of sapwood rings may vary between 15 and 50 years, depending on the position in and the age of the tree. Thus the year of the last ring dated could be misleading to the construction date and be underestimated by an unknown number, possibly 60 years. Sapwood may be found on at least some of the timbers in the dendrochronological survey and the site master chronology will lead to a more reliable date than an individual core.

RADIOCARBON DATING

Whereas tree-ring dating is limited in this country to oak structures, radiocarbon dating may be used for any wood species and, indeed, for any other organic based materials found in buildings such as: wattle and daub; straw used for insulation; hair used in plaster; leather wall hangings and, perhaps surprisingly, mortar. The range of radiocarbon dating reaches back to 60,000 years. For the last few thousand years it can have a precision of a few decades and may, in certain circumstances, be comparable with tree-ring dates. The method was conceived by American Professor Willard F Libby of Chicago in 1947. The laboratory at Cambridge here in England was among the first six to be set up anywhere in the world. There are now several radiocarbon dating laboratories in Britain including those at Belfast, Cambridge, East Kilbride, Oxford and Swansea as well as a commercial unit near Harwell.

Radiocarbon dating is based on the element carbon, the basis of all life on earth. The atoms of this element are of three different types or 'isotopes'. They are identical chemically but have slightly different physical properties, particularly in mass. The isotopes are respectively 12, 13 and 14 times as heavy as the common hydrogen atom (the base unit by which the weight of other elements is measured). The isotopes C-12 and C-13 are stable and make up the bulk of the element but the C-14 isotope, which is mildly radioactive, is extremely rare. The instability of radiocarbon results in half of it disappearing in 5,730 years (its 'half-life'). This instability is the basis of the dating method. All creatures have the same concentration of radiocarbon in their cells while they remain alive. This level is maintained constant by a sequence of events affecting the food web. It starts with photosynthesis in green leaves of plants, whereby atmospheric water vapour and carbon dioxide, containing the radiocarbon, are combined in the presence of sunlight to produce sugar. The plant biological process converts this to the myriad of substances required for life. These substances are shared via the food network to all animals including man. For our purposes it may be assumed that the amount of radiocarbon in the atmosphere is constant over time.

Once the creature dies the food chain is broken and the concentration of radiocarbon in the cells falls away. By measuring the residual C-14 concentration in the material the date of its death may be calculated. In the case of tree-rings the food chain is effectively broken at the end of the growing season and the radiocarbon concentration immediately begins to fall. Thus, in principle, the age of each growth ring may be measured. In practice, the measurements may resolve differences of about 20 or 30 years.

Samples from a building for radiocarbon dating should be taken with care and due regard to provenance. For timber specimens, samples should be obtained as near to the bark as possible, as for dendrochronology. Samples such as leather, cloth, food residues or straw represent a year's growth and so a point in time. Thatch, whether straw or rush will date the last repair and not necessarily the construction date. Mortar is made by heating limestone to over 850°C to form quicklime. When slaked and used as mortar between layers of bricks it dries by absorption of contemporary carbon dioxide from the air and so may be used to date this event.

THERMOLUMINESCENCE DATING

Thermoluminescence, or TL, was first used in the 1950s for the measurement of radiation exposure, and underwent a period of difficulties before being applied to dating; the first dates it produced being too young. Upon resolution of the technical problems the method was used for dating pottery and burnt flints from archaeological sites with a precision of about 7–10 per cent. Subsequently it has been used in the investigation of recent geological formations reaching back to half a million years. In its most common form it may shed light on the age of fired clay and quartz-based materials but approaching the present no closer than about a thousand years. A modern variant on the technique is able to date far more recent fired clay material.

TL depends upon minute levels of background radiation in the clay matrix, a tiny fraction of which is absorbed and stored as a charge at imperfections in the crystal lattice of quartz inclusions. The firing of pottery removes the inherited geological TL and sets the dating clock to zero. In the laboratory grains of quartz are extracted from the pottery and heated in light-tight apparatus at a constant rate to around 400°C . Superimposed upon the red-hot glow, a tiny flash of light is produced as the stored energy is released (hence 'thermoluminescence') and the flash is recorded by computer. The quantity of light produced is proportional to the length of time since it was last fired. Unfortunately, problems remain since all samples do not have the same sensitivity to the radiation and background radiation levels vary. Furthermore, the results are sensitive to water content. Thus many measurements must be made in order to obtain a date.

Recently this method has been improved. The flash of light is released by scanning the sample with an energetic green laser beam and light-emitting diodes are used as detectors. This form of the method, known as 'optically stimulated luminescence dating', enables objects which are not more than a few hundred years old to be dated to within a few decades. Hence it is far more useful than the original TL technique in dating buildings. The requirement remains that the sample should have undergone some heating event to set the clock to zero. It also requires that a dosimeter be left undisturbed in situ at the site for some months in order to discover the natural radioactivity permeating the samples. These must be inorganic and contain some light transmitting materials. Pottery artefacts and certain bricks might be suitable specimens, and often TL provides the only way to distinguish medieval or Tudor bricks and chimney pots from Victorian reproductions.

There are several laboratories capable of this sort of measurement in this country which include the Geology Department, Aberwystwyth; the British Museum; the Godwin Institute, Cambridge; the Department of Archaeology, Durham; Environmental Sciences; the Institute of Archaeology, University College, London; Research Laboratory for Archaeology and the History of Art, Oxford.

There are thus now various ways in which chronologists are able to help conservators, historians and architects in their endeavours in studies of dating our heritage.

PROF ROY SWITSUR, a pioneer of radiocarbon dating, established The Dendrochronology Laboratory in 1976 in the Godwin Institute for Quaternary Research, Cambridge University. He is the Research Director of the Cambridge Radiocarbon Dating Research Laboratory (see Cambridge Dating Unit opposite)

DENDROCHRONOLOGISTS

▶ AOC ARCHAEOLOGY GROUP

Edgefield Industrial Estate, Edgefield Road, Loanhead, Midlothian EH20 9SY
Tel 0131 555 4425 Fax 0131 555 4426
E-mail annec@aocscot.co.uk
Website www.aocarchaeology.com

DENDROCHRONOLOGY: AOC Archaeology Group provides the only dendrochronological service based in Scotland and has over a decade's experience in the analysis of material from standing buildings, archaeological excavations and objects in museum collections. The company can also age living trees, a valuable tool in the analysis of historic and designed landscapes. Although working primarily with oak, AOC has experience of other species such as pine, ash and yew. Providing a full service from assessment and sampling to analysis and reporting, clients include Historic Scotland, the National Trust for Scotland, the Tayside Building Preservation Trust, as well as private clients.

▶ BELFAST TREE-RING LABORATORY

Palaeoecology Centre, Queen's University Belfast, Belfast BT7 1NN
Tel 02890 335143 Fax 02890 335354

DENDROCHRONOLOGY: Holds exciting possibilities as a dating method. It is the only method which can provide exact calendrical dates. However, the extreme accuracy is tempered by its applicability. This method has three basic requirements, (1) timbers must be oak, (2) they must be long lived i.e. have many growth rings, and (3) they must be derived from an area for which an oak chronology exists. Fortunately, all three of these conditions exist in Britain. Oak has always been a preferred building timber and as a consequence turns up regularly in buildings. However, the method does not offer a guarantee of success, especially with single samples. As a result, the charge is for the work involved in attempting to date samples.

▶ CAMBRIDGE DATING UNIT

Environmental Sciences Research Centre, Anglia Polytechnic University, East Road, Cambridge CB1 1PT
Tel 01223 363271 x2594 E-mail vrs1@anglia.ac.uk

RADIOCARBON DATING AND DENDROCHRONOLOGY SPECIALISTS: Cambridge Dating Unit undertakes tree ring dating surveys of buildings containing oak timbers. Radiocarbon dating of materials such as timber, fabric, leather, bone, plaster, mortar etc. There is access to resources and staff across the university for studies including palaeontology, petrology and sedimentology. Cambridge Dating Unit has a strong policy of research and publication. Studies have included listed buildings, churches, boats and vernacular structures. Contracts are with English Heritage, Cambridge Archaeological Unit, Oxford Archaeological Unit, Cambridge colleges, local authorities and private individuals.

▶ HERITAGE & ARCHAEOLOGY RESEARCH PRACTICE (HARP)

University of Wales, Lampeter, Ceredigion SA48 7ED
Tel 01570 424730 Fax 01570 421240
E-mail harp@lamp.ac.uk

DENDROCHRONOLOGY: A broad range of specialist services is available through HARP from the fully equipped dendrochronology laboratory situated on the Lampeter campus. Historic buildings can be accurately dated through a programme of on-site assessment and sampling followed by laboratory-based analysis. Reports can be produced to the client's specification. A comprehensive service is available for the analysis of archaeological wood assemblages from wetland and underwater excavations including species identification, technological studies, archive and publication illustration, tree-ring dating and analysis. Recent corporate clients include Cambrian Archaeological Projects Ltd, Clwyd-Powys Archaeological Trust, English Heritage and the National Museum of Wales.

▶ INSTITUTE OF ARCHAEOLOGY, UNIVERSITY COLLEGE LONDON

31-34 Gordon Square, London WC1H 0PY
Tel 020 7679 1540 E-mail martin.bridge@ucl.ac.uk
Website http://members.aol.com/MarBrdg/
Contact Dr Martin Bridge

DENDROCHRONOLOGY: Dr Bridge has been working in dendrochronology since 1979 and has extensive experience of tree-ring dating buildings, other artefacts and tree-ring studies on living trees. He has published widely, both in popular and academic sources, and lectures on dendrochronology and other related areas. Most of the work is undertaken in southern Britain and northwest France, but work has also been done in New Zealand, Newfoundland and Ireland. Work undertaken includes assessment of the suitability of the material for dendrochronological study, sampling, analysis and reporting. Clients include English Heritage, county councils, architects and private individuals.

▶ OXFORD DENDROCHRONOLOGY LABORATORY

Mill Farm, Mapledurham, Oxon RG4 7TX
Tel 0118 972 4074 Fax 0118 972 4404 Workshop 0118 972 3682
E-mail daniel.miles@rlaha.ox.ac.uk
Website www.dendrochronology.com
Contact Daniel Miles or Michael Worthington

DENDROCHRONOLOGY LABORATORY: Specialising in the dating of buildings throughout the south and west of England, the Midlands, and Wales. Ongoing research includes development of minimal and non-destructive sampling techniques allowing the dating of ephemeral objects such as doors, furniture, and other artefacts. Intensive projects currently conducted in West Sussex, Surrey, Hampshire, Somerset, Oxfordshire, Shropshire, Wales, and the Channel Islands. Potential sites subjected to rigorous preliminary assessment resulting in high success rate. Over the past 15 years 350 sites encompassing 600 phases have been dated and published in Vernacular Architecture; all results obtained by the laboratory together with full architectural descriptions are accessible on the website. Staff are trained in archaeology, architectural history, and timber conservation, ensuring the best interpretation and most efficient and economical sampling of historic structures.

▶ SHEFFIELD DENDROCHRONOLOGY LABORATORY

Research School of Archaeology and Archaeological Science, University of Sheffield, West Court, 2 Mappin Street, Sheffield S1 4DT
Tel 0114 222 5107 Fax 0114 276 3146
E-mail dendro@sheffield.ac.uk

DENDROCHRONOLOGY: The Sheffield Laboratory staff have extensive experience of dating projects, both large and small, on buildings, historical objects such as furniture and paintings, archaeological assemblages of waterlogged timbers of all periods from the Neolithic to the present, and modern woodlands. For building conservation projects they can provide free advice on the likely suitability of timbers in any building and can undertake assessment, sampling, and reporting of results to agreed standards and deadlines anywhere in England and Wales. Sheffield Dendrochronology Laboratory can work with specialist building recording teams, or provide a complete solution for PPG15 and other statutory requirements.

▶ THE UNIVERSITY OF NOTTINGHAM TREE-RING DATING LABORATORY

University of Nottingham, University Park, Nottingham NG7 2RD
Tel 0115 951 4837 Fax 0115 951 4812
E-mail robert.laxton@nottingham.ac.uk

DENDROCHRONOLOGY: The laboratory has dated oak-framed buildings for over 20 years, including roofs in cathedrals, for example Lincoln and Ely; churches; other large structures and many vernacular buildings in most areas of England for national and local authorities as well as local societies and private individuals. The dates of the buildings range from the medieval period to the 18th century. Initially the oak timbers are inspected to assess the probability of dating them and if satisfactory, the overall cost is estimated. Sampling is by coring, there is no structural damage and each small hole is plugged afterwards to make it unobtrusive.

CONSERVATION PLANNING

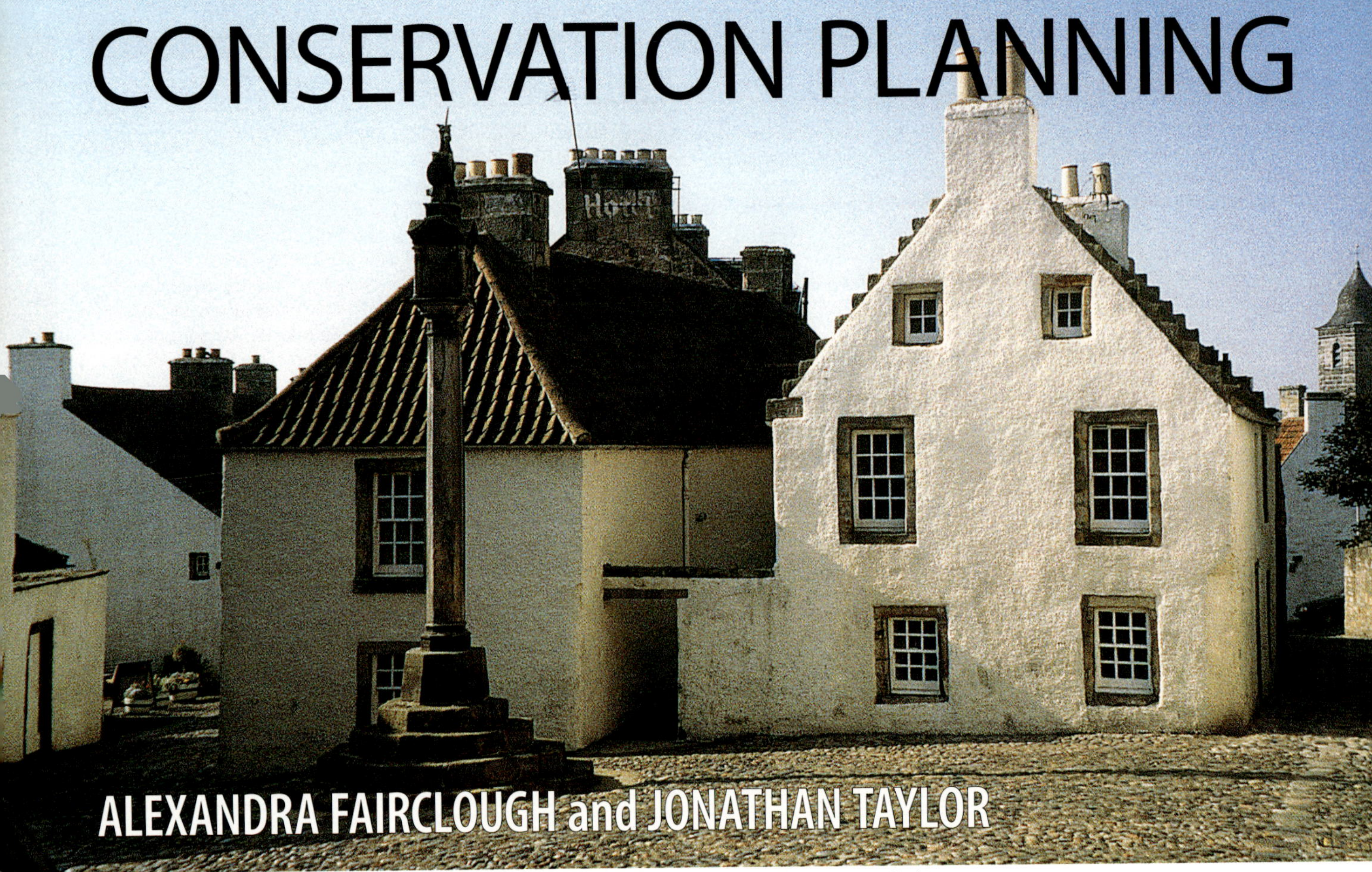

ALEXANDRA FAIRCLOUGH and JONATHAN TAYLOR

A CONSERVATION AREA AT CULROSS, FIFE *Four out of five conservation areas in Scotland are designated as being 'of outstanding quality', like this one. The designation is used primarily for targeting grant aid. (Simon Montgomery)*

Control over the development of new buildings and the alteration of historic buildings in the UK has evolved over the past 50 years as a reactive and a preventative system of control. It is preventative in that most development requires permission from the local planning authority by law, and the application procedure leads to unacceptable applications either being refused or being modified and improved. It is reactive in that breaches of planning can be put right by enforcement procedures and specific cases can result in criminal prosecution.

Underpinning this system is a public law mechanism which is embodied in a variety of statutes, regulations, rules, policy guidance notes, circulars and technical advice notes, and which is implemented at a local level by public authorities vested with powers to control the development and use of land. Local authorities achieve this through three different elements; the development plan, the control of development and a centralised system of administration.

An application may be decided by a planning committee or delegated to the planning officer if the case is not controversial. In deciding whether or not to approve a planning application, the local authority is required to take into account its own policies within the development plan and 'any material considerations' – that is to say, anything else which is particularly relevant. This allows the authority wide discretionary powers on the basis that the development plan is not the only consideration in planning applications.

The development plan sets out the policies of the local authority rather than prescriptive rules. The plan is reviewed periodically to take into account national policy and local requirements. There are also areas of specialised heritage protection and control within the planning arena. These include listed buildings, conservation areas, trees, churches and ancient monuments.

Planning control is rarely proactive although recent governmental policy attempts to promote development (urban regeneration for example). The planning system is more negative than positive and sometimes unaccountable, in that the decision-making process is influenced by Government policy rather than strict rules or guidance.

This system is political in so far as decisions are made by a committee of local politicians. Thus political accountability is substituted for legal accountability and issues of policy are negotiated between local politicians, national politicians and the Executive arm of the Government – that is the Houses of Parliament and, specifically, the Secretaries of State. Thus there exists a system of central supervision based on pragmatism rather than legal principle.

CONSERVATION

The impact of planning law on the historic environment is acute. The existence of a separate planning act is indicative of the importance of historic buildings and structures to our national identity and cultural heritage. In England and Wales this is the *Planning (Listed Buildings and Conservation Areas) Act 1990*. In Scotland it is the *Planning (Listed Buildings and Conservation Areas) (Scotland) Act 1997*. The provisions of these two acts can work independently and in parallel with the main planning act, the *Town and Country Planning Act 1990*. There are also many associated regulations, circulars and guidance notes.

In England, *Planning Policy Guidance Note 15: Planning and the Historic Environment (PPG15)* is particularly useful for both the conservation practitioner and the historic building owner. Equivalent documents for Wales are the *Welsh Office Circulars 61/96* and *1/98* (their full titles are given at the bottom of the page) and for Scotland it is the *Memorandum of Guidance on Listed Buildings and Conservation Areas*. These regulations, circulars and planning policy guidance notes are produced by central government. They are recognised as important sources of government policy which assist local planning authorities in their decision-making, and are used to interpret planning law.

A wide variety of information has to be assimilated by an applicant or the professional when proposing to alter, extend or demolish a listed building, and owners of historic buildings are best guided by the conservation officer, whose experience and expert knowledge can be invaluable. If possible, establish a relationship with the conservation officer prior to the submission of an application and submit as much information as you can at this stage. It is often necessary to include a full and detailed justification for the proposals to assist in the decision-making process.

Often there will be issues and proposals

CATHEDRAL COMMUNICATIONS LIMITED

that will be contrary to both the guidance given by the conservation officer and in PPG15 or its equivalent Circular or Memorandum. If an applicant prefers to proceed against the advice of the conservation officer, then the alternative is to lodge an appeal against the subsequent refusal with either the Planning Inspectorate if the property is in England or Wales, or with the Scottish Official Inquiry Reporters Office if in Scotland.

The appeal system enables an aggrieved applicant or those subject to enforcement, to appeal the decision of the planning authority. A planning inspector is generally the arbiter of such adversity, although government ministers are the final decision-makers.

Until recently the rights of the property owner were not protected by any constitutional rules. The incorporation of the *Human Rights Act 1998* may lead to change but this will be on a case by case basis and will be inevitably slow. The impact of the Act is discussed further in the article on page 22.

LISTED BUILDINGS

Some 440,000 buildings are listed on the statutory register of buildings of 'special architectural or historic interest' in Britain. As some list entries include several buildings at the same address, the total number of listed buildings is larger – perhaps 600,000 – amounting to almost two per cent of our total housing stock. The listings are graded according to the architectural or historic importance of each building, Grade I being the most important in England and Wales and Grade A in Scotland. The grade generally reflects the age and rarity of the building, but many other factors are also taken into account, such as technological innovation, townscape value or connection with a particular historical event.

LISTED BUILDINGS BY GRADE

	ENGLAND	WALES	SCOTLAND
	Grade I 2%	Grade I 2%	Category A 7%
	Grade II* 2%	Grade II* 6%	Category B 57%
	Grade II 96%	Grade II 91%	Category C(s) 36%
		Others 1%	
TOTAL	370,505	25,430	45,498

(5 April 2000)

All alterations and extensions to a listed building require listed building consent if they 'affect its character as a building of special architectural or historic interest', irrespective of its grade (although some churches are exempt – see below). This includes alterations inside the building and may also include many aspects of repair. The only real difference between grades is that whilst listed building consent applications are considered by the local authority alone, those affecting buildings of the most important grades (I, II* and A) are also considered by either Cadw, English Heritage or Historic Scotland. In Wales, all applications to which the local planning authority are minded to grant consent are notified to the National Assembly of Wales (ie Cadw); unless the works affect the interior only of a Grade II listed building which has not been grant aided or is not subject to an undecided grant aid application, under Section 4 of the Historic Buildings and Ancient Monuments Act 1953.

Listed building consent is also required for alterations to any object or structure which lies within the grounds or 'curtilage' of a listed building and which was constructed before 1 July 1948. This may be taken to include garden walls, sundials, dovecotes and other such objects and structures, as well as buildings which are ancillary to the principal building.

ECCLESIASTICAL EXEMPTION

Until recently all churches and chapels in England, Northern Ireland and Wales were exempt from listed building and conservation area controls. The exemption remains intact in Northern Ireland, but in England and Wales, 'ecclesiastical exemption' as it is known, has been restricted to churches and chapels of the six denominations deemed to be operating an acceptable internal system of control, provided that the building remains in use as a place of worship. These include the buildings of the Church of England, the Church in Wales, the Methodists, the Roman Catholics, the United Reformed and those Baptist churches where the Baptist Union acts in the capacity of trustee.

Ecclesiastical exemption in Scotland, which was limited to listed building consent only, was also reassessed and a pilot scheme was introduced on 1 January 1999 for a trial period of three years. Under this scheme a congregation wishing to carry out works to the exterior of a listed church building is required to submit details of the proposal to the planning authority and Historic Scotland and obtain their agreement before undertaking the work. For the moment works to the interior of a listed church building remain exempt.

CONSERVATION AREAS

There are almost 10,000 conservation areas in the UK, designated by the local authorities. In these areas the demolition of a building requires conservation area consent. Unauthorised demolition is a criminal offence.

The meaning of 'demolition', at least as far as Great Britain is concerned, was clarified by the case of *Shimizu v Westminster City Council* in 1997. The result was that the 'demolition' of part of an unlisted building, such as a chimney stack or a front porch, is now considered as an alteration, not demolition, and so does not require conservation area consent. However, alterations such as this may require planning permission.

The alterations which require planning permission are complicated by 'permitted development rights', the effect of which is that certain small extensions and other alterations are granted planning permission automatically where they affect a house which is occupied as a 'single family dwelling' – that is to say, it is lived in by one family only, not subdivided to form flats. Within a conservation area these 'permitted development' rights are more limited, and exclude for example certain types of cladding, the insertion of dormer windows and satellite dishes, all of which therefore require planning applications. In Scotland, changes in a roof covering are also excluded.

Permitted development rights for a prescribed range of developments may also be withdrawn by the local authority under an Article 4 direction. This enables the local authority to control certain types of alteration which do so much damage to the character of conservation areas, such as the alteration or removal of doors and windows.

No separate application is required where an unlisted building lies within a conservation area, but the policies of the local authority should be carefully noted as local authorities are required to pay special attention to 'the desirability of preserving or enhancing the character or appearance of that area' when considering an application for planning permission.

PARKS AND GARDENS

Alterations to parks and gardens generally do not require statutory consent unless they involve development work requiring a planning application or affecting a tree covered by a tree preservation order (TPO). However, local planning authorities are encouraged to include policies in their development plans for the protection of designed landscapes and to protect 'registered' historic parks and gardens. In England and Wales being 'registered' means inclusion on the Register of Parks and Gardens of special historic interest. In Scotland the register is called the Inventory of Gardens and Designed Landscapes.

There are also statutory procedures for consultation where development is likely to affect a registered historic park or garden. The consultation varies slightly between the regions of the UK.

Where an application affects a park or garden, Historic Scotland can request the Planning Division of the Scottish Office to call in the application under a 'notification directive' before the planning authority issues its decision.

The consent of the local planning authority is required to cut down, top, or lop a tree which is protected by a TPO. The principal exception to this is where a tree is dying, dead or dangerous, in which case notice should be given to the local authority before carrying out the work. Within a conservation area anyone proposing to cut down, top or lop a tree is also required to give the local planning authority six weeks notice, giving the authority the opportunity to consider whether a TPO should be made.

FURTHER INFORMATION

General historic building issues

Mynors, Charles; *Listed Buildings, Conservation Areas and Monuments, Third Edition.* Sweet and Maxwell, 1999

Government guidance:

England: *Planning Policy Guidance Note 15: Planning and the Historic Environment*

Wales: *61/96 Planning and the Historic Environment: Historic Buildings and Conservation Areas*

1/98 Planning and the Historic Environment: Directions by the Secretary of State for Wales

Scotland: *Memorandum of Guidance on Listed Buildings and Conservation Areas*

This article was prepared by **ALEXANDRA FAIRCLOUGH** (IHBC Law and Practice) and **JONATHAN TAYLOR** (BCD Editor) with contributions from **GRAHAM REED** (Historic Scotland) and **DAVID ROWE** (Cadw).

THE HUMAN RIGHTS ACT 1998 – The death of planning or the birth of a fairer system?

ALEXANDRA FAIRCLOUGH

The incorporation of the European Convention of Human Rights into UK legislation has caused uncertainty in our planning system. Cases currently on appeal to the House of Lords could lead to a radical overhaul of the present system, not least to separate the decision making process from political influence, and to establish the rights of objectors.

The *Human Rights Act 1998* (HRA) came into force on 2nd October 2000. The act incorporates into domestic law many of the provisions of the *European Convention of Human Rights* (ECHR). The UK Government was heavily involved in the drafting of the ECHR, and it was one of the first to sign it and *the* first to ratify it in 1951. Furthermore, since 1966, the UK has accepted the rights of individuals to petition the Strasbourg authorities in respect of alleged breaches of the Convention. Yet those rights have not themselves been part of, nor actionable within our legal system. The reason argued repeatedly was that there was no conflict between any of the provisions of the Convention and UK domestic law.

The main forum for the protection of human rights in Europe is the Council of Europe. The Council of Europe has a court of Human Rights in Strasbourg. The citizens of individual member states can either apply to the Council of Europe alleging infringement of their human rights, or they can petition the Court directly if they fulfil certain criteria relating to standing, eligibility and time limitation. In the UK this has rarely happened; the cost – in time and money – has deterred many people from taking cases involving potential breaches of human rights to Strasbourg.

The Convention has been binding on UK law for almost half a century. However, it has not had direct effect within national boundaries. So if you complained of a violation of a Convention right, you had to exhaust the domestic court system first before you could apply to the European Court of Human Rights. In the meantime, the domestic courts would continue to apply the alleged violating, domestic law in the same way.

This position has been gradually changing on the basis that the law of the European Union is part of domestic law. The European Courts of Justice (ECJ) in Luxembourg has held in recent decisions that member states must respect human rights. Therefore European law has overlapped with the law of the ECHR.

Also, prior to the HRA coming into force, the UK courts accepted human rights points on the basis that Convention rights were said to have influenced previous decisions of the UK courts, and that all subsequent decisions were already bound by these ones.

In essence it is unlawful for a 'public authority' such as a court, tribunal or local planning authority, to act in breach of a Convention right, unless it is necessary to do so to 'give effect to legislation' – that is to say that by obeying the Convention they would be in breach of another law. Such a breach may be justified on the grounds that it protects the amenity of the community.

A victim of such an act has a right of action or a defence where the public authority is acting unlawfully. Where the authority is acting in breach of a Convention right, but is giving effect to legislation, the victim can seek a declaration from the courts that the legislation is incompatible with the ECHR.

However, there are clear exceptions: an interference with a right may be justified if it is in accordance with the law and is necessary to a democratic society.

Several cases considered by the European Court of Human Rights have provided the following conditions under which an interference may be justified:

1. There is some specific, accessible and precise legal rule justifying the interference
2. The interference serves one of the aims set out in the qualification to the relevant article
3. The interference is necessary in a democratic society; namely that there is a pressing social need for the interference and the interference is proportionate to the aim pursued.

It is for the public authority in question to justify any interference in all the above respects and the burden of proof is on the public authority once interference has been established.

Public authorities, which includes local planning authorities by definition, are prohibited from acting in a way which is incompatible with any of the human rights described by the Convention (Section 6 (1)), unless legislation makes this unavoidable.

If an authority acts in a way which is incompatible, then separate proceedings can be brought against it under Section 7 (1). Therefore the Act creates new rights of action and grounds of appeal whether civil or criminal by a 'victim' of the unlawful act.

THE CONVENTION RIGHTS

The rights defined by the ECHR are set out in Schedule 1 of the *Human Rights Act 1998*. Those rights most likely to affect planning and historic buildings include:

Procedural guarantees

- Article 6 (the procedural right to a fair trial)
- Article 14 (the prohibition of discrimination)

Substantive rights

- Article 8 (the substantive right of respect for a person's home)
- Protocol 1 Article 1 (the substantive right of peaceful enjoyment of one's possessions which include one's home and other land)

The two 'substantive rights' listed above will enable those affected by the planning process to reinforce their objections by stating that to allow such a development to proceed or such an enforcement notice to stand would infringe their human rights. The two 'procedural guarantees' will ensure that all 'victims' are given the chance of a fair hearing.

The full implications of the impact of the HRA are unknown. However, at the very least, if an interference is established, then it is necessary for the 'public authority to justify that incursion'.

Article 6 (the right to a fair trial)

Article 6 relates entirely to procedure and it applies wherever there is a determination of a person's 'civil rights'. These rights encompass property rights, thus affecting planning and conservation law.

This article could be of significance to the planning process in that it enables a complainant, whether developer or objector, to argue that he or she has not had a fair hearing. Article 6 will only extend to an objector if he or she is directly affected by any development proposals. However, the complainant must prove that his rights are also affected. If this is so then he also must be given a fair hearing. This could lead to major changes in the way that planning committees are operated.

The developer's position has already been tested in *Bryan v UK*. Here it was held that a developer could challenge an enforcement notice as a breach of Article 6.

The planning appeal system is presently under scrutiny on the basis that the adjudication of a planning inquiry by a planning inspector is not considered independent or impartial. In *County Properties v Scottish Ministers*, it was established that there was a breach of Article 6 where a listed building was called in by the Scottish executive agency, Historic Scotland. This case illustrates how elements of the planning consultation process can infringe upon the *European Convention on Human Rights*.

Similarly, appeals heard by the Secretary of State could also be open to challenge on the grounds that the Secretary of State is in effect both policy maker and judge. More recently the *Alconbury et alia* cases have illustrated that the Secretary of State's position as final decision maker and policy maker was contrary to Article 6. These cases are currently on appeal to the House of Lords. If the Lords' uphold the first decision this could result in a radical overhaul of the present planning system, perhaps by creating an independent tribunal or environmental court.

The objector to the grant of permission has had limited rights until now. Under the *Local Government Act 1972*, objectors are often unable to present their case at a planning committee and are hindered by a time restraint. Often the planning officer's report may only make a passing reference to the objections, and once permission has been granted, there is no opportunity for third parties to appeal against planning permission unless there has been an error in the legal process of the decision.

If an appeal against the process of the decision is made, the success of any 'judicial review' which ensues will depend on whether there has been bias or procedural error in the decision-making process, or the decision is *ultra vires*. The judicial review process is itself complicated by the fact that planning committees do not have to give reasons for approval. Therefore no appeal on the merits of an approval ever occurs.

In the recent planning case *Ortenberg v Austria* the European Court of Human Rights found that a third party did have the protection of Article 6 (1) where the grant of planning permission might adversely affect the value of property. Third party property values may not be considered as material considerations within the planning process, nonetheless, a person's civil rights could possibly be infringed depending on how the courts interpret Article 6.

Article 6 may allow third parties to lobby the Minister responsible (in England the Secretary of State) to persuade him to call in controversial applications. If an application is not called in there may be a right to challenge under Article 6.

The right to a fair hearing may induce changes to the way a planning committee makes decisions, for example oral hearings by objectors and cross-examination.

Article 8 (the right to respect for private and family life)

Article 8 gives everyone the 'right to respect' *for* his or her home but not a right *to* a home. Ultimately the courts will have to determine how far 'respect' is to be interpreted.

It must be necessary to safeguard a democratic society in the interests of national security, public safety(highway safety etc.) or the economic well-being of the country (recreation or amenity): for example; for the prevention of disorder or crime, for the protection of health or morals and for the protection of the rights or freedoms of others. This last element in particular provides plenty of scope for planning policy which overrides the freedom of the individual in the interests of the public.

In a recent case, *Britton v SOS,* the courts reappraised the purpose of the law and concluded that the protection of the countryside falls within the interests of Article 8 (2). 'Private and family life' therefore encompasses not only the home but also the surroundings. Arguably, this could mean that Article 8 (2) would also apply where a listed building or a conservation area is affected, enabling people to demand respect for the special interest of the conservation area in which they live or nearby listed buildings as a human right.

First Protocol Article 1 (the protection of property)

In many cases there is likely to be a significant overlap between Article 8 and First Protocol Article 1. However, this right is wider than Article 8 in the sense that it applies to the peaceful enjoyment of all of a person's possessions and not merely to his home. This could include land, curtilage property, fixtures and fittings.

The grant or refusal of planning permission, listed building consent or conservation area consent will frequently affect the lives, homes and property of others. Notably the applicants and the owners and occupiers of neighbouring properties, all of whom have the right to respect for their home and a right for the peaceful enjoyment of their possessions.

In practice it is likely that the interests of the community and those of the applicant will be balanced. It will be necessary for the local planning authority, the planning inspectorate and the courts to ensure this balance is fair.

Planning Policy

It is only in exceptional cases that personal circumstances may be relevant to planning decisions. However, the Convention puts the rights of the individual first on the basis that the rights of the individual are paramount unless there is justification in the public interest. Planning policy always puts the public interest before the rights of the individual. This may lead to changes in planning policy and in the determination of planning applications.

CONCLUSIONS

The full impact of incorporating the *European Convention of Human Rights* into domestic law is unknown. However, it must be emphasised that challenges can only be made against a public authority. The intention behind the incorporation was to provide as much protection as possible for the rights of individuals against the misuse of power by the state, within the framework of the act that preserves parliamentary sovereignty. Challenges regarding Article 6 are currently under consideration. It is anticipated that there will be an initial surge of challenges until the legislature or judiciary has fully established the position of Articles 8 and Protocol 1 Article 1 within the planning arena.

The impact of the *Human Rights Act* is causing uncertainty including parliamentary debate on issues such as the creation of an environmental court and third-party rights of appeal. Changes that may occur could include the following:

- the creation of an independent inspectorate, tribunal or environmental court
- rights of appeal against the grant of planning permission, conservation area consent or listed building consent by third parties whose homes or property may be affected
- rights of appeal against the designation of listed buildings and conservation areas
- the withdrawal of the right of a local planning authority to determine some of its own applications.

The most important cases are under consideration by the House of Lords and its decision, which was expected in April 2001, may result in a new approach to the administration of planning law.

FURTHER INFORMATION

Human Rights Act 1998 – Website **www.hmso.gov.uk/acts**

For further information on the Institute of Historic Building Conservation please view **www.ihbc.org.uk**.

For further information on the European Convention on Human Rights please view **http://conventions.coe.int**, the website for the Treaty Office of the Council of Europe

ALEXANDRA FAIRCLOUGH MA (Arch), BSc (Hons) MRTPI, IHBC, a conservation officer at Macclesfield Borough Council, is currently Law and Practice Officer for the Institute of Historic Building Conservation and is Chairman to a Committee of eight. The Committee has recently revived the Law and Practice Roadshows which tour the country providing information on legal issues relating to conservation. She has recently completed a law degree at Manchester Metropolitan University and is a member of the Chester Diocesan Advisory Committee for the care of churches. She can be contacted on law@ihbc.org.uk.

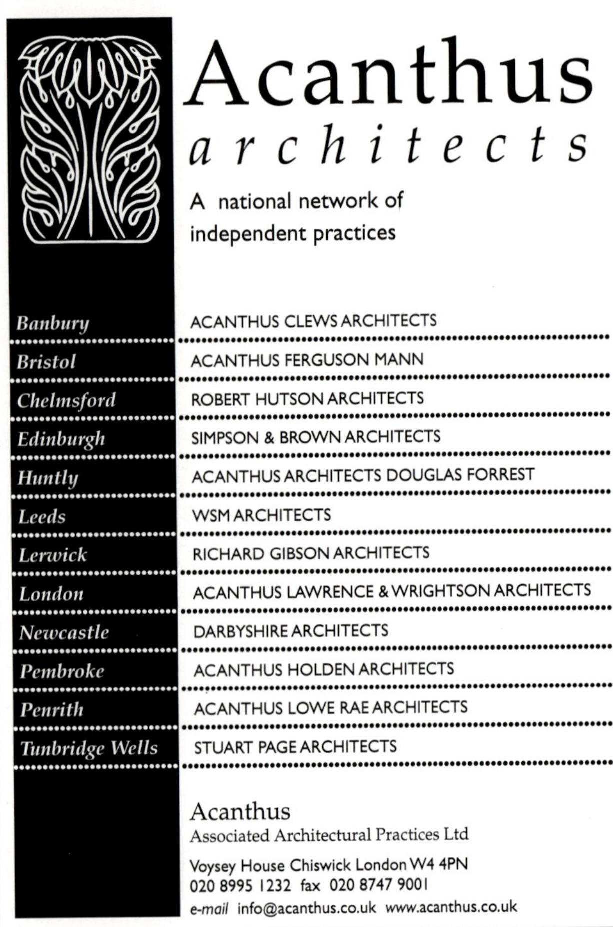

Acanthus
architects
A national network of
independent practices
Banbury ACANTHUS CLEWS ARCHITECTS
Bristol ACANTHUS FERGUSON MANN
Chelmsford ROBERT HUTSON ARCHITECTS
Edinburgh SIMPSON & BROWN ARCHITECTS
Huntly ACANTHUS ARCHITECTS DOUGLAS FORREST
Leeds WSM ARCHITECTS
Lerwick RICHARD GIBSON ARCHITECTS
London ACANTHUS LAWRENCE & WRIGHTSON ARCHITECTS
Newcastle DARBYSHIRE ARCHITECTS
Pembroke ACANTHUS HOLDEN ARCHITECTS
Penrith ACANTHUS LOWE RAE ARCHITECTS
Tunbridge Wells STUART PAGE ARCHITECTS
Acanthus
Associated Architectural Practices Ltd
Voysey House Chiswick London W4 4PN
020 8995 1232 fax 020 8747 9001
e-mail info@acanthus.co.uk www.acanthus.co.uk

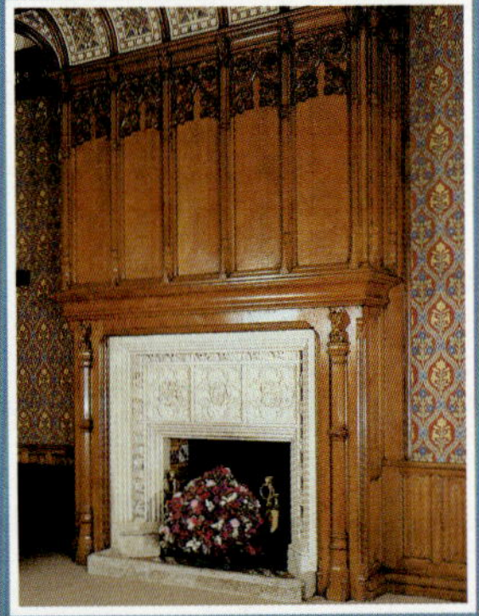
The Pugin Library, Bilton Grange

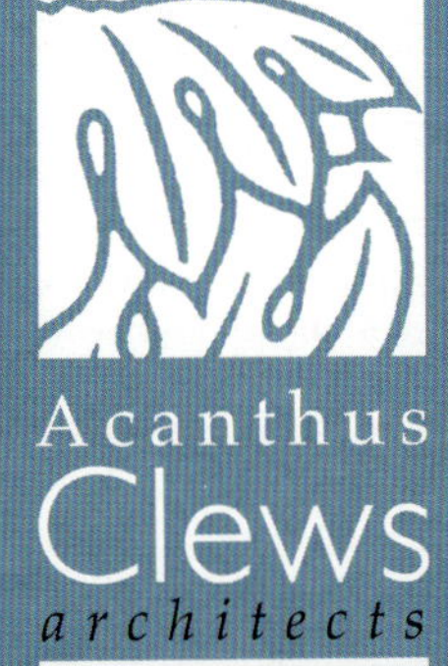

▶ A & Q (CONSTRUCTION MANAGEMENT) LIMITED

**The Hop Pole, 32 Pitfield Street, London N1 6EU
Tel 020 8880 0850 Fax 020 8880 0707
E-mail andrew.jackson10@virgin.net**

CHARTERED ARCHITECTS, QUANTITY SURVEYORS AND CONSERVATION SURVEYORS: The practice specialises in the repair and conservation of buildings and interiors offering expertise in stone, brick, timber, lime plaster, render and all types of traditional roofing. They also advise on funding and all aspects of finance. Projects vary from full repair contracts to preparation of costed feasibility studies. Clients include The Church of England, City of Westminster and several registered social landlords. Projects include conservation of church buildings and refurbishment of listed/conservation area housing estates of architectural interest.

▶ A R P LORIMER AND ASSOCIATES

**11 Wellington Square, Ayr, Scotland KA7 1ET
Tel 01292 289777 Fax 01292 288896
E-mail architects@arpl.co.uk Website www.arpl.co.uk**

CONSERVATION ARCHITECTS: The firm has a long tradition of conservation works to historic buildings and scheduled ancient monuments. A R P Lorimer is currently handling a wide variety of projects throughout Scotland, from Orkney to Galloway and throughout the Western Isles involving structures which date from the 12th to 19th centuries. Clients include the Church of Scotland, NTS, BPTs, local and national authorities along with private individuals. The firm has in-house quantity surveyors and planning supervisors, and maintains close working relationships with specialist engineers, archaeologists and other conservators. Patrick Lorimer has attained RIAS Conservation Accreditation.

▶ ALLEN TOD ARCHITECTS

**The Studio, 32 The Calls, Leeds LS2 7EW
Tel 0113 244 9973 Fax 0113 242 3687**

ARCHITECTS: Formed in 1977, Allen Tod Architects has established an enviable record in conservation, the re-use of historic buildings and the design of new buildings in conservation areas. The firm's work encompasses nationally the whole field of conservation from the very highest quality conservation repair work to the rescue and imaginative re-use of historic 'wrecks'. In particular it has developed a rare level of expertise in formulating feasibility studies and other projects that successfully combine conservation and architectural intervention with funding packages and grant assistance from a wide variety of sources; skills appreciated by clients including English Heritage, the National Trust, The Civic Trust and the Government of Japan.

▶ AMBO ARCHITECTS

**377 Kennington Road, London SE11 4PT
Tel 020 7735 2888 Fax 020 7735 2966 E-mail paul@amboarchitects.demon.co.uk
Contact Paul Baker AA Dipl GradDiplCons(AA) RIBA
▸ Forum House, Caledonian Road, Chichester, West Sussex PO19 2EN
Tel 01243 782254 Fax 01243 785019 E-mail jane@ambo-architects.demon.co.uk
Contact Jane Jones-Warner BA (Hons) DipArch RIBA FRSA**

DESIGN AND CONSERVATION ARCHITECTS: Affiliated practices, based in Chichester and London, specialising in repairs, alterations and additions to historic buildings. Recent projects include conservation work to listed college buildings in West London and Suffolk, new-build houses in Surrey and Hampshire conservation areas, church repairs and inspections, refurbishments and extensions to listed private properties.

▶ ANDERSON AND GLENN

**Yew Tree Nurseries, Frampton West, Boston, Lincolnshire PE20 1RQ
Tel 01205 724047 Fax 01205 723792
E-mail glenngaa@aol.com
Contact Mary Anderson BSc AADipCons AABC IHBC RIBA**

See also: profile entry in Horticultural Consultants section, page 39.

ARCHITECTS

▶ ANTHONY BLACKLAY & ASSOCIATES

120 Hospital Street, Nantwich, Cheshire CW5 5RY
Tel 01270 610050 Fax 01270 610273

ARCHITECTS, HISTORIC BUILDINGS AND LANDSCAPE CONSULTANTS: The practice has special expertise in all aspects of the conservation of historic buildings and landscapes, parks and gardens. As architects, they also deal with the alteration, conversion and extension of existing buildings and the design of new buildings in an historic context. As urban design and landscape consultants they also deal with conservation area environmental enhancement schemes and similar projects. Projects include the restoration of Biddulph Grange Garden, and the Apprentice House at Styal for the National Trust; both awarded Europa Nostra Diplomas of Merit. Clients include the National Trust, The Churches Conservation Trust, English Heritage, and local authorities, churches and private individuals.

▶ ANTHONY SHORT AND PARTNERS

34 Church Street, Ashbourne, Derbyshire DE6 1AE
Tel 01335 342345 Fax 01335 300624
E-mail asap@ashbourne1.demon.co.uk

CONSERVATION ARCHITECTS: Established in 1966, Anthony Short and Partners specialises in conservative repair and reuse of historic buildings. The firm's client list includes private individuals, church and cathedral bodies, trusts, societies, country estates and local authorities. It is currently appointed architects for some 80 churches, several grade I or II*, in three dioceses, and has successfully completed numerous church re-orderings and repairs. Secular work includes sympathetic listed building repair and refurbishment, conversions and extensions, surveys and feasibility and conservation studies. CAD facilities are used where appropriate. Client satisfaction and attention to detail are the highest priorities.

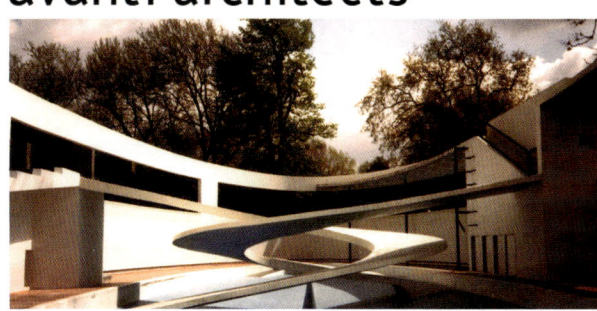

▶ ARCHITECTON

The Wool Hall, 12 St Thomas Street, Bristol BS1 6JJ
Tel 0117 910 5200 Fax 0117 926 0221

CHARTERED ARCHITECTS, DESIGN AND CONSERVATION: The practice undertakes projects of new work, adaptive reuse and conservation. Schemes range from delicate specialist repairs to complex new buildings. Work includes interior design, landscaping and fittings design with careful attention to detail at all levels. Architecton works through the whole South West region and has special interest in sustainable solutions and earth buildings. Clients include local authorities, private clients, commercial organisations, the National Trust, The Churches Conservation Trust, English Heritage and various trusts and housing associations.

▶ ARROL & SNELL LTD

St Mary's Hall, St Mary's Court, Shrewsbury SY1 1EG
Tel 01743 241111 Fax 01743 241142
E-mail andrew.arrol@which.net

ARCHITECTS: A highly experienced practice responsible for many complex and demanding conservation projects involving high quality repair and sensitive new work. Clients include English Heritage, Landmark Trust, Vivat Trust, Chester Cathedral, churches in three dioceses, numerous local authorities, private and corporate owners. Services include grant and funding applications, feasibility studies, condition surveys, conservation plans, building archaeology and research, landscape and garden design.

▶ AVANTI ARCHITECTS LIMITED

1 Torriano Mews, London NW5 2RZ
Tel 020 7284 1616 Fax 020 7284 1555
E-mail aa@avantiarchitects.co.uk

CONSERVING 20TH CENTURY BUILDINGS: *See also: display entry above.*

▶ BAILY•GARNER

146-148 Eltham Hill, London SE9 5DY Tel 020 8294 1000 Fax 020 8294 1320
Contact Lisa J Brooks BSc (Hons) DipBldgCons ARICS

BUILDING SURVEYORS, ARCHITECTS, QUANTITY SURVEYORS AND PROJECT MANAGERS: *See also: profile entry in Surveyors section, page 45.*

▶ BARLOW SCHOFIELD PARTNERSHIP

41 Church Street, Helston, Cornwall TR13 8GT
Tel 01326 563395 Fax 01326 564445

CHARTERED DESIGN AND CONSERVATION ARCHITECTS: The practice is involved in all aspects of building conservation work, including repairs, alterations, extensions, design of new buildings in the context of historic buildings, and conversions to new uses. In addition to design services, the practice also undertakes condition surveys, feasibility studies and grant applications, and advises local authorities on conservation matters including development control. Conservation projects are led by a partner with an MA in Architectural Conservation, whose experience covers all grades of listed buildings together with Scheduled Ancient Monuments. Clients include the National Trust, Cornwall Buildings Preservation Trust, MoD, local authorities and private individuals.

▶ BENJAMIN TINDALL ARCHITECTS

17 Victoria Terrace, Edinburgh EH1 2JL
Tel 0131 220 3366 Fax 0131 220 3535
E-mail BenjaminTindallArchitects@btinternet.com

DESIGN AND CONSERVATION ARCHITECTS: A medium-sized practice working in a traditional, professional manner. Projects are undertaken throughout the UK and cover town and country, small and large, houses and offices, arts and commerce. Repairs get to the root of the problem using best methods and workmanship. Alterations are carried out with appropriately imaginative and sympathetic style. Conservation plans are prepared with a deep and well-researched appreciation. The practice actively supports the SPAB in Scotland, the National Trust for Scotland and the RIAS.

▶ BLAMPIED AND PARTNERS LTD

A member of the Areen Design Group of Companies
Areen House, 282 King Street, London W6 0SJ
Tel 020 8563 9175 Fax 020 8563 9176
E-mail arlette@blampied.co.uk
Website www.blampied.co.uk

CHARTERED ARCHITECTS, PROJECT MANAGERS AND PLANNING SUPERVISORS: The practice has 40 years experience of working with clients on refurbishment, alteration and conversion of listed buildings and non-listed buildings in conservation areas. Particularly experienced in hotel and residential projects and most recently involved in the sympathetic conversion of office premises back to single occupancy houses liaising closely with English Heritage and other authorities.

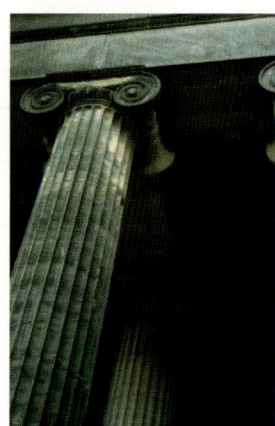

▶ BROCK CARMICHAEL ARCHITECTS

Federation House, Hope Street, Liverpool L1 9BS
Tel 0151 709 1087 Fax 0151 709 6418
E-mail office@bcalpool.demon.co.uk
Contact David Watkins, DipArch, AADiplCons, RIBA

ARCHITECTS AND PLANNERS: Brock Carmichael Architects is an award winning practice in the conservation, refurbishment, extension and adaptation of historic and listed buildings, and the appropriate and imaginative design and integration of new buildings within sensitive and historic sites. Operating on a wide geographical basis, a comprehensive range of services is provided including building surveys and analysis, feasibility studies, masterplanning, full architectural and planning services and interior design. Clients include the National Trust, Leeds City Council, Huddersfield MBC, Stockport MBC, Pilkington Properties and Liverpool Anglican Cathedral.

▶ THE BROWN MATTHEWS PARTNERSHIP

11A High Street, Warwick CV34 4AS
Tel 01926 495141 Fax 01926 410134
E-mail brownmatthews@btinternet.com

CHARTERED ARCHITECTS AND HISTORIC BUILDING CONSULTANTS: The practice has extensive experience in the specialised field of conservation and repair of historic buildings and structures. It is known for its sensitive work and innovative design with private and public listed buildings of all types. The practice's portfolio includes many scheduled monuments, listed buildings, churches and country houses of national and international significance for example: Shire Hall, Warwick; Tregothnan House, Cornwall; Kings High School, Warwick; Penshurst Place, Kent; Christchurch Spire, Coventry.

▶ PEREGRINE BRYANT

The Courtyard, Fulham Palace, Bishop's Avenue, London SW6 6EA
Tel 020 7384 2111 Fax 020 7384 2112

ARCHITECTURE AND BUILDING CONSERVATION: Peregrine Bryant has been running his own practice since 1994. The firm concentrates on historic buildings and conservation but also carries out new building work. Clients include the National Trust, the Landmark Trust and the Crown Estate as well as private owners of historic buildings.

ARCHITECTS

▶ BUILDING DESIGN PARTNERSHIP

PO Box 4WD, 16 Gresse Street, London W1A 4WD
Tel 020 7462 8000 Fax 020 7462 6342 Contact Tim Leach
▶ Sunlight House, PO Box 85, Quay Street, Manchester M60 3JA
Tel 0161 834 8441 Fax 0161 832 4280 Contact Ken Moth
▶ Also at Sheffield, Glasgow, Belfast, Dublin

HISTORIC BUILDING CONSULTANTS, ARCHITECTS, ENGINEERS, COST CONSULTANTS: BDP is multi-specialist, its reputation is based on a wide range of skills and broad knowledge of many building types. We understand the complexities and sensitivities involved in the conservation and the creative adaptive re-use of historic buildings, and its experience ranges from the repair of individual buildings to the regeneration of derelict areas in our historic cities. Completed projects include the award-winning Royal Opera House and National Maritime Museum, London, and the Elizabethan Town House, Plas Mawr, Conwy, together with the Round Tower, Windsor Castle and the Museum of Science and Industry, Manchester. BDP's portfolio of current projects includes the Royal Albert Hall and the Royal College of Music, London, Liverpool Ropewalks and the Historic Dockyard, Chatham.

▶ BYROM CLARK ROBERTS

117 Portland Street, Manchester M1 6EH
Tel 0161 236 9601 Fax 0161 236 8675
▶ Jubilee House, West Bar Green, Sheffield S1 2BT
Tel 0114 275 7878 Fax 0114 272 8954

ARCHITECTS, SURVEYORS AND ENGINEERS: Byrom Clark Roberts' Historic Buildings Division offers comprehensive consultancy services for historic buildings and conservation projects. The firm has experience in complicated issues of 'appropriate' re-use as well as traditional maintenance and repair techniques for stone and timber framed structures. They are members of the SPAB, IHBC and EASA and are church architects for over 80 parishes in four dioceses. Byrom Clark Roberts advises on grant aid and sources of funding together with VAT to listed buildings. Their work conforms with BS 7913 1988 'Principles of the conservation of historic buildings'. Contact Ian Lucas in Manchester or Alex Roberts in Sheffield.

▶ CARDEN & GODFREY ARCHITECTS

9 Broad Court, Long Acre, London WC2B 5PY
Tel 020 7240 0444 Fax 020 7836 2244 E-mail vcab@cardenandgodfrey.demon.co.uk
Partners: Ian Stewart Dip Arch Dip Cons (AA) RIBA
Richard Andrews MA Dip Arch RIBA
Ian Angus Dip Arch RIBA
Russell Taylor D Arch Dip Cons (AA) RIBA IHBC FRSA

ARCHITECTS: Specialists in all aspects of historic architecture: conservation, repairs, new buildings in sensitive sites, sympathetic alterations and additions, interior design and landscape design. A sound technical knowledge with a scholarly approach to historic detail and innovative design. Clients include the National Trust, English Heritage, colleges, churches, commercial developers and private clients, on projects ranging from small to large.

▶ CHARLES KNOWLES DESIGN

80-82 Chiswick High Road, London W4 1SY
Tel 020 8742 8322 Fax 020 8742 8655

CHARTERED ARCHITECTS: CKD were established in 1984 to concentrate on high quality architectural design. From pure conservation, through refurbishment and additions to listed and historic properties, their work also includes contemporary design which is often in an historical context. In all instances they believe that intelligent planning, good design, sound construction and the greatest attention to detail produce permanent, stylish and timeless solutions. The practice has built up a reputation for work on country houses and historic houses on the old London family owned estates. Awards and commendations include most recently the 1998 Runner up Structural Brick Award.

▶ CHEDBURN DESIGN & CONSERVATION

Bath Brewery, Toll Bridge Road, Bath BA1 7DE
Tel 01225 859999 Fax 01225 859343 Website www.chedburn.co.uk

CONSERVATION ARCHITECTS: This small practice, with technical and CAD backup, specialises in works to historic buildings throughout the country. Their experience covers private houses to churches, public buildings to monuments, including works for the National Trust and Court Service.

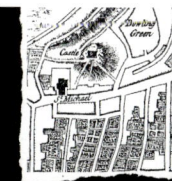

▶ CHRIS ROMAIN ARCHITECTURE

Griffin Mews, 22 High Street, Fordingbridge, Hants SP6 1AX
Tel 01425 650980 Fax 01425 650978
E-mail chris.romain@virgin.net Website www.chrisromain.co.uk

ARCHITECTS: Over 25 years experience, specialising in ecclesiastical repairs and conservation and public buildings. Traditional design, drawing services (CAD), church inspections, contract administration, design and build. The geographical area covered includes Devon to Sussex and Dorset to Worcestershire.

▶ CHRISTOPHER RAYNER ARCHITECTS

Apple Cross House, 52 The Rise, Sevenoaks, Kent TN13 1RN
Tel 01732 461806 Fax 01732 461824
Principal Christopher Rayner BA MArch (California) RIBA

CHARTERED ARCHITECTS: Christopher Rayner Architects is a small architectural practice specialising in all aspects of work to churches and other historic buildings. Projects have included conservation, repairs and sympathetic alterations/extensions to Wealden hallhouses, post-medieval domestic buildings, barns, churches and buildings at risk.

▶ CLAGUE

62 Burgate, Canterbury, Kent CT1 2BH
Tel 01227 762060 Fax 01227 762149

CONSERVATION ARCHITECTS: Medium sized consultancy established in 1936, with a branch office in Ashford. Clague provide a full range of architectural and landscape design services alongside their highly experienced conservation specialists. Clague operate an in-house quality management system. Clients range from private individuals to large corporations and they are appointed to many parish churches, whom they advise in all aspects of remodelling, alteration and maintenance. Have carried out work for the Dean and Chapter of Canterbury Cathedral and The King's School, Canterbury in respect of repair and adaptation of their historic town centre buildings. The practice specialises in the repair and alteration of historic country houses and hotels.

ARCHITECTS

▶ CoDA CONSERVATION

No 2 Harewood Yard, Harewood, Leeds LS17 9LF
Tel 0113 288 6766 Fax 0113 288 6765

CHARTERED ARCHITECTS AND ENGINEERS: The practice offers combined architect/engineer expertise in restoring and adapting redundant old buildings for new uses, and in designing new work in historic contexts, using traditional materials and skills.

▶ DAVID ASHTON HILL ARCHITECTS

18 Tavistock Terrace, Islington, London N19 4DB
Tel 020 7272 8399 Fax 020 7281 8369
E-mail ashton.hill@clara.net
Website www.ashton.hill.clara.net
Principal David Ashton Hill AA Dip (Cons) RIBA FRSA

ARCHITECTS: This practice specialises in the conservation and rehabilitation of historic buildings of all three listed grades, on rural and urban sites. Experience in this field over 20 years includes consultancy in conservation areas and CAPS, preparation of action plans and feasibility studies for repair and improvement work to historic buildings and churches, implementation of grant aided work for local authorities and consultation and assessment of funded work. The practice offers a full architectural service for private clients and public bodies, quinquennial surveys and management plans, design for the disabled, landscaping, garden and furniture design. They work in a multi-disciplinary team with community artists and other specialist disciplines.

▶ DAVID BROWN AND PARTNERS

51 High Street, Hampton, Middlesex TW12 2SX
Tel 020 8941 2112 Fax 020 8941 1742

ARCHITECTS: The practice has received nine awards since 1981 for careful works to existing structures and new buildings in sensitive architectural situations. Recent award winning projects include new buildings associated with The Great Barn, Harmondsworth, Middlesex, a scheduled ancient monument within a listed site; the construction and re-ordering for Kingston University of Dorich House, Kingston, a listed three storey sculptor's studio built in 1937, and the refurbishment of the listed Castle Stables, Sunbury on Thames, to create two houses.

▶ DAVID GIBSON ARCHITECTS

35 Britannia Row, London N1 8QH
Tel 020 7226 2207 Fax 020 7226 6920
E-mail DGibArch@aol.com
Website www.DGibArch.co.uk

ARCHITECTS: David Gibson Architects is a practice committed to the art of architecture and the design of good buildings. Specialising in work to listed buildings and buildings in conservation areas, it is guided by the philosophy of sympathetic interaction between good modern interventions and existing structures. The practice has an established reputation with the national and regional heritage bodies and amenity societies and its work includes the refurbishment of listed buildings such as St Luke's Church, Battersea, Ovington Square, Vestry House Museum and Sainsbury's, Streatham. The practice is able to advise on town planning, technical, space planning and aesthetic issues for listed and other buildings.

▶ DAVID HARVEY ARCHITECTS

Abergavenny, Monmouthshire NP7 5SE
Tel 01873 857232 Fax 01873 859344
E-mail dharchs@aol.com

CHARTERED ARCHITECTS: Established in 1980, the practice has extensive experience of conservation work in South Wales and the Marches, including condition surveys, quinquennial inspections, the restoration, repair and adaptation of Listed Buildings, Industrial Archeological projects and the design of new buildings in sensitive locations. Recent work includes Pontywaun Historic Town Scheme, repairs to Great Castle House, Monmouth, Llanyrafon Farm Conservation Plan and a feasibility study for the Hendref Building Preservation Trust in Blaenavon. A member of SPAB, the practice offers their clients a committed, efficient and personal service.

▶ DAVID LEWIS ASSOCIATES

Delf View House, Church Street, Eyam, Sheffield S32 5QH
Tel 01433 630030 Fax 01433 631972
Contact David Lewis B Arch MA (Architectural Building Conservation) RIBA

CHARTERED ARCHITECTS AND CONSERVATION SPECIALISTS: The practice, established 1978, has an enviable reputation for the careful restoration and conservation of historic buildings, sympathetic conversions, and for new designs and extensions in historic contexts. Current work includes the conservation of a Grade I listed half-timbered country house, the conversion of an early 20th century synagogue to a Baptist church and conference centre and the extension of a 19th century school at one of Derbyshire's great country estates along with quinquennial inspections of churches dating from 1350. Expert witness services are provided for planning appeals, public enquiries and building disputes. Clients include the National Trust and churches, along with private and commercial owners of listed buildings.

▶ DAVID PITTS CHARTERED ARCHITECTS

12a The Waits, St Ives, Cambs PE27 5BY
Tel 01480 466213 Fax 01480 493330
E-mail dpittsriba@aol.com
▶ PO Box 8, Oundle, Northants PE8 5JQ
Tel 01780 470170 Fax 01780 470800

CHARTERED ARCHITECTS: Established since 1976, the practice specialises in the conservation and repair of historic buildings, including agricultural, horticultural, residential and ecclesiastical fabric, their upgrading, refurbishment and alteration as necessitated by the client brief and current usage.

▶ DAVIES SUTTON ARCHITECTURE LIMITED

30 Cowbridge Road, Pontyclun, Glamorgan CF72 9EE
Tel 01443 225205 Fax 01443 238965
E-mail office@davies-sutton.co.uk
Website www.davies-sutton.co.uk

CONSERVATION ARCHITECTS: Specialising in the care and conservation of historic buildings and churches to preserve their patina and character, Davies Sutton Architecture aspires to being Wales' leading historic buildings practice, providing sensitive and practical conservation, and new buildings in a sensitive context. The practice has a commitment to quality, attention to detail, good management and continuous learning. Members and field workers for SPAB, providing a highly personal and dedicated service.

▶ DEAN & CHEASON – ARCHITECTS

4 Gwenllian Terrace, Trefforest, Pontypridd, Wales CF37 1DN
Tel/Fax 01443 208046 or 029 2070 0949
E-mail rspencdean@aol.com

ARCHITECTURE AND BUILDING CONSERVATION: Award winning practice having extensive experience of sensitive conservation, refurbishment and adaptation of historic buildings and structures. Work ranges from a 13th century castle at Llantrisant to industrial archaeology and non-conformist chapels. The practice offers a comprehensive, efficient and personal service including detailed surveys, funding plans, liaison with Cadw and acts as Planning Supervisors. Members of the Institute of Historic Building Conservation, SPAB, Ancient Monuments Society and Agoriad Community Architects.

▶ DEREK ROGERS ASSOCIATES

Church Square, 48 High Street, Tring, Hertfordshire HP23 5AG
Tel 01442 824298 Fax 01442 890616

CHARTERED ARCHITECTS: Founded in 1980, the practice has extensive experience in repairs, alterations and extensions to historic buildings, be they listed or simply of local vernacular construction. New buildings, designed in an historic context, is a major area of their work as is condition survey or measured survey and recording of standing structures. Buildings and street scene improvement in conservation areas, using CAD, linked to electronic survey when appropriate, is as much a part of their work as is energy-conscious and historically sympathetic design. Apart from the historic aspects of their work, their design skills encompass all aspects of housing design, arcadia, education and conference and management training buildings, medical and commercial projects.

ARCHITECTS

▶ DONALD INSALL ASSOCIATES LTD

19 West Eaton Place, Eaton Square, London SW1X 8LT
Tel 020 7245 9888 Fax 020 7235 4370
Website www.insall-lon.co.uk
▶ with branch offices in Canterbury, Shrewsbury, Chester, Cambridge and Bath

ARCHITECTS AND PLANNING CONSULTANTS: The practice has over 40 years experience in the care and adaptation of historic buildings and towns, as well as the design of new buildings for sensitive sites. Its projects have now been recognised in more than 90 design and construction awards. London and branch offices provide a comprehensive service across a wide geographical area, including building surveys and site analysis, conservation plans, development and feasibility studies, grant aid, planning and listed building consent negotiations, fire protection and security advice, services integration in old buildings, all basic architectural services, and contemporary and historic interior design.

▶ DREW-EDWARDS KEENE

150 Upper New Walk, Leicester, LE1 7QA
Tel 0116 254 5015 Fax 0116 254 8019
E-mail rjwooddek@aol.com

CONSERVATION ARCHITECTS: Drew-Edwards Keene specialise in conservation work but also undertake refurbishment and new-build projects, throughout the North, Midlands, central and east England. They assist churches and private clients with feasibility studies, building appraisals, planning advise, grant applications, design, contract administration, and as a Planning Supervisor.

▶ JAMES DUNNETT MA, Dipl Arch (Cantab), RIBA

142 Barnsbury Road, London N1 0ER
Tel 020 7833 3451 Fax 020 7833 2126

ARCHITECT: Experienced in the conservation, extension, and improvement of 18th, 19th and 20th century houses, and listed social housing. Winner of Kensington and Chelsea Environment Award for Restoration and Conversion 1994; SPAB member; committee member of the Twentieth Century Society and DOCOMOMO-UK.

▶ ELAINE RIGBY ARCHITECTS

33 Chapel Street, Appleby-in-Westmorland, Cumbria CA16 6QR
Tel/Fax 017683 52572
E-mail elaine.rigby@virgin.net
Contact Elaine Rigby, MA(Conservation Studies York) DIP ARCH IHBC RIBA

ARCHITECTS, CONSULTING ENGINEERS AND LANDSCAPE CONSULTANTS: Experts on the repair and conservation of historic buildings, structures and landscapes, particularly medieval and 17th century vernacular buildings, ruins and churches. Elaine Rigby Architects applies a strong conservation philosophy of conservative repair using traditional materials, and has considerable expertise in the use of lime and stonework consolidation.

▶ FEILDEN + MAWSON

36 Grosvenor Gardens, London SW1W 0EB
Tel 020 7730 8880 Fax 020 7730 8881
E-mail london@feildenandmawson.com
▶ 1 Ferry Road, Norwich NR1 1SU
Tel 01603 629571 Fax 01603 633569
E-mail norwich@feildenandmawson.com
▶ Horningsea Road, Fen Ditton, Cambridge CB5 8SZ
Tel 01223 294017 Fax 01223 293458
E-mail cambridge@feildenandmawson.com

ARCHITECTS, PLANNERS AND PROJECT MANAGERS: Leaders in the conservation field since 1957. The firm has won awards in conservation work with new uses for old buildings and the design of modern buildings. Working on over 20 Grade I listed buildings and scheduled ancient monuments, clients include: Somerset House Trust, Stanhope Plc, Court Service, Historic Royal Palaces, PACE, Cabinet Office, Crown Estate, British Academy, University of Sussex, English Heritage, National Trust, University of London, Chatham House, Royal Opera House Developments and Railtrack.

▶ GIBBON, LAWSON, MCKEE LIMITED

41A Thistle Street Lane South West, Edinburgh EH2 1EW
Tel 0131 225 4235 Fax 0131 220 0499

ARCHITECTS: GLM combines the disciplines of architecture with those of building surveying and project management. Good sensible building surveyors' skills are combined in this practice with architects' design flair, the architecture department being headed up by David Roulston, with 25 years of experience behind him. His sensitivity towards vernacular architecture is matched by a lively and ingenious design sense and a robust determination to satisfy the client's requirements. The office operates the full range of Computer Aided Design facilities. The combination disciplines is a particular strength of the practice in undertaking sensitive building conservation. *See also: profile entries in Surveyors section, page 46 and Fire Protection Consultants section, page 182.*

▶ GILES QUARME & ASSOCIATES

41 Cardigan Street, London SE11 5PF
Tel 020 7582 0748 Fax 020 7587 0576
E-mail mail@quarme.com
Website www.quarme.com

CONSERVATION ARCHITECTS AND PLANNING CONSULTANTS: The team is led by Giles Quarme, RIBA FRSA, an architect and art historian with considerable planning experience acting as expert witness at public inquiries. The practice has worked on a wide variety of listed and historic buildings which range from churches and hospitals to art galleries and offices. It prides itself on providing the same care and attention to detail in the repair of a small terraced house as for a large country mansion. It aims to provide radical solutions which combine new ideas with traditional cost conscious conservation methods. They are currently restoring the Indian High Commissioner's residence and have recently converted the stables at Althorp for the Princess Diana Memorial Museum and have been advising Foster and Partners on their millennium design for the British Museum. *See also: display entry on page 31.*

ARCHITECTS

▶ GILMORE HANKEY KIRKE LTD

GHK House, Heckfield Place, 526 Fulham Road, London SW6 5NR
Tel 020 7736 8212 Fax 020 7736 0784
E-mail architects@ghkint.com
Website www.ghkint.com/architects

ARCHITECTS, PLANNERS AND CONSERVATION SPECIALISTS: Gilmore Hankey Kirke (GHK), established in 1973, with offices in London and Plymouth, brings a high degree of professionalism and enthusiasm to every project. GHK does not believe in standard solutions; each project is examined from first principles to achieve the optimum solution. In dealing with listed buildings the company adopts a pragmatic approach within the framework of an understanding of the cultural significance of the historic fabric and location with which it is dealing. This allows for creative reuse and innovation whilst respecting the past. GHK's work on the restoration of Lulworth Castle, Dorset for English Heritage received an RICS 2000 Conservation Award.

▶ GRAHAM HOLLAND ASSOCIATES

4 King Street, Knutsford, Cheshire WA16 6DL
Tel 01565 651066 Fax 01565 755265
▸ Plas Draw, Ruthin, Denbighshire LL15 1RT
Tel 01824 704709 Fax 01824 704912

ARCHITECTS AND HISTORIC BUILDING CONSULTANTS: The practice's policy is for sensitive design and conservation. Formed in 1986, now with three associates, technical and secretarial staff, principally in the care and repair of historic buildings and new buildings in historic settings. Graham Holland is a member of the Chester and Liverpool DACs and Cathedral Architect at Bangor. Clients include the Historic Chapels Trust, National Trust, Duchy of Lancaster, Church of England, Church in Wales, Methodist, URC and Unitarian churches, and private estates. Work has included: Hawarden, Gwydir and Halton Castles, Ecclesfield St Mary, Todmorden Unitarian Chapel & Lodge, Balderstone St Mary and Tutbury Priory.

▶ GRAY, MARSHALL & ASSOCIATES

23 Stafford Street, Edinburgh EH3 7BJ
Tel 0131 225 2123 Fax 0131 225 8345

CHARTERED ARCHITECTS: Established in 1972, the practice has a proven track record and particular interest in conservation work. Recent work includes Heriot's Hospital and Victoria Terrace, Edinburgh, and the Dower House, Corstorphine. Working throughout Scotland, with clients who include building preservation trusts, churches, Edinburgh World Heritage Trust, housing associations and private owners, Gray, Marshall & Associates has considerable experience of listed buildings, including sympathetic adaptation and reuse, new buildings on sensitive sites, complex funding packages, phased restoration projects, church quinquennials and conservation plans.

▶ H G P GREENTREE ALLCHURCH EVANS LIMITED (HGP)

Furzehall Farm, Wickham Road, Fareham, Hampshire PO16 7JH
Tel 01329 283225 Fax 01329 237004
E-mail hgp-architects.co.uk

ARCHITECTS AND TOWN PLANNERS: Founded in 1968, HGP offers innovative design and rapid response to clients' requirements. HGP's experience includes major new-build and historic projects nationally, winning the RICS/The Times Conservation Award for their own 17/18th century Grade II listed offices. Other projects include conversion of Grade II/Grade II* Eastney Barracks, Portsmouth to 270 residential units; restoration of Portsmouth Dockyard's scheduled monument Storehouse 11 into the Royal Naval Museum; conversion of listed Fort Picklecombe, Plymouth into 121 flats; and masterplanning of Gunwharf Quays, Portsmouth with repairs to scheduled monuments and listed buildings including £2.5 million repairs to Vulcan Building. HGP's work includes conservation of churches, large historic houses and commercial buildings, with English Heritage, employing archaeologists, historians and specialist contractors.

▶ HAIGH ARCHITECTS

29 Lowther Street, Kendal, Cumbria LA9 4DH
Tel 01539 720560 Fax 01539 723570
E-mail rh@haigharchitects.co.uk
Website www haigharchitects.co.uk
Contacts EM Bottomley, B Arch, ARIBA CR Haigh, BA, Dip Arch, RIBA

DESIGN AND CONSERVATION ARCHITECTS: The practice specialises in restoring and adapting old redundant buildings for new uses in historic contexts, using traditional materials and skills to blend the new with the old. A sensitive approach is taken to the repair, restoration and conservation of historic buildings, from country houses to traditional cottages, churches to museums. It has received several Civic Trust Awards and Commendations. Clients include private owners of historic properties, the National Trust, Churches Conservation Trust, Diocese of Carlisle, English Heritage and civic society building preservation trusts.

▶ THE HALPERN PARTNERSHIP LTD

The Royle Studios, 41 Wenlock Road, London N1 7SH
Tel 020 7251 0781 Fax 020 7251 9204
E-mail info@halpern.com Website www.halpern.com

ARCHITECTURE, PLANNING, URBAN DESIGN, HERITAGE: Halpern combines architecture, town planning, urban design and conservation expertise to create improved urban environments. Still one of the few practices to have architects and town planners working literally side by side, Halpern has over 50 years experience of realising its designs as successful buildings. The firm offers a comprehensive accredited and qualified conservation service which has secured national awards. Work includes new development in conservation areas, and repair, conversion and extension of listed buildings, reconciling important conservation principles with commercial viability. Listed building work in Westminster and the City of London includes cinemas and theatres in the West End, and office, hotel, residential and mixed-use projects.

▶ HAWKES EDWARDS & CAVE

1 Old Town, Stratford-upon-Avon, Warwickshire CV37 6BG
Tel 01789 298877 Website www.hawkesedwards.com

ARCHITECTS: Over 45 years of experience in dealing with listed historic buildings and churches, where a sensitive, skilled and creative approach is required, both in repair and alteration. They act, amongst others, for the Churches Conservation Trust and the Ancient Monuments Society.

▶ THE HERITAGE PRACTICE

27 Grange Road, Saltford, Nr Bristol BS31 3AH
Tel/Fax 01225 400066

ARCHITECTS: A specialist practice working on historic buildings, waterways and garden structures. Current commissions include: Environment Agency, British Waterways and Duchy of Cornwall.

▶ HOK INTERNATIONAL LIMITED

(incorporating Cecil Denny Highton)
216 Oxford Street, London W1R 1AH
Tel 020 7636 2006 Fax 020 7636 1987

ARCHITECTS: HOK Conservation and Cultural Heritage aims to create environments that keep intact our history and values by preserving fine buildings and maintaining the historic urban texture; respect historic buildings as irreplaceable parts of a region's cultural heritage; and retain a building's integrity whilst providing sensible solutions for new occupants. These aims are guiding current projects; a new research centre beside London's famous Natural History Museum; three listed buildings in Whitehall forming the Cabinet Office; conserving fine interiors at the Ministry of Defence; writing a Guide to the care and development of Railtrack's architectural heritage; repairing and conserving five lodge buildings and proposals for a visitor centre at Ashton Court in Bristol; repair and conservation of the cemetery at Nunhead.

ARCHITECTS

▶ JAMES BROTHERHOOD & ASSOCIATES LIMITED

Golly Farm, Golly, Burton, Rossett, Wrexham LL12 0AL
Tel 01244 579000 Fax 01244 571133
E-mail jba-architects.co.uk
Website www.jba-architects.co.uk

CONSERVATION ARCHITECTS: Energetic and successful practice with the resources to undertake development projects throughout the British Isles. Specialists in the adaptation and care of historic buildings and in the integration of new buildings in conservation areas. Based on over 25 years experience, they achieve practical and high quality solutions, acceptable to clients, local authorities, English Heritage, Cadw and the Historic Monuments Commission. Services include building and land surveys, feasibility and development studies, planning negotiations, lottery and grant advice, interior design, building control matters, CDM and fire and security advice. The firm has received top Europa Nostra, Conservation and Civic awards and its work is included in HRH The Prince of Wales', 'A Vision of Britain'. Full CAD facilities.

▶ JOHN D CLARKE & PARTNERS

2 West Terrace, Eastbourne, East Sussex BN21 4QX
Tel 01323 411506 Fax 01323 410064
E-mail JDC_ARCH@compuserve.com

CHARTERED ARCHITECTS AND HISTORIC BUILDING CONSULTANTS: The practice was established in 1909 and one of the first projects was the restoration of the 13th Century Lamb Inn at Eastbourne. The practice has developed a reputation for sensitive conservation and repair of historic buildings as part of a portfolio which includes work for local authorities, building societies, hotels and churches. The practice looks after Harveys Brewery in Lewes winning a Civic Trust Commendation for its Brewery Tower extension, and has won a Friend of Lewes Award for the restoration of a large Jacobean fire damaged house. The practice deals with a great variety of historic work from the very large, down to the restoration of a Victorian cast iron drinking fountain.

▶ JOHN STARK & CRICKMAY PARTNERSHIP

13 & 14 Princes Street, Dorchester, Dorset DT1 1TW
Tel 01305 262636 Fax 01305 260960
E-mail jscp@johnstark.co.uk
Website www.johnstark.co.uk

CONSERVATION ARCHITECTS: Established over 150 years ago the practice has a long history in the conservation, repair and remodelling of listed buildings and ancient monuments. Notable projects include the repair of Ince Castle, Cornwall after its total devastation by fire in 1988, ongoing works of repair and remodelling at Wardour Castle, Wiltshire in conjunction with English Heritage, and the conservation of the remains of the Roman town house in Dorchester for which the practice received the Dorset Archaeological Award 2000. Other recent conservation awards were received in 1990, 1992 and 1994. The practice acts for over 90 churches in Dorset and the surrounding counties.

▶ JOHNSTON AND WRIGHT

15 Castle Street, Carlisle, Cumbria CA3 8TD
Tel 01228 525161 Fax 01228 515559
E-mail jw@jwarchitects.co.uk

CHARTERED ARCHITECTS: Founded in 1885 and with over 50 awards for good design, Johnston & Wright have an extensive track record in new design, alterations/repair and regeneration of historic buildings and their surroundings. Clients include, private owners of historic properties, National Trust, English Heritage, churches, commercial developers and local authorities.

▶ JONATHAN RHIND ARCHITECTS

The Old Rectory, Shirwell, Barnstaple, Devon EX31 4JU
Tel 01271 850416 Fax 01271 850445
E-mail jonathan@jonathan-rhind.co.uk
Website www.jonathan-rhind.co.uk

CONSERVATION ARCHITECTS: The practice specialises in the careful repair and sensitive alteration or extension of historic buildings from barns, cottages and farmhouses to churches, mansions and castles, with an imaginative approach to new design and a keen eye for detail. Also, for new uses to existing historic buildings and new buildings within sensitive settings.

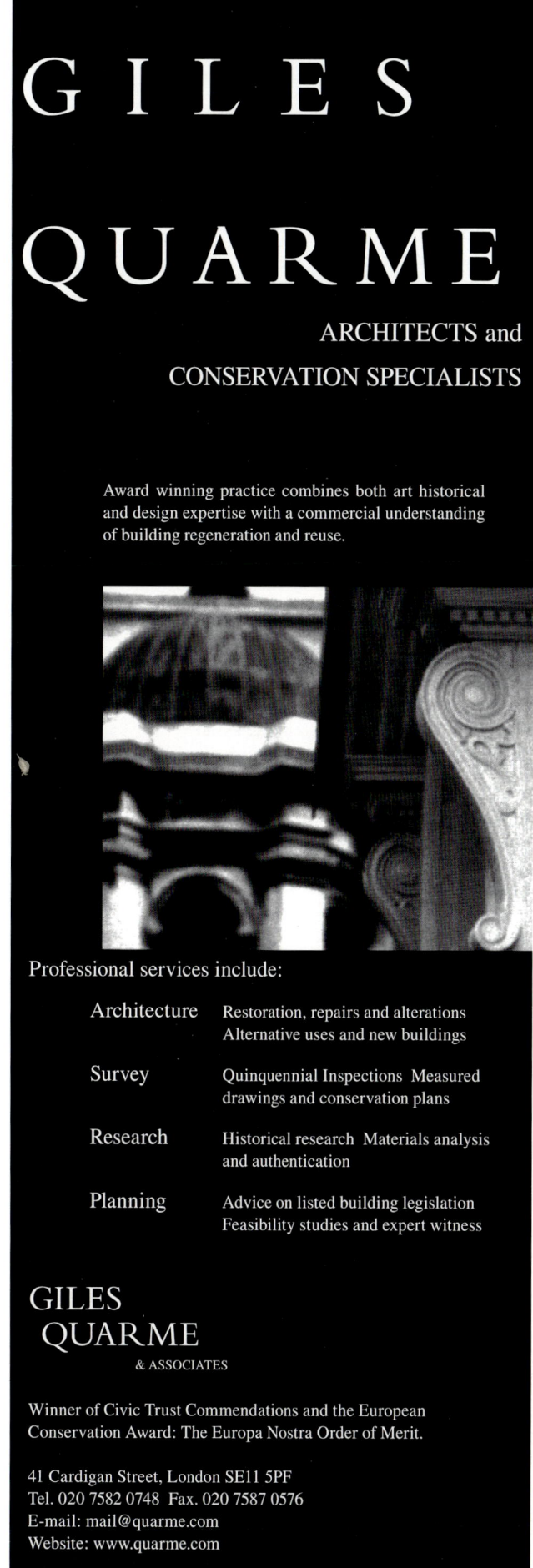

▶ JULIAN HARRAP ARCHITECTS

95 Kingsland Road, London E2 8AG
Tel 020 7729 5111 Fax 020 7739 8306

DESIGN AND CONSERVATION ARCHITECTS: Julian Harrap Architects are a medium sized specialist practice offering a full range of architectural services for the repair and restoration of historic buildings, estates and landscapes and for new buildings in historic settings. They work with buildings of all ages from scheduled ancient monuments to 20th century icons. All projects large or small are planned within a careful management structure. Established in 1975 they have a reputation for scholarly conservation and attention to fine detail. Clients include the National Trust, English Heritage, The Royal Academy of Arts, Sir John Soane's Museum, and many notable churches. The practice has received many national and international awards.

▶ LATHAM ARCHITECTS

St Michael's, Queen Street, Derby DE1 3SU
Tel 01332 365777 Fax 01332 290314

DESIGN AND CONSERVATION ARCHITECTS, PLANNING CONSULTANTS, LANDSCAPE ARCHITECTS: A highly experienced firm, working with scheduled monuments, historic and ecclesiastical buildings, including structural and condition surveys, inspections, conservation and repair work and creative-reuse of redundant buildings. Winners of many conservation awards and increasingly involved in the wider planning issues of conservation and finding uses for old buildings, together with designing modern buildings in sensitive areas. Clients include The Royal Household, Duchees of Lancaster and Cornwall, MOD, English Heritage, National Trust, local authorities and many small agencies, including historic building preservation trusts.

▶ LAW & DUNBAR-NASMITH

16 Dublin Street, Edinburgh EH1 3RE
Tel 0131 556 8631 Fax 0131 556 8945 E-mail architects@ldne.co.uk
▶ 29 St Leonards Road, Forres IV36 1EN
Tel 01309 673221 Fax 01309 676397 E-mail architects@ldnf.co.uk

ARCHITECTS: Established in 1957, the practice has enjoyed a long association with some of the most historically important buildings of Scotland, and recent projects are spread across the whole of the UK. Their experience encompasses cathedrals, churches, law courts, castles, country houses, theatres, museums and galleries, ancient monuments and the many small buildings which characterise both towns and countryside. They are skilled in the historical analysis of buildings and sites, and in the preparation of conservation plans. Their current client list includes Historic Scotland, the National Trust for Scotland, the Phoenix Trust and several other historic building preservation trusts.

▶ LINCOLN & CAMPBELL ASSOCIATES LTD

11 Carlton Street, Edinburgh EH4 1NE
Tel 0131 332 4888 Fax 0131 332 3384
▶ 47 Roderick Road, London NW3 2NP
Tel 020 7485 7442 Fax 020 7267 9779
E-mail orancampbell@lincoln-campbell.co.uk

ARCHITECTURE AND CONSERVATION: Oran Campbell and Richard Lincoln trained under Philip Jebb RIBA in the 1970s. Each has 30 years experience of historic buildings repair and conservation. Recent workload has included historic buildings, new institutions, houses and garden design in Scotland, England, Spain, Australia and the United States. New or scholarly traditional design in detail, always aiming for harmony with setting and landscape. Other languages: Spanish and Italian.

▶ JULIAN R A LIVINGSTONE CHARTERED ARCHITECT

BAHons DipArch(Leic) GradDiplConservation(AA) RIBA
Dahlia Cottage, Vicarage Lane, Upper Swanmore, Hampshire SO32 2QT
Tel/Fax 01489 893399

CHARTERED ARCHITECT AND HISTORIC BUILDINGS CONSULTANT: Specialising in the sensitive repair, conversion, extension and conservation of country and town houses in private ownership. Julian Livingstone concentrates on domestic buildings, producing designs with an emphasis on imaginative solutions, within the constraints of historic buildings and materials, to delight, excite and stimulate. He advises clients on the most appropriate methods of caring for their buildings and has received several awards for sympathetic and imaginative conversion of agricultural buildings to residential use. Julian Livingstone serves on the Portsmouth Diocesan Advisory Committee, is a member of SPAB, ASCHB and an associate member of IHBC.

▶ LLOYD EVANS PRICHARD

5, The Parsonage, Manchester M3 2HS
Tel 0161 834 6251 Fax 0161 832 1785
E-mail post@lep-architect.co.uk

CHARTERED ARCHITECTS: Lloyd Evans Prichard are general practitioners with proven skills in conservation and restoration. Backed by the broad experience of their professional staff they are able to combine expertise in historic building work and conservation plans with an extensive commercial background to tailor services offered to individual client requirements, including CDM and project management. Their successful approach to conservation work, under the guidance of Director John Prichard, Surveyor to the Diocese of Manchester, has led to commissions from: English Heritage; the Holy Name Church, Manchester; Capesthorne Hall; St Michael's, Ashton-under-Lyne, John Rylands Library, Manchester, Manchester City Council at Heaton Hall and Park and Carlisle Castle.

▶ T C R MacMILLAN-SCOTT

11 Lansdowne Road, Alton, Hants GU34 2HB
Tel/Fax 01420 549233

CHARTERED ARCHITECT

ARCHITECTS

▶ MANSFIELD THOMAS & PARTNERS

81 Albany Street, Regent's Park, London NW1 4BT
Tel 020 7224 4446 Fax 020 7935 8991
▶ **Little Heath Farm, Berkhamsted, Hertfordshire HP4 2RY Tel/Fax 01442 864951**

ARCHITECTS: This small practice offers a full range of conservation, architectural, town planning, surveying, quantity surveying, landscaping and historical investigation with the benefit of over 50 years experience. Committed to achieving high quality design standards for each client's complex needs – to the best solution for each commission - combining artistic expression disciplined to the brief, to 'value for money' and to the programme. Recent projects include Crown Estate quinquennial surveys, restoration, refurbishment and repairs to many Grade I listed buildings including Ockwell's Manor, Maidenhead; Upton Court, Slough; Childwicksbury, St Albans; Cumberland and Chester Terraces and York Terrace East, Regent's Park; 7 Palace Green and 15 Kensington Palace Gardens; and Trevor Hall and Bettisfield Park, North Wales. (Associated practice – John Moore and Associates, Barnet, Hertfordshire).

▶ MARTIN STANCLIFFE ARCHITECTS

29 Marygate, York YO30 7WH
Tel 01904 644001 Fax 01904 623462 E-mail post@msarchitects.co.uk

ARCHITECTS AND CONSERVATION CONSULTANTS: The practice has 20 years experience of caring for, conserving and restoring a diverse range of nationally important historic buildings including cathedrals, churches, country houses, colleges, museums and scheduled ancient monuments. Extensive remodelling and extensions to listed and historic buildings especially in sensitive locations and adaptive re-use of old and redundant buildings, form part of the practice's portfolio together with feasibility studies and specialist advice to a range of clients. These include the National Trust, English Heritage, building preservation trusts, The Landmark Trust, local and county authorities, cathedrals, churches, country estates and private owners of historic properties.

▶ MICHAEL DRURY ARCHITECTS

St Ann's Gate, The Close, Salisbury, Wiltshire SP1 2EB
Tel 01722 555200 Fax 01722 555201
E-mail stannsgt@globalnet.co.uk

ARCHITECTS, CONSERVATION AND DESIGN: Michael Drury Architects specialise in high quality architectural design within historic settings, ranging from pure conservation, new interventions to listed buildings and completely new design work. Whatever the project, their approach to each one is specifically tailored to its situation. In some instances, the new work will be designed to integrate seamlessly with the existing setting. In other cases, the new interventions will clearly spring from today. Current projects include alterations and new extensions to listed private properties, repair and reordering of parish churches, and major conservation and enhancement programmes for cathedrals.

▶ NICHOLAS GROVES-RAINES ARCHITECTS LTD

Liberton House, 73 Liberton Drive, Edinburgh EH16 6NP
Tel 0131 467 7777 Fax 0131 467 7774
E-mail mail@nicholas-groves-raines-architects.co.uk

ARCHITECTS: A rot ridden tenement, a vandalised mansion, an Adam style house collapsed above first floor level and ruined castles are all architectural challenges of the highest order which, over the last few years the practice has undertaken. Projects which others might find daunting have been embraced, nurtured and brought to completion, including several historic churches for which new uses have been found. The award winning practice has developed a range and variety of skills which whilst respecting our architectural heritage are very much of the present and looking to the future.

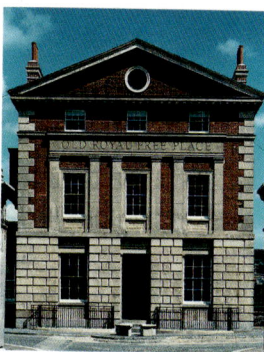

▶ PWP ARCHITECTS

Newnham House, 61 South Street, Havant, Hants P09 1BZ
Tel 023 9248 2494 Fax 023 9248 1152
E-mail design@pwp-architects.com
Contact David Blunden BA (Hons) BArch (Hons) RIBA

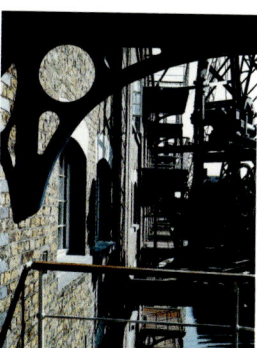

ARCHITECTS, SURVEYORS, PLANNING CONSULTANTS: Established in 1921 the practice handles the conservation, restoration and re-development of historic buildings in sites of special landscape interest, working with a broad range of conservators, landscape architects, archaeologists and environmentalists. PWP's clients use the firm repeatedly because it meticulously balances budget and time constraints against essential and urgent conservation needs. The practice has special expertise adapting and extending Grade I and Grade II buildings at risk, and negotiating viable new uses with English Heritage and the amenity societies. The practice carries out developments across the country and undertakes feasibility studies and concept designs for UK based clients in France.

▶ MICHAEL PEARCE ARIBA MRTPI FRSA IHBC

The Lodge, 52 Hollows Close, Salisbury SP2 8JX
Tel/Fax 01722 334355

CONSERVATION CONSULTANT: Clients include English Heritage, local planning authorities, landed estates, and other owners. He has advised on the future of a number of important historic buildings and sites, and has appeared at many public local inquiries. His contacts extend throughout the country, and he woks closely with other leading professional firms.

▶ PETER CODLING ARCHITECTS

7 The Old Church, St Matthews Road, Norwich, Norfolk NR1 1SP
E-mail pcodling@globalnet.co.uk
Tel 01603 660408 Fax 01603 630339

ARCHITECTS: Church repairs, re-ordering and extensions; quinquennial reports. Repair and conversion of buildings of all ages and types. Housing for individual clients and special needs groups.

▶ PETER YIANGOU ASSOCIATES

Puckham Barn, Whittington, nr Cheltenham, Glos GL54 4EX
Tel 01242 821031 Fax 01242 820193
E-mail admin@yiangou.com Website www.yiangou.com

ARCHITECTS AND SURVEYORS: PYA has covered the Cotswolds and surrounding counties for 20 years building up in the process considerable experience with country properties, conservation areas, listed buildings and natural stone construction. Projects include new country and manor houses in stone, indoor pools, extensions and repair of listed buildings, new Oxford College buildings, small quality housing developments, museums, corporate HQ in traditional buildings; please see website for examples. Full CAD capability with eight full time staff.

▶ PIDDUCK & WHITTAKER

39 St Johns Hill, Shrewsbury, Shropshire SY1 1JQ
Tel 01743 360015 Fax 01743 360041
E-mail pw.architect@btinternet.com

ARCHITECTS, SURVEYORS AND SPECIALISTS IN HISTORIC BUILDING CONSERVATION: Pidduck & Whittaker specialises in the conservation of historic buildings, together with the design of new buildings in sensitive locations. The practice is responsible for a number of churches in both the Hereford and Lichfield dioceses. Clients have included the National Trust as well as a number of private and commercial organisations.

▶ POLLARD THOMAS & EDWARDS ARCHITECTS

Diespeker Wharf, 38 Graham Street, London N1 8JX
Tel 020 7336 7777 Fax 020 7336 0770
E-mail name.surname@ptea.co.uk Website www.ptea.co.uk

ARCHITECTS: Pollard Thomas and Edwards combine conservation and restoration experience with urban regeneration expertise. Their masterplan for London's Shad Thames area paved the way for the regeneration of that whole quarter between Tower Bridge and St Saviours Dock. They also won Europa Nostra conservation awards for two buildings in this area, New Concordia Wharf and the Anchor Brewhouse and several others. Pollard Thomas and Edwards work with listed buildings and in conservation areas has included dwellings for housing associations as well as landmark historic buildings. *See also: display entry in this section, page 33.*

▶ RICHARD GRIFFITHS ARCHITECTS

14/16 Cowcross Street, London EC1M 6DG
Tel 020 7251 6334 Fax 020 7490 2251

ARCHITECTS: The practice combines the highest standards of conservation work with the highest quality of new building in a historic context, and has been awarded Civic Trust and Europa Nostra awards. Richard Griffiths Architects also carry out conservation and development plans for historic buildings in support of lottery bids. The practice has worked extensively on buildings owned by the National Trust, English Heritage, the Church, and local authorities, providing access and use by the community. The practice is responsible for the £5 million Millennium Project at Southwark Cathedral and for the new glazed courtyard at Lambeth Palace.

▶ ROBERT SEYMOUR CONSERVATION

The Merchants House, 10 High Street, Totnes, Devon TQ9 5RY
Tel 01803 865568 Fax 01803 834722

CONSERVATION ARCHITECTS AND HISTORIC BUILDING CONSULTANTS: The practice has over 25 years experience carrying out sympathetic, appropriate repairs to a wide range of historic and listed buildings. It has strong links with English Heritage and SPAB, working with private clients, local authorities, charitable trusts, almshouse associations, churches and other groups. With other offices in Dartmouth and London, the firm's architects, qualified and experienced in building conservation, are able to carry out detailed surveys, evaluations and repair programmes. Robert Seymour Conservation combines technical expertise with innovative and sensitive design by working with a network of specialists as well as providing sensitive solutions to new or historic building projects, often in sensitive urban conservation areas, throughout the south of England.

▶ ROBIN KENT ARCHITECTURE & CONSERVATION

Newtown Street, Duns, Berwickshire TD11 3AS
Tel 01361 884401 Fax 01361 884402 Mobile 07051 884401
E-mail rk@robinkent.com Website robinkent.com
Contact Robin Kent BAHons DiplArch(Oxford) MACons(York) RIBA ARIAS IHBC

ARCHITECTURE AND CONSERVATION CONSULTANCY: Founded in 1981 and established in Berwickshire in 1997, specialising in building design in sensitive locations, ancient monument and listed building conservation work, conservation area character appraisals, historic building evaluations, building surveys and investigations, disability access, research and education. RIAS conservation accredited to highest level.

▶ ROGER JOYCE ASSOCIATES

39 Bouverie Square, Folkestone, Kent CT20 1BA
Tel 01303 246400 Fax 01303 246455
Contact Roger A Joyce DipArch (Cant) DipConservation (AA) RIBA

ARCHITECT: Workload includes repair and refurbishment of historic buildings and design for new uses. The practice has won awards for conversions of redundant agricultural buildings and new build. The principal is a Diocesan approved inspecting architect. Works include Grade I listed churches, Institutional clients including further education colleges, Blue Circle, Union Rail, Kent Trust for Nature Conservation. Roger Joyce serves on the Executive Committee of Kent Building Preservation Trust, amenity groups, the Education Committee of ICOMOS UK, member of SPAB, EASA, and IHBC and also lectures at the Lille School of Architecture.

ARCHITECTS

▶ ROGER MEARS ARCHITECTS

**2 Compton Terrace, London N1 2UN Tel 020 7359 8222 Fax 020 7354 5208
E-mail rma@rogermears.com Website www.rogermears.com**

ARCHITECTS: Founded in 1980, the practice has built up a reputation for sensitive work to historic and domestic buildings, guided by the principles of the SPAB. Past work includes alterations and repairs to listed houses in London, Dorset, Wiltshire and Oxfordshire. Among them are Tudor House, Cheyne Walk, listed Grade II* and formerly Rossetti's house and studio, and a terrace of Grade I listed houses in Newington Green, London, dating from 1658. Current work includes refurbishment of houses in London, a Grade II* house in Richmond designed by Sir William Chambers, and a Sir George Gilbert Scott church in Forest Gate.

▶ EDWARD SARGENT CONSERVATION ARCHITECT

**Heritage House, 79-80 High Street, Gravesend, Kent DA11 0BH
Tel 01474 535221 Fax 01474 564857**

ARCHITECT: Small architectural practice specialising in repairs and alterations to historic buildings and the design of new buildings in historic environments.

▶ SIMPSON AND BROWN ARCHITECTS

**St Ninian's Manse, Quayside Street, Edinburgh EH6 6EJ
Tel 0131 555 4678 Fax 0131 553 4576
E-mail admin@simpsonandbrown.co.uk Website simpsonandbrown.co.uk**

ARCHITECTS AND HISTORIC BUILDING CONSULTANTS: Established in1977, S&B aspires to be one of Scotland's leading historic buildings practices and regularly undertakes work in England. The firm has a particular interest in the conversion and reuse of historic buildings and is committed to simple wholesome new building design, fusing sound ecological principles and ethics with experience of traditional materials and techniques. S&B's work ranges from straightforward but high quality repairs to the most careful conservation and scholarly restoration work. Associated firm, Addyman and Kay, undertakes archaeology and building investigations. S&B is committed to conscientious conservation, to green and beautiful buildings and to caring for the needs and interests of its clients.

▶ SPENCE & DOWER ARCHITECTS

**Column Yard, Cambo, Morpeth, Northumberland NE61 4AY
Tel 01670 774448 Fax 01670 774446**

CHARTERED ARCHITECTS: The practice, established in 1946, has been responsible for work to both grand and modest historic houses, castles, churches and historic industrial buildings in the North of England. Grade I Elsdon Tower major re-ordering, the consolidation of Harbottle Castle, Woodhouses Bastle and Low Cleughs Bastle all within the Northumberland National Park. The consolidation of extensive standing structures at Nenthead Leadmines near Alston, Cumbria has led to other work for the North Pennines Heritage Trust. Robin Dower is responsible for the inspection and maintenance of a number of churches in Newcastle Diocese and is Chairman of Durham Cathedral Fabric Advisory Committee.

▶ STAINBURN TAYLOR ARCHITECTS

**Bideford House, Church Lane, Ledbury HR8 1DW
Tel 01531 634848 Fax 01531 633273
E-mail architects@stainburn-taylor.co.uk**

ARCHITECTS: A practice with a wide range of work encompassing historic ecclesiastical and secular projects which range from a cathedral through to private housing. The practice has developed considerable expertise in the diagnosis and conservation of buildings. Clients include English Heritage, the National Trust, Ironbridge Gorge Museum Trust together with numerous churches in Herefordshire, Worcestershire, Shropshire and Gloucester cathedral.

▶ STUART PAGE ARCHITECTS

**Forge House, The Green, Langton Green, Tunbridge Wells, Kent TN3 0JB
Tel 01892 862548 Fax 01892 863919 E-mail stuart.page@member.riba.org**

ARCHITECTS AND INTERIOR DESIGNERS: Historic buildings require economic and appropriate uses to ensure their survival. Stuart Page Architects undertake architectural and interior design projects for new buildings and conservation and repair of historic and listed buildings. They believe the architect's role to be especially important when working in conservation areas or with historic buildings to ensure sympathetic buildings integrated with their surroundings and that satisfy the client's brief. Current projects include work for the National Trust, Historic Royal Palaces, English Heritage and private owners of historic buildings.

▶ ANTHONY SWAINE FSA, FRIBA, FASI, IHBC

**The Bastion Tower, 16 Pound Lane, Canterbury, Kent CT1 2BZ
Tel 01227 462680 Fax 01227 472743**

CONSERVATION ARCHITECT AND CONSULTANT: Private practice: restoration of historic buildings, conservation areas and conservation in general. Advisor: Thanet District Council and other authorities for historic buildings. Member of Council and Technical Panel of Ancient Monuments Society. Representative for International Council on Monuments and Sites. Patron: Venice in Peril. Author: *Faversham Conserved* and *Margate Old Town*. Architectural consultant for Historic Churches Preservation Trust and responsible during last war for care of Canterbury Cathedral. Past part-time teacher – History of Architecture and Construction. Past part-time listing, Ministry of Housing, Local Government. Past member Churches Conservation Trust, member of Friends of Friendless Churches, lecturing.

▶ THOMAS FORD & PARTNERS

**177 Kirkdale, Sydenham, London SE26 4QH
Tel 020 8659 3250 Fax 020 8659 3146 E-mail tfp@thomasford.co.uk
Partners: Paul Sharrock, BSc, DipArch(UCL), RIBA
Daniel Golberg, MPhil(Nottm), BArch(Nottm), RIBA, ACI Arb, FRSA
Clive England, BA Hons, DipArch(Sheffield), RIBA**

CHARTERED ARCHITECTS AND SURVEYORS: Established in 1926, the practice has extensive experience of historic building projects up to £14 million including churches, domestic buildings, museums, palaces, military buildings and structures. Work includes feasibility studies, quinquennial inspections, conservation, repair, extensions, remodelling and new buildings in historic settings. The practice's portfolio includes many scheduled monuments, Grade I and Grade II* buildings of national and international significance. Clients include English Heritage, National Trust, Historic Royal Palaces Agency, Royal Household, Ministry of Defence and numerous museums and churches.

ARCHITECTS

▶ THORNBURROW HOLDSWORTH
The Eden Centre, 47 City Road, Cambridge CB1 1DP
Tel 01223 460475 Fax 01223 464142 E-mail kthornburrow@carltd.com
CHARTERED ARCHITECTS: Architectural practice specialising in the conservation, restoration and extension of historic buildings. The practice has completed works in London, Kent, and throughout East Anglia. Current works include a house dating from the 14th century, two barn conversions, and a listed Cambridge college. The practice works closely with Cambridge Architectural Research Ltd on environmental matters in historic buildings, conservation plans and researches into conservation issues.

▶ TIM BENTON ARCHITECT
33 Northgate, Sleaford, Lincs NG34 7BX
Tel 01529 304524 Fax 01529 306981
HISTORIC BUILDINGS ARCHITECTS: The practice is involved in a wide range of work, a high percentage of which concerns historic buildings. From small domestic repairs to the £2 million repair, extension and adaptation of a listed building to alternative use, the firm offers a personal professional service. Against a backbone of church work which includes quinquennial Inspections and English Heritage grant aid involvement upon programmes of work, the practice offers innovative solutions to the challenging problems posed by interesting buildings.

▶ THE VICTOR FARRAR PARTNERSHIP
57 St Peters Street, Bedford MK40 2PR
Tel 01234 353012 Fax 01234 363473
CHARTERED ARCHITECTS AND SURVEYORS: Established by Victor Farrar in 1962, the firm has proven expertise in all facets of ancient building conservation, repair and sympathetic extension. Ecclesiastical work, stone and timber frame repairs are specialities.

▶ W R DUNN & CO
27 Front Street, Acomb, York YO24 3BW
Tel 01904 784421 Fax 01904 784679
CHARTERED BUILDING SURVEYORS, ARCHITECTS AND HISTORIC BUILDINGS CONSULTANTS: *See also: profile entry in Surveyors section, page 48.*

▶ MICHAEL WALTON MA Dip Arch (Cantab), RIBA
Thriplow House, Thriplow, Cambridgeshire SG8 7RD
Tel/Fax 01763 208887
ARCHITECT: Michael Walton specialises in sympathetic alterations, extensions and repairs to listed buildings and interiors, primarily in East Anglia.

▶ WATSON BERTRAM & FELL
5 Gay Street, Bath, Somerset BA1 2PH
Tel 01225 337273 Fax 01225 448537 E-mail wbfbath@compuserve.com
ARCHITECTS AND SURVEYORS: Watson Bertram and Fell specialise in the restoration and alteration of listed buildings or new buildings in conservation areas. Projects include Abbotsbury, Dorset – estate policy for restoration of village (European Heritage Award 1975); Cliveden, Berkshire (Grade I) conversion into hotel; Royal Crescent, Bath (Grade I) conversion of two houses into the Royal Crescent Hotel and construction of new annex (Civic Trust Commendation 1986); the Sloane Club, London, extension in conservation area (RBKC Highly Commended Conservation Award).

▶ THE WHITWORTH CO-PARTNERSHIP
18 Hatter Street, Bury St Edmunds, Suffolk IP33 1NE
Tel 01284 760421 Fax 01284 704734
CHARTERED ARCHITECTS AND SURVEYORS: The original practice was founded in 1963, and reconstituted in its present form in 1985. The firm handles a wide range of work, a high proportion of which is to do with historic buildings, ranging from structural surveys and analysis of building defects for private clients, to the conservation and repair of major historic buildings including churches of all denominations. It aims to approach each project with care and sensitivity, bringing together a range of expertise most appropriate for each project. Of the three partners, Philip Orchard, a Lethaby scholar, and Matthew Stearn are chartered architects and Tony Redman, is a chartered building surveyor accredited by RICS for the conservation of historic buildings. The practice additionally has CAD experience which is used when appropriate.

VALUE ADDED TAX

Implications for historic buildings, including changes announced in May 2001

ROGER WOOD

Value Added Tax (VAT) is a significant additional cost that has to be borne by those responsible for the preservation and conservation of our architectural heritage. VAT is charged on building work of all descriptions at the standard rate (currently 17.5 per cent) unless either it can be 'zero-rated' or it can be charged at the new reduced rate of five per cent.

This new reduced VAT rate, which was introduced with effect from 12 May 2001, applies only to building work carried out in the following circumstances:

- renovation or alteration of a single dwelling which has been empty for three years or more
- conversion of a dwelling, or changing the number of dwellings (eg from two houses into one)
- conversion of a dwelling into a multiple occupation dwelling (ie flats)
- conversion of a dwelling or multiple occupation dwelling to a building intended for 'relevant residential purposes' *(see table of definitions opposite)*.

This new legislation is very complex and of course untested. It is therefore essential that the rules are considered in full when determining whether the new five per cent rate can be applied.

A new scheme is being introduced shortly to ease the VAT burden when carrying out repairs to listed places of worship. However, rather than reducing the VAT rate, there will be a grant available from Customs and Excise which will compensate for a large proportion of the VAT charged on the work. The intention is to reduce the VAT cost in this way so that the 'effective VAT rate' becomes five per cent. When introduced, it will be backdated to 1 April 2001, so it is important that those involved in current projects keep detailed records including all invoices received.

Zero-rating is not granted automatically, and the burden is placed on the owner of the building to prove that the works undertaken are not liable to VAT at the standard rate. Unfortunately, there is no general relief for historic and listed buildings. The relief is dependent on the individual circumstances of the particular project, as illustrated in the flow charts opposite.

Perhaps the most important point to bear in mind before embarking on any project is to ensure that the appropriate planning and listed building consent is granted before any work on the building is undertaken. Whilst this may seem obvious, HM Customs and Excise will not grant any zero-rating relief if the planning and listed building consents are retrospectively granted.

ROGER WOOD is a tax partner with PricewaterhouseCoopers – ***see page 38 for contact details.***

CATHEDRAL COMMUNICATIONS LIMITED

VAT on repairs to churches and other places of worship is now, in effect, five per cent

SOME USEFUL DEFINITIONS

Approved alterations:
Alterations which cannot be carried out without listed building consent and have received such consent – these do not include repairs or 'incidental alterations' which are carried out as a result of the need to repair the structure of the building
Protected building:
A listed building or scheduled monument

Dwelling:
House, flat or similar

Long lease:
A lease in excess of 21 years

Option to tax:
The option can be exercised by anybody to enable VAT to be charged on most property transactions which would otherwise be exempt, excluding transactions involving dwellings and certain other types of property

Qualifying use:
Use of the building for a relevant charitable use (see below) or a relevant residential use (see below)

Relevant charitable use:
Use by a charity for non-business activities or to provide social or recreational activities for a local community

Relevant residential use:
A home providing residential accommodation, student accomodation or personal care, including army accomodation, hospices or any institution which is the sole or main residence of 90 per cent of its residents

Substantial reconstruction:
No more than the external walls of the building remain, and/or the total cost of the approved alterations (excluding repairs) represents 60 per cent or more of the total cost of the work.

LISTED BUILDING ALTERATIONS AND RECONSTRUCTION WORK

Work to a 'protected building' (which includes listed buildings and scheduled monuments) can only be zero-rated if it is an 'approved alteration' or a 'substantial reconstruction'. In addition, the refurbished building must be used as either a dwelling, for 'relevant residential purpose' (for example, as a residential home of some kind such as a hospice, nursing home or similar) or for a 'relevant charitable purpose' (for example, as a non-business charity use such as a church). Professional fees can never be zero-rated.

For example, VAT will be charged at zero rate on materials and services used in the alteration of a qualifying listed building, such as an owner-occupied house or a church. In this case only costs related to the alteration work will be eligible. However, if the building is largely demolished, a developer will be able to recover all the VAT incurred on the redevelopment (including repairs) when he sells it.

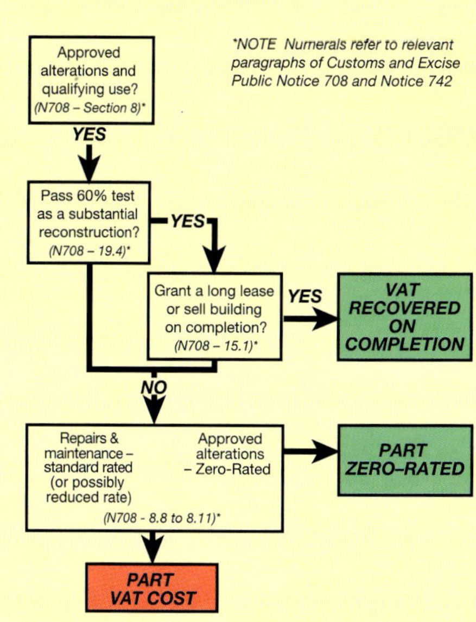

CONVERSIONS

A more general relief from VAT is provided to any non-residential building which is converted into a *dwelling* or for use solely for a *relevant residential purpose.* But to qualify for the relief, the building must be sold or leased for more than 21 years once the conversion has been completed. For example, a disused warehouse (listed or not) could be converted to flats and their sale (or long lease) could be zero-rated, thus allowing the VAT incurred on the conversion, alteration and professional fees to be recovered. There are similar rules for housing associations which allow them to request that contractors zero-rate work done for them in converting non-residential buildings to dwellings or residential homes.

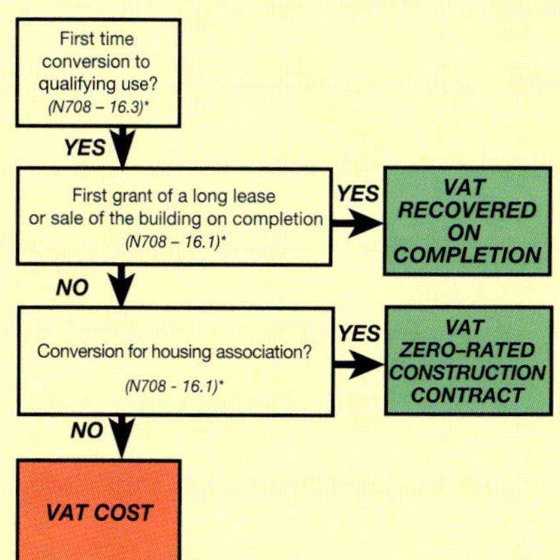

BUILDINGS IN ORIGINAL OWNERSHIP

Many old buildings are still in the ownership of the organisation that originally constructed them (such as churches, universities and hospitals). Provided a long lease can be granted, or the building can be sold, perhaps to a subsidiary company, and the building is used for a *qualifying purpose,* the work may be zero-rated, allowing recovery of the tax incurred on completion of the refurbishment.

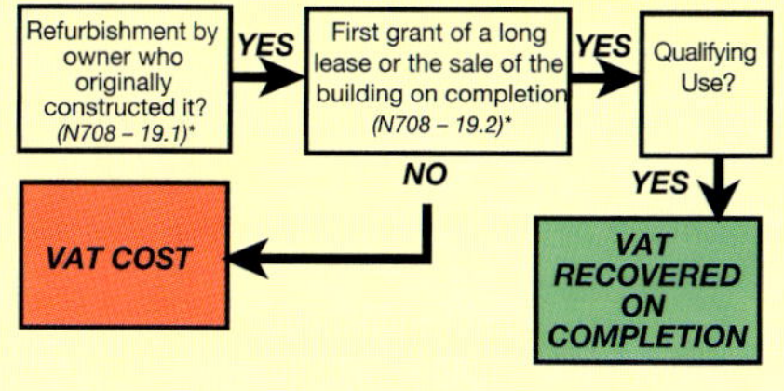

'OPTION TO TAX': PROPERTY DEVELOPMENT FOR BUSINESS USE

If the restored building is going to be sold or put to business use, then the *'option to tax'* can be made. Under such an option, VAT is charged on the sale price or lease rents, and this allows any VAT incurred on the restoration to be reclaimed.

The Government has legislated that in certain circumstances HM Customs and Excise can direct that the option to tax be disapplied for any particular transaction. The effect of such a ruling would be that VAT would not have to be charged on the sale or lease of the property, but any VAT paid on the restoration would not be recoverable.

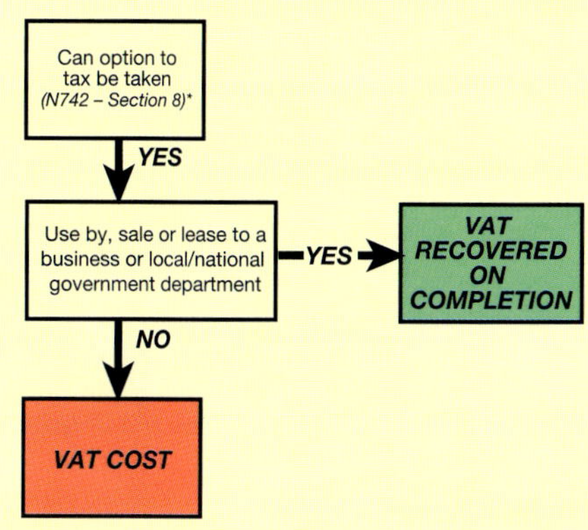

PLANNING CONSULTANTS

▶ ABERCROMBIE PLANNING AND CONSERVATION

21 Holmesdale Avenue, East Sheen, London SW14 7BQ
Tel/Fax 020 8392 2188 E-mail j1abercrombie@netscapeonline.co.uk

CONSERVATION CONSULTANT AND CHARTERED SURVEYOR: Expert planning advice tailored to development in the historic environment. The practice aims to reconcile commercial and conservation objectives through a thorough understanding of conservation legislation, philosophy, practise and case law, coupled with extensive commercial experience on major projects. Site appraisals to identify conservation considerations; historic building research; conservation area assessments and design; feasibility studies, grant aid and funding; effective liaison between applicant/decision maker/ amenity groups; planning, listed building and conservation area consent applications and appeals; expert witness. Clients include local authorities, commercial and residential developers and landowners and amenity groups. Contact Eleanor Abercrombie BSc(Hons) DipBldgCons MRICS.

▶ ANDREW MARTIN ASSOCIATES

Croxton's Mill, Little Waltham, Chelmsford, Essex CM3 3PJ
Tel 01245 361611 Fax 01245 362423
E-mail ama@amaplanning.com Website www.amaplanning.com
Contact Andrew Martin

CHARTERED TOWN PLANNERS AND URBAN DESIGNERS: AMA prides itself on its highly successful track record, carefully guiding client's proposals through the complex planning system, not only to obtain a successful development outcome, but also to ensure quality and enhancement in both urban and rural environments. AMA provides a genuinely personal service and a wide range of professional expertise. The firm is committed to a quality management system based upon the principles of ISO 9000. With its established state-of-the-art studio, experienced technicians, designers and illustrators AMA produces a variety of high quality graphics and visualisations in all media to support planning submissions.

▶ ROTHERMEL THOMAS

14-16 Cowcross Street, Smithfield, London EC1M 6DG
Tel 020 7490 4255 Fax 020 7490 1251
E-mail inquiries@rothermelthomas.co.uk

CHARTERED ARCHITECTS AND TOWN PLANNERS: Rothermel Thomas (RT), celebrating ten years in practice, have recently been appointed Historic Buildings Architects for English Partnerships' regeneration of The Royal Arsenal, Woolwich. RT have acted as architects for 6 Lothbury, St Mary's Harrow, Upper Montagu Street, Skinners' Hall, Telford Bridge St Katharine-by-the-Tower, St Ethelburga's, and 6 Grove Terrace, and have produced three reports on 'The Conservation and Management of Salisbury Cathedral Close' for English Heritage, the Cathedrals Fabric Commission for England, and Dean and Chapter. James Thomas, the principal, is very experienced at historic buildings planning inquiries, including Chepstow Castle, 57 Smithfield, Charterhouse, Letchworth Garden City, Bedford Park, and Coventry Street, Westminster.

▶ T P B PLANNING

One America Street, London SE1 0NE
Tel 020 7208 2002 Fax 020 7208 2023
E-mail mlowndes@tpbennett.co.uk Website www.tpbennett.co.uk
Contact Michael Lowndes

CONSERVATION ARCHITECTS, CHARTERED TOWN PLANNERS AND URBAN DESIGNERS: Specialist conservation area and historic building consultants providing the full range of services for the historic built environment. Conservation area proposals; historic building assessments; planning, listed building and conservation area consent applications, negotiations and appeals; expert witness. Restoration, interventions and contextual design.

LEGAL SERVICES

▶ WILSONS

Steynings House, Fisherton Street, Salisbury, Wiltshire SP2 7RJ
Tel 01722 412412 Fax 01722 333021 E-mail prf@wilsons-solicitors.co.uk
Contact Peter R FitzGerald

SOLICITORS: Wilsons provides a specialist service in heritage law. They act for some 70 agricultural estates including a great number of heritage properties and houses open to the public, as well as some 25 heritage maintenance funds and many collections of heritage exempt chattels. Their clients range from Cumberland to Cornwall.

VAT CONSULTANTS

▶ HIGHMEAD VAT CONSULTANCY

Barrons Cuckoo, Portfield Gate, Haverfordwest, Pembs SA62 3LL
Tel 01437 762278 Fax 01437 769124
E-mail fjgolden@hotmail.com

VAT CONSULTANCY: *See also: Landfill Tax Environmental Credit Scheme, page 54.*

▶ PRICEWATERHOUSECOOPERS

Churchill House, Churchill Way, Cardiff CF1 4XQ
Tel 029 2023 7000 Fax 029 2080 2404
Website www.pwcglobal.com

VAT CONSULTANCY: *See also: Value Added Tax, page 36.*

▶ THE VAT CONSULTANCY

Laurel House, Station Approach, Alresford, Hants, SO24 9JH
Tel 01962 735350 Fax 01962 735352
Website www.thevatconsultancy.com

VAT CONSULTANCY: An independent consultancy specialising in providing first class advice on VAT and other indirect taxes to professional and corporate clients alike. The Vat Consultancy specialises in advising professionals, developers and land owners on the best way to minimise the impact of VAT on any building project, be it listed or otherwise. The new rules on the lower rates of VAT for construction services and the extension of zero-rating will be of benefit to house owners.

HERITAGE CONSULTANTS

▶ BYROM CLARK ROBERTS
117 Portland Street, Manchester M1 6EH Tel 0161 236 9601 Fax 0161 236 8675
▶ Jubilee House, West Bar Green, Sheffield S1 2BT Tel 0114 275 7879 Fax 0114 272 8954
ARCHITECTS, SURVEYORS AND ENGINEERS: *See also: profile entry in Architects section, page 27.*

▶ INGRAM CONSULTANCY LTD
Netley House, Gomshall, Guildford, Surrey GU5 9QA
Tel 01483 205170 Fax 01483 205175 E-mail enquiries@ingram-consultancy.co.uk
Website www.ingram-consultancy.co.uk
ARCHITECTS, SURVEYORS AND CONSERVATION CONSULTANTS: Ingram Consultancy is a co-ordinated practice of architecture, surveying, archaeology and practical conservation. The firm carries out condition surveys of historic buildings and archaeological sites with recommendations on conservation and repair, conservation and management planning for historic sites, historic building assessments, building recording, investigation of materials failures, and training and exemplar works. Ingram Consultancy designs and delivers Building Conservation Masterclasses for West Dean College. Clients include English Heritage, The Royal Household, The States of Guernsey, local authorities, and architectural and surveying practices.

▶ RESURGAM®
Netley House, Gomshall, Surrey GU5 9QA
Tel 01483 203221 Fax 01483 202911 E-mail ei@handr.co.uk Website www.handr.co.uk
Contact Christopher Marsh RIBA
ARCHITECTURAL, ARCHAEOLOGICAL AND TECHNICAL CONSERVATION CONSULTANTS: Resurgam, a division of H+R Environmental Investigations Limited, consists of a group of experts and scientists. Resurgam carries out research, condition surveys and analysis of traditional buildings and sites, combining extensive architectural and construction experience with innovative investigative technology. Stone, mortar, plaster and decorative finishes are specialities. Remedial specifications, schedules, bills of quantities and tender procurement are provided. Resurgam provides conservation consultancy and management for refurbishment projects. Clients include The Royal Household, English Heritage, National Trust, UNESCO, Crown Estate, architects, surveyors and property managers. *See also: Hutton+Rostron profile entry in Damp & Timber Decay section, page 176.*

▶ STATS CONSULTANCY (STATS Limited, founded 1974)
Porterswood House, Porters Wood, St Albans, Herts AL3 6PQ
Tel 01727 833261 Fax 01727 835682 E-mail ian.sims@stats.co.uk
Contact Dr Ian Sims
SPECIALIST ENGINEERING, MATERIALS AND ENVIRONMENTAL CONSULTANTS: *See also: profile entry in Materials Analysis section, page 45.*

▶ TAYWOOD ENGINEERING
345 Ruislip Road, Southall, Middlesex UB1 2QX
Tel 020 8575 4283 Fax 020 8575 4044
HERITAGE CONSULTANTS: Preservation, repair and restoration of historic buildings and sensitive sites. *See also: display entry in Building Contractors section, page 68.*

LANDSCAPE ARCHITECTS

▶ ANTHONY BLACKLAY & ASSOCIATES
120 Hospital Street, Nantwich, Cheshire CW5 5RY
Tel 01270 610050 Fax 01270 610273
HISTORIC LANDSCAPE, PARKS AND GARDENS CONSULTANTS: *See also: profile entry in Architects section, page 25.*

▶ JULIAN HARRAP ARCHITECTS
95 Kingsland Road, London E2 8AG
Tel 020 7729 5111 Fax 020 7739 8306
DESIGN AND CONSERVATION ARCHITECTS: *See also: profile entry in Architects section, page 32.*

HORTICULTURAL CONSULTANTS

▶ ANDERSON AND GLENN
Yew Tree Nurseries, Frampton West, Boston, Lincolnshire PE20 1RQ
Tel 01205 724047 Fax 01205 723792 E-mail glenngaa@aol.com
Contact Mary Anderson BSc AADipCons AABC IHBC RIBA
ARCHITECTURE AND GARDENS: Anderson and Glenn are specialists in topiary, formal garden design and historic buildings. This multi-disciplinary practice of horticulturist/garden historian and conservation architect offers a unique combination of services for owners. Anderson and Glenn prepare feasibility studies, management plans, advisory reports, design proposals and implementation on both historic sites and buildings. Recent projects include works to a wide variety of listed buildings, conservation advice to local authorities, garden designs and advice to owners of sensitive and registered sites.

ENVIRONMENTAL CONSULTANTS

▶ STATS CONSULTANCY (STATS Limited, founded 1974)
Porterswood House, Porters Wood, St Albans, Herts AL3 6PQ
Tel 01727 833261 Fax 01727 835682
E-mail ian.sims@stats.co.uk
Contact Dr Ian Sims
SPECIALIST ENGINEERING, MATERIALS AND ENVIRONMENTAL CONSULTANTS: *See also: profile entry in Materials Analysis section, page 45.*

EXHUMATION OF HUMAN REMAINS

▶ CHERISHED LAND LIMITED
Robinson House, Robinson Road, Crawley, West Sussex RH11 7AD
Tel 01306 627321 Fax 01306 621357
E-mail information@cherishedland.com
Website www.cherishedland.com
EXHUMATION SPECIALISTS: Cherished Land provides a comprehensive exhumation service for professionals contemplating the disturbance of human remains, advising on every aspect of this complex and emotive subject including public relations, legal, ecclesiastical and health requirements, budget costings, site clearance and final disposition. Evaluation and advice provided without charge or obligation.

STRUCTURAL ENGINEERS

▶ ADRIAN COX ASSOCIATES

The Studio, 3 Bayham Road, Sevenoaks, Kent TN13 3XA
Tel 01732 462640 Fax 01732 740893
E-mail engs@adriancox.co.uk Website www.adriancox.co.uk

CONSULTING CIVIL AND STRUCTURAL ENGINEERS: A small practice of engineers with a proven track record of work to ancient buildings. Adrian Cox Associates are committed to achieving the best for the building using experience and modern analysis methods, coupled with traditional repair techniques and materials as appropriate. The practice is also dedicated to achieving the best for their clients by working efficiently and using plain English. Projects have included numerous historic buildings including houses, churches and schools.

▶ AUSTIN TRUEMAN ASSOCIATES

30 Britton Street, London EC1M 5TF
Tel 020 7490 2885 Fax 020 7250 3834
▶ 8 Spicer Street, At Albans, Herts AL3 4PQ
Tel 01727 858752 Fax 01727 852376
E-mail engineers@austintrueman.co.uk

CIVIL AND STRUCTURAL ENGINEERS: Established in 1973 Austin Trueman Associates specialise in the renovation and repair of existing buildings especially those with an architectural and or historic interest. The firm operates a philosophy of sympathetic engineering through a unique blend of technical flair and creative problem solving. Projects include Brocket House, Knebworth House, Hatfield House, Somerset House, The Royal Naval College and The Palace of Westminster. Offices in London and St Albans and associated offices in Jersey, Paris and Prague. Member firm of the Association of Consulting Engineers and British Consultants Bureau.

▶ BILL HARVEY ASSOCIATES

85 Pennsylvania Road, Exeter EX4 6DW
Tel 01392 499934 Fax 0870 458 0295
E-mail bill@obvis.com

STRUCTURAL ENGINEER: Experience ranges from domestic scale to large bridges. Assessment and conservation of masonry and timber structures. Specialist analysis of complex masonry and other structures; usually able to prove the structure is stable. Clients include Historic Scotland, English Heritage, the Landmark Trust, numerous local authorities.

▶ BLACKETT-ORD CONSULTING ENGINEERS

33 Chapel Street, Appleby-in-Westmorland, Cumbria CA16 6QR
Tel/Fax 017683 52572
Contact Charles Blackett-Ord, CEng FICE FConsE and
Elaine Rigby, DipArch, MA (Con Studies, York) RIBA

CIVIL AND STRUCTURAL ENGINEERS AND ARCHITECTS: The practice works throughout the country on repair and conservation of historic buildings and ancient monuments, including listed railway viaducts, 18th century grand houses, churches, medieval buildings and ruins. The Lambley Viaduct repairs won the RICS Award for Craftsmanship in Building Conservation in 1997. The use of traditional materials both in repairs and new work is encouraged and the practice has considerable expertise, in particular, in the use of lime mortars and renders.

▶ THE BUDGEN PARTNERSHIP

56 Lisson Street, London NW1 5DF
Tel 020 7224 8887 Fax 020 7224 8883
E-mail budgenp@aol.com
Website www.budgenpartnership.com

STRUCTURAL ENGINEERING CONSULTANTS: The Budgen Partnership was founded in 1960 acting for clients in all sectors. They have earned a significant reputation for works to historic buildings. Involvement with many fine buildings over the years has engendered a close working relationship with English Heritage. Important works include the Foreign Office, Royal Albert Hall, Parliament Hill Mansions, St Pancras Chambers and many smaller projects. They have received awards and commendations in recognition of design excellence in both new building and conservation work.

▶ BYROM CLARK ROBERTS

117 Portland Street, Manchester M1 6EH
Tel 0161 236 9601 Fax 0161 236 8675
▶ Jubilee House, West Bar Green, Sheffield S1 2BT
Tel 0114 275 7879 Fax 0114 272 8954

ARCHITECTS, SURVEYORS AND ENGINEERS: *See also: profile entry in Architects section, page 27.*

▶ CAMERON TAYLOR BEDFORD

Lorne Close, London NW8 7JJ
Tel 020 7262 7744 Fax 020 7724 0917
E-mail c.richardson@camerontaylor.co.uk

CONSERVATION ENGINEERS: Specialists in the survey, repair and development of buildings, large and small. Expert advice for planning inquiries and litigation. Clients include the National Trust, the Church Commissioners, and the Crown Estate. CTB's dedicated conservation team is known to English Heritage, and led by Clive Richardson, Engineer to Westminster Abbey. Offices also at Solihull and Leeds.

▶ DAVID NARRO ASSOCIATES

36 Argyle Place, Edinburgh EH9 1JT
Tel 0131 229 5553 Fax 0131 229 5090
E-mail mail@davidnarro.co.uk Website www.davidnarro.co.uk

CONSULTING STRUCTURAL AND CIVIL ENGINEERS: David Narro Associates was established with the aim of providing a high quality service from committed and experienced staff. A desire to extend the experience of the practice has led to expertise in fields such as conservation of ancient monuments and repair and sensitive alteration of listed buildings. Their skills are shown to best advantage when designing innovative solutions to engineering problems. The challenges posed by old buildings requiring sensitive non-invasive strengthening or alteration are met with a flexible and inventive approach. The initial contact can be the most fruitful and this is always entrusted to senior personnel.

▶ DEMAUS BUILDING DIAGNOSTICS LTD

Stagbatch Farm, Leominster, Herefordshire HR6 9DA
Tel 01568 615662 E-mail info@demaus.co.uk
Website www.demaus.co.uk

STRUCTURAL TIMBER TESTING AND BUILDING DIAGNOSTICS: Demaus Building Diagnostics specialises in the detection and assessment of decay, weakness and fire damage in structural timber using non-destructive techniques. *See also: profile entry in Non-destructive Investigations section, page 50.*

▶ ELLIOTT & COMPANY, STRUCTURAL ENGINEERS

51 Niddry Street, Edinburgh EH1 1LG
Tel 0131 558 9797 Fax 0131 558 9696 E-mail structures@ecoeng.co.uk

STRUCTURAL ENGINEERS: Award winning small specialised practice offering bespoke service on the conservation and alterations of historic structures and ancient monuments throughout Scotland. Extensive knowledge of traditional building methods ensures a considered assessment of the existing structure and highlights any original or imposed defects to be addressed with minimal intervention. Close liaison with other members of the design team ensure that implications of structural alterations and repairs are considered in their wider context. Projects include Stanley Mills (Historic Scotland and The Phoenix Trust), Newhailes House (National Trust for Scotland), Highland Tolbooth (Edinburgh Festival Centre) and Lady Victoria Colliery (Scottish Mining Museum Trust).

▶ ELLIS & MOORE

9th Floor, Hill House, Highgate Hill, London N19 5NA
Tel 020 7281 4821 Fax 020 7263 6613

CIVIL AND STRUCTURAL ENGINEERS: The practice is involved in the repair and refurbishment of listed buildings. Services include structural investigations, organising testing and preparing reports identifying defects with recommendations on remedial works. Lime mortars and renders are used in repair work where possible.

STRUCTURAL ENGINEERS

▶ GIFFORD AND PARTNERS

Carlton House, Ringwood Road, Woodlands, Southampton SO40 7HT
Tel 023 8081 7500 Fax 023 8081 7600

CONSERVATION ENGINEERS AND ARCHAEOLOGISTS: Award winning specialist expertise in structural, mechanical and electrical services engineering and archaeology for historic buildings. This unique blend of multi-disciplinary engineering and archaeological services allows Gifford and Partners to provide a fully integrated service to their clients. Numerous commissions on buildings of national importance have been secured with clients in the Public and Private sectors including English Heritage, National Trust, Historic Royal Palaces Agency, the Royal Household, Ministry of Defence, church PCCs, along with a number of key developers and private owners. Specialisms include: post fire repair, fire safety services, environmental control, electronic security and surveillance, energy conservation and public health engineering.

▶ MANN WILLIAMS

4 Palace Yard Mews, Bath BA1 2NH
Tel 01225 464419 Fax 01225 448651

CONSULTING STRUCTURAL AND CIVIL ENGINEERS: Working nationally on historic structures and ancient monuments since 1986, Mann Williams has built an impressive portfolio of projects for clients such as the National Trust and Landmark Trust. Projects range from Cathedrals at Winchester, Exeter and St David's, plus Avebury Stones and Corfe Castle, through to the Bath Spa Roman complex and significant timber framed structures. The firm's innovative approach provides uniquely tailored solutions combining accepted guidance from historic bodies with the individual requirements of both building and client.

▶ THE MORTON PARTNERSHIP LTD

The Old Cavalier, 89 Dunbridge Street, Bethnal Green, London E2 6JJ
Tel 020 7729 4459 Fax 020 7729 4458
E-mail london@themortonpartnership.co.uk
▸ Arcadia House, 19 Market Place, Halesworth, Suffolk IP19 8BB
Tel 01986 875651 Fax 01986 875085
E-mail halesworth@themortonpartnership.co.uk
Website www.themortonpartnership.co.uk

STRUCTURAL AND CIVIL ENGINEERS: Brian Morton founded this practice in 1966. It is now almost completely involved in minimum repair solutions to preserve historic buildings. Current work includes: work to Canterbury Cathedral, Pell Wall Hall by Sir John Soane, the new tower at Bury St Edmonds Cathedral, work to many parish churches, barns and domestic buildings, work for the Crown Estate in Regents Park, and National Trust properties. The practice carries out a considerable amount of work for local authorities, and is well known to all the national amenity groups. Work to small buildings is an important part of its work. Services include preliminary advice, structural surveys and presentation of the most cost effective solution to the proper repair of historic buildings and structures.

▶ IAN RUSSELL CEng MICE MIStructE

Shulbrede Priory, Lynchmere, Haslemere, Surrey GU27 3NQ
Tel 01428 653049 Fax 01428 645068

CIVIL AND STRUCTURAL ENGINEERING CONSULTANCY WITH BUILDING CONSERVATION: Specialising since 1983 in repairs and alterations to historic buildings and structures where using sympathetic materials and methods is important. Past projects: Ashton Court Mansion, Bristol; Royal Naval Hospital, Plymouth; offices, City Road EC1; Michelham Priory, Sussex. Currently engaged on extensions to Council House, Chichester and URC chapel, Elstead Surrey.

▶ FRED TANDY

9 Cannock Drive, Heaton Mersey, Stockport, Cheshire SK4 3JB
Tel 0161 432 4416 Fax 0161 442 1283

▶ JOHN WARDLE BSc, CEng, MIStructE

5 Lotus Road, Biggin Hill, Kent TN16 3JL
Tel 01959 540696

CONSULTANT STRUCTURAL ENGINEER

PROJECT MANAGEMENT

▶ BYROM CLARK ROBERTS

117 Portland Street, Manchester M1 6EH
Tel 0161 236 9601 Fax 0161 236 8675
▸ Jubilee House, West Bar Green, Sheffield S1 2BT
Tel 0114 275 7879 Fax 0114 272 8954

ARCHITECTS, SURVEYORS AND ENGINEERS: *See also: profile entry in Architects section, page 27.*

▶ GIBBON, LAWSON, MCKEE LIMITED

41A Thistle Street Lane South West
Edinburgh EH2 1EW
Tel 0131 225 4235 Fax 0131 220 0499

CONSERVATION PROJECT MANAGEMENT: *See also: profile entry in Architects section, page 29*

▶ GIFFORD AND PARTNERS

Carlton House, Ringwood Road, Woodlands, Southampton SO40 7HT
Tel 023 8081 7500 Fax 023 8081 7600

PROJECT MANAGEMENT: *See also: profile entry on this page.*

▶ W R DUNN & CO

27 Front Street, Acomb, York YO24 3BW
Tel 01904 784421 Fax 01904 784679

CHARTERED BUILDING SURVEYORS, ARCHITECTS AND HISTORIC BUILDINGS CONSULTANTS: *See also: profile entry in Surveyors section, page 48.*

See also:
Services Engineers, page 188
Fire Protection, page 182

CATHODIC PROTECTION OF IRON AND STEEL

Recent Applications to Heritage Buildings

DAVID FARRELL, KEVIN DAVIES and IAIN McCAIG

Metal dowels and cramps were often built into traditional masonry structures to secure stones which might otherwise be prone to movement or displacement, for example, copings, parapets and cornices. They were also widely used in ordinary ashlar walls, which usually consisted of a relatively thin facing of finely dressed and narrow jointed stonework, with a core or backing of rubble or brick. Cramps would be incorporated to tie the facing back to the core. Dowels and cramps were also embedded in the facing itself to help maintain its structural integrity.

In 18th and 19th century buildings, dowels and cramps were usually made from wrought iron which is susceptible to corrosion if exposed to air and moisture. In ashlar masonry of this period it is common to find vertical joints not filled with mortar to their full depth. When the shallow bead of mortar at the surface decays or cracks, rainwater is able to penetrate freely. The narrowness of the joints makes effective repointing very difficult, so water penetration continues, causing the embedded cramps to rust. The expanding rust eventually exerts such pressure on the stone that the stone cracks or spalls. The conventional remedy involves major surgery to remove the cramps, to replace them with non-corroding phosphor bronze or stainless steel and then to repair the wounded stonework.

Some 19th and 20th century masonry-clad buildings incorporate steel frames which are also liable to corrosion. Again, conventional treatments can be highly invasive involving large-scale opening up to expose and treat the effected components. Cathodic protection offers an alternative approach to the treatment of rusting iron and steelwork buried in masonry and stone.

The Whitchurch almshouses before repairs (top), showing spalling stonework along the base of the front wall caused by the rusting metal cramps, and (below) a schematic diagram showing the SACP system used to protect the cramps

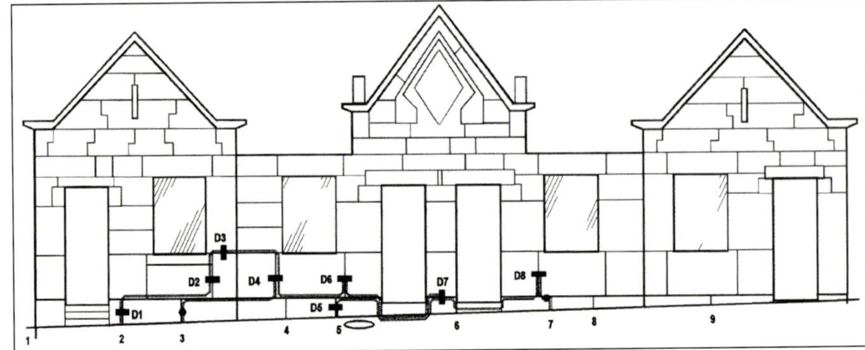

CATHODIC PROTECTION TECHNIQUES

Cathodic protection (CP) encompasses a range of techniques used to suppress corrosion of metal structures and components. CP is not a new process: in 1824 Sir Humphrey Davy presented a series of papers to the Royal Society describing how CP could be used to prevent the corrosion of copper sheathing in the wooden hulls of British naval vessels. Since then it has been applied in many areas, including marine applications and for the preservation of buried underground structures such as oil pipelines and tanks. CP technology has, over the past 20 years, been applied to reinforced concrete to protect steel reinforcements from corrosion and more recently, it has also been applied to iron and steel embedded in brick, masonry and stone in heritage buildings.

CP systems work on the principle that corrosion is an electrochemical reaction in which one part of a piece of iron or steel acts as an anode while adjacent metal acts as a cathode. At the anode corrosion occurs as iron gives up electrons and forms soluble iron ions ($Fe \rightarrow Fe^{2+} + 2e$). At the cathode the electrons released by the corrosion process combine with water and oxygen to form hydroxide ions ($\frac{1}{2} O_2 + H_2O + 2e \rightarrow 2OH^-$). In CP systems the metal to be protected is forced to act as the cathode, as on this side of the reaction the surface of the metal is unaffected by the reaction, preventing further corrosion. When used to protect structural iron and steel this is achieved by applying small DC electric currents, via the building material. This supplies a constant stream of electrons to satisfy the cathodic reaction. The anodic (corrosive) reaction then becomes suppressed. There are two methods of achieving this, either sacrificial anode cathodic protection (SACP) or impressed current cathodic protection (ICCP).

SACP systems use sacrificial anodes (zinc, aluminium or magnesium) which are placed in close proximity to the corroding metalwork and electrically connected to it. As the sacrificial anode corrodes, it generates a current that passes through the building material to provide protection to the embedded metalwork. The current is ionically conducted by means of pore water contained within the building material. These systems are capable of protecting small metal components such as embedded iron cramps or restraints set into walls, floors or roofs of a building.

ICCP systems use transformer rectifiers, normally mains powered, to provide the DC current to the iron or steel being protected. These systems use corrosion resistant anodes, fixed close to the metalwork, to provide part of the current pathway. ICCP systems are more complex than the SACP systems, but are suitable for providing CP to much larger areas of embedded steel such as I beams, supports and columns and where the stone or masonry has inherently higher electrical resistance.

WHITCHURCH ALMSHOUSES – A SACRIFICIAL ANODE CP APPLICATION

In 1999, an SACP system was installed to protect rusting iron cramps in the stone façade of four Grade II listed almshouses in Whitchurch, Shropshire. This was the first known application of its kind in the UK. The stones formed an interlocked frontage with iron cramps fitted between adjacent blocks. Water ingress beneath the eaves and below window sills had permeated into stonework joints and had allowed the iron cramps to corrode, especially in the vertical joints around windows and doors. The expanding corrosion product had introduced internal stresses into the stonework which had resulted in cracking and spalling of some of the stones.

Conservation policy (see PPG15 for example) favours the 'minimum intervention' principle, whereby any repairs should preserve the historically important features of the architecture. For this structure, a novel repair technique using SACP was adopted. Damaged stones located on the outer edges of the façade were replaced with new stones fitted with stainless steel cramps. For the remaining, as yet undamaged stones, an SACP system was installed to control further corrosion of the iron cramps.

To provide the cathodic protection, six magnesium anodes were buried in the pavement in front of the cottages and these were connected directly through to the iron cramps in a 'ring circuit' arrangement. Electrical connections were made to the cramps using a 'keyhole surgery' technique to minimise damage to the stones; the titanium connection wires were sunk into the mortar joints; and the current from the sacrificial anodes was conducted through the stone façade thus completing the CP's 'electrical circuit'. This allowed the cramps to be polarised from an external source and thus protected from further corrosion.

A corroding cramp causing masonry to fracture at the Grade II listed almshouses

The Wellington Arch, London and a detail (below) showing the corrosion of the iron beams supporting the quadriga

THE WELLINGTON ARCH – AN IMPRESSED CURRENT CP APPLICATION

A number of heritage structures within the London area have been fitted with ICCP over the past few years, including an external staircase at Kenwood House, to protect the embedded steel and the Inigo Jones Gateway at Chiswick House, to protect iron cramps.

An ICCP system has recently been installed within the Grade I listed Wellington Arch, Hyde Park Corner. The arch was built in the late 1820s using Portland stone and the roof slab was formed from concrete, supported by steel I beams. The arch has recently undergone extensive renovation, including the repair of the steel and concrete structure which supports the bronze sculpture of the quadriga. During the inspection it was discovered that some of the key steel I beams within the roof had suffered significant corrosion. This was partly due to rainwater penetrating the roof structure. The worst corrosion however, was where the beams had been in direct contact with the Portland stone.

Portland stone, having a neutral pH, offers no corrosion protection for steel. This may be compared to concrete, which is alkaline and, when in intimate contact with steel, helps to passivate the surface, inhibiting further corrosion. In fact, as a sedimentary rock formed in a marine environment, Portland stone may contain significant concentrations of chlorides or sulphates (salts) which, in the presence of moisture, can accelerate the corrosion process.

Corrosion of the arch's I beams had several consequences.

- The corrosion products physically occupied a much greater volume than the original steel and they began to push against adjacent surfaces. The corroding I beams had been physically displaced upwards by 20mm in places, due to the formation of corrosion products underneath the beams and this had resulted in cracking of some of the stonework. This process is known as corrosion jacking.
- The corrosion products had caused delamination of some areas of concrete on the roof slab allowing further corrosion of the now unprotected steel.
- Substantial thinning of the steel I beams caused concern about the future integrity of the whole support structure if significant corrosion continued.

THE APPROACH FOR RENOVATION

English Heritage did not wish to replace the beams as this would have been both expensive and disruptive, possibly requiring lifting the quadriga. Besides, conservation is about retaining as much as possible of the original structure and conserving its current condition. Ultrasonic thickness readings of the steel beams showed that their remaining thickness was sufficient to support the roof slab and quadriga, provided that ongoing corrosion could be controlled.

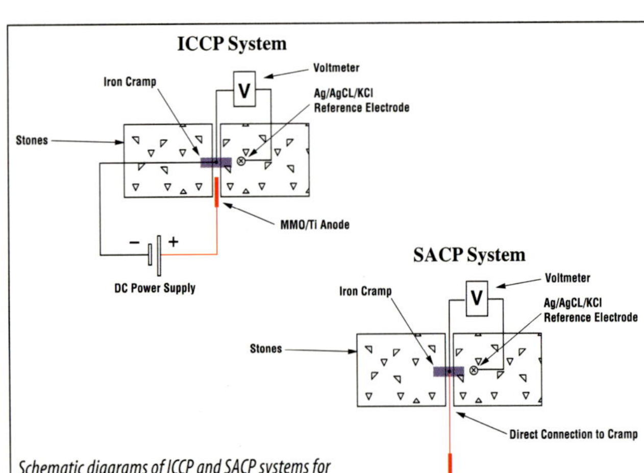

Schematic diagrams of ICCP and SACP systems for corroding iron embedded in stone

INSTALLATION OF THE ICCP SYSTEM

To provide efficient long-term CP performance, titanium expanded mesh ribbon anodes, with a mixed metal oxide (MMO) coating, were cast into the top surface of the concrete slab above the steel frame structure. This ribbon anode system provides maximum anode surface area for efficient transfer of the electrical current through the structure's concrete sections.

Small, discreet MMO coated titanium rod anodes were used to protect the steel beams where they were laid over the Portland stone. These anodes were embedded into 12mm diameter, 300mm deep holes at centres of 300mm in the mortar pointing of a brick course under the stone, using a conductive backfill to provide electrical continuity. These discreet anodes provide current into the depth of a wall to protect embedded steelwork.

System wiring from the beams and anodes was installed within the internal roof space back to an instrumentation cabinet. The cabinet houses the system electronics and computer for system control and monitoring.

BENEFITS FROM USING THE CP APPROACH

One of CP's principal advantages in the protection of embedded metalwork is that it provides corrosion protection without changing the immediate physical environment: there may still be now, or in the future, damp concrete or stone adjacent to the metal which would previously have allowed corrosion to continue. Cathodic protection provides the electrochemical conditions to control this corrosion process.

The implication of this is that there is no need to gain full access to the structure by removing the surrounding material and the structure can remain largely intact. All that is required is to install the necessary cables and anodes that form part of the CP system. Usually, these can be installed in such a way as to have little or no impact on the structure's visual appearance. In the case of the Whitchurch almshouses, this avoided dismantling of the stone façade and removal of the expanded corroded cramps, a project with uncertain consequences. On the Wellington Arch, major structural upheaval was avoided by adopting a CP approach.

RECOMMENDED READING

J Morgan, *Cathodic Protection* [second edition], National Association of Corrosion Engineers, ISBN 0-915567-28-8, 1993

K Blackney and B Martin, *The Application of Cathodic Protection to Historic Buildings*, English Heritage Research Transactions, ISBN 1 873936 62 1, Vol 1, April 1998

DAVID FARRELL PhD, MICorr is a corrosion engineer and a director of Rowan Technologies Ltd, a consultancy and research and development company specialising in the conservation of metals, stones and concrete in historic buildings and structures. Website www.rowantech.dial.pipex.com.

KEVIN DAVIES BSc, MICorr is a specialist corrosion engineer and has been responsible for the design, supervision, installation, commissioning and monitoring of over 50 cathodic protection systems for civil engineering structures and buildings.

IAIN McCAIG DipArch has specialised in building conservation for over 25 years, much of this time spent with the GLC's Historic Buildings Division and English Heritage. He is now Buildings at Risk Officer for North Shropshire District Council.

QUANTITY SURVEYORS

▶ BAILY•GARNER

146-148 Eltham Hill, London SE9 5DY
Tel 020 8294 1000 Fax 020 8294 1320
Contact Lisa J Brooks BSc (Hons) DipBldgCons ARICS

BUILDING SURVEYORS, ARCHITECTS, QUANTITY SURVEYORS AND PROJECT MANAGERS: *See also: profile entry in Surveyors section, page 45.*

▶ BARE LEANING & BARE

2 Bath Street, Bath, Somerset BA1 1SA
Tel 01225 461704 Fax 01225 447650
E-mail blbbath@btinternet.com
▸ Exeter office – Tel 01392 272245 Fax 01392 412089
E-mail blbexeter@btinternet.com

QUANTITY SURVEYORS: The partners and staff at Bath and Exeter have many years experience in the repair, alteration and conservation of listed buildings, churches and cathedrals throughout the UK. A full range of quantity surveying services is offered to clients for work being undertaken by contractors, conservators or direct labour. Applications for grant aid are prepared and cost advice given on maintenance programmes, quinquennial reports and VAT liability. Clients include English Heritage, the National Trust, Landmark Trust, cathedrals, Anglican, Roman Catholic and non-conformist churches, preservation trusts, local authorities, and private owners.

▶ KING SUMNERS PARTNERSHIP

1 Gloster Court, Segensworth West, Fareham, Hampshire PO15 5SH
Tel 01489 578811 Fax 01489 577123
E-mail KSP.fareham@virgin.net
Contact Tony Sumners

QUANTITY SURVEYORS: Founded in 1959 KSP operates from offices in Hampshire and Kent and offers quantity surveying, project management and planning supervisor services throughout the southern half of England. Current appointments in Lincolnshire, Kent, Devon and Warwickshire indicate the geographical range, and current clients include English Heritage, churches and private estates. The practice philosophy is to provide a high quality, cost effective, value for money professional service to its clients.

▶ MILDRED, HOWELLS & CO

Royal Colonnade, 14 Great George Street, Bristol BS1 5RH
Tel 0117 929 2894 Fax 0117 925 4356

CHARTERED QUANTITY SURVEYORS AND COST CONSULTANTS: Mildred, Howells & Co has many years experience in the repair, alteration and conservation of a wide variety of listed buildings, undertaking work from their Bristol, Swansea and Saltash offices. A full range of QS services is offered including cost advice, preparation of suitable tender documents, advice on procurement methods, assistance with grant applications, advice on VAT liability, maintenance programmes etc. Their clients include English Heritage, churches, local authorities, preservation trusts and private owners.

▶ MURDOCH GREEN KENSALLS (a member of Babtie Group)

1 Grand Parade, Brighton, E Sussex BN2 2QB
Tel 01273 676766 Fax 01273 696977
E-mail mgk.brighton@babtie.com
▸ Other offices at London, Croydon, Manchester and Glasgow
Contact Andrew Kirk FRICS Dip Bldg Cons

CHARTERED QUANTITY SURVEYORS/PROJECT MANAGERS: The firm has a well founded specialisation in the conservation, repair and alteration of ecclesiastical and secular listed buildings and structures, scheduled ancient monuments, historic parks and gardens and World Heritage sites. Comprehensive project services have ranged from consolidation of Roman and medieval remains to repairs and alterations of internationally important historic buildings. Clients include English Heritage, Historic Royal Palaces, Royal Household, Royal Parks, Parliamentary Works Directorate, Crown Estate, national museums, ecclesiastical authorities, preservation trusts, local authorities, private firms and individuals.

QUANTITY SURVEYORS

▶ PRESS & STARKEY

9-12 Stonehills House, Welwyn Garden City, Hertfordshire AL8 6NH
Tel 01707 325408 Fax 01707 338333
E-mail press_&_starkey@compuserve.com

CHARTERED QUANTITY SURVEYORS, HISTORIC CONSERVATION COST CONSULTANTS: Established in 1960 with offices in Welwyn Garden City and Maidstone the practice serves clients in both the public and private sectors in all aspects of construction. A wealth of experience has been gained in the conservation, consolidation and presentation of historic buildings providing the full range of QS services and advice on the valuation for fire insurance, detailed costing of quinquennial and other maintenance plans and reports and Value Added Tax. Clients include English Heritage, Cadw, National Trust, private estates, preservation trusts and churches. The practice provides a professional service tailored to the specific requirements of each client in compliance with BS/EN/ISO9002.

▶ SHAMBROOKS

10 Clayfield Mews, Newcomen Road, Tunbridge Wells, Kent TN4 9PA
Tel 01892 540399 Fax 01892 540416

CHARTERED QUANTITY SURVEYORS, BUILDING COST CONSULTANTS AND HISTORIC BUILDING CONSULTANTS: Founded in 1942, Shambrooks have considerable experience of conservation and repair, and of installing services in historic buildings. Clients include English Heritage, Royal Parks, Canterbury Cathedral. Services include grant applications, costing of listed building work and quinquennial maintenance, and Planning Supervision.

MATERIALS ANALYSIS

▶ HIRST CONSERVATION

Laughton, Sleaford, Lincolnshire NG34 0HE
Tel 01529 497449 Fax 01529 497518

ANALYSIS OF PLASTER AND PAINT LAYERS: *See also: display entry on the inside front cover and profile entry in Building Contractors section, page 62.*

▶ LISA OESTREICHER

Jubilee House, High Street, Tisbury, Wiltshire SP3 6HA
Tel 01747 871717 Fax 01747 871718

ARCHITECTURAL PAINT ANALYSIS: *See also: profile entry in Historic Paint Analysis section, page 200.*

▶ STATS CONSULTANCY (STATS Limited, founded 1974)

Porterswood House, Porters Wood, St Albans, Herts AL3 6PQ
Tel 01727 833261 Fax 01727 835682
E-mail ian.sims@stats.co.uk
Contact Dr Ian Sims

SPECIALIST ENGINEERING, MATERIALS AND ENVIRONMENTAL CONSULTANTS: A professional service in building and construction materials throughout the UK and overseas, including bricks, concrete, mortars, plasters, roofing slates, natural stone, terracotta, timber, tiles and other traditional fabrics. In addition to experienced advice on materials selection, condition and conservation, STATS investigates actual or suspected deterioration. In-house UKAS-accredited laboratories conduct materials identifications and assessments using examinations, analyses and tests. Sites include Buckingham Palace and the British Museum.

SURVEYORS

▶ BAILEY PARTNERSHIP

Amhurst House, 22 London Road, Riverhead, Sevenoaks, Kent TN13 2BW
Tel 01732 455522 Fax 01732 460607
▶ Plymouth office Tel 01752 229259 Fax 01752 224280

HISTORIC BUILDINGS CONSULTANCY: This specialist consultancy has been created in recognition of the requirement to apply the appropriate level of knowledge and experience to the needs of historic buildings, and is soundly based upon many years of experience in dealing sympathetically with listed buildings and scheduled ancient monuments. A comprehensive range of professional services is available covering aspects of surveying, repair and conservation, adaptation and management. The consultancy is headed by Richard Sutch BSc Dip Bldg Cons ARICS, who is also a commissioned surveyor to English Heritage.

▶ BAILY•GARNER

146-148 Eltham Hill, London SE9 5DY
Tel 020 8294 1000 Fax 020 8294 1320
Contact Lisa J Brooks BSc (Hons) DipBldgCons ARICS

BUILDING SURVEYORS, ARCHITECTS, QUANTITY SURVEYORS AND PROJECT MANAGERS: Baily•Garner were established in 1976 and have a commendable track record in the care and conservation of historic buildings. As a multi-disciplinary practice they are able to offer clients a comprehensive range of services from the preparation of feasibility reports, surveys, detailed design and specification, contract administration, through to party wall matters, listed building consent and planning applications. Their multi-disciplinary approach enables them to maintain a much closer control of projects both in terms of adherence to programme and cost whilst providing clients with a 'one stop' service.

▶ BONIFACE ASSOCIATES (part of The Whitworth Co-Partnership)

4 Simon Campion Court, High Street, Epping, Essex CM16 4AU
Tel 01992 579555 Fax 01992 571386
E-mail boniface@ukgateway.net
Contact Stephen L Boniface DipBldgCons (RICS), FRICS, MAE, IHBC

HISTORIC BUILDINGS CONSULTANCY: RICS Building Conservation Accredited. Services include: architectural (alteration, adaptation, extension and new-build); defect analysis; planning; feasibility studies; conservation plans; project management; maintenance advice; insurance advice and claim resolution; purchaser/vendor surveys; valuations; party walls; landlord and tenant issues; expert witness reports. Experienced with various building types, all listing grades, conservation areas and scheduled monuments. Work undertaken throughout London, Home Counties, East Midlands and East Anglia. Dedicated to the sympathetic handling of historic buildings and providing sound independent advice. *See also: The Whitworth Co-Partnership profile entry, page 36.*

▶ BOSENCE & CO

Oxenham Farm, Sigford, Newton Abbot, Devon TQ12 6LF
Tel/Fax 01626 821609
Contact Oliver Bosence MA DipBldgCons(RICS)

SPECIALIST PROJECT MANAGER AND CONSERVATION CONSULTANT: Extensive sitework experience with traditional buildings in the South West, especially masonry repair, slatework and lime renders; vernacular to polite; practical guidance on specifications, programming, project and site management; defects analysis and reports; lectures and demonstrations of repair techniques and conservation craft skills.

▶ BYROM CLARK ROBERTS

117 Portland Street, Manchester M1 6EH
Tel 0161 236 9601 Fax 0161 236 8675
▶ Jubilee House, West Bar Green, Sheffield S1 2BT
Tel 0114 275 7879 Fax 0114 272 8954

ARCHITECTS, SURVEYORS AND ENGINEERS: *See also: profile entry in Architects section, page 27.*

▶ DAVIS BLACKBURN (CLEVEDON)

30a Old Church Road, Clevedon, North Somerset BS21 6LY
Tel 01275 877024 Fax 01275 342617
E-mail info@davisblackburn.co.uk
Website www.davisblackburn.co.uk

CHARTERED BUILDING SURVEYORS AND BUILDING DESIGN CONSULTANTS: Founded in 1976, Davis Blackburn has established a reputation for sympathetic refurbishment, extension and alteration of older buildings. The practice offers a comprehensive in house service including feasibility studies, measured surveys, architectural design, detailed specification, tender and supervision of works and also full and limited building surveys combining comprehensive reporting with defect analysis backed up by practical advice on remedial works. Above all, the firm provides a friendly, cost effective, professional service.

▶ ALAN DICKINSON, ARICS

1 The Grove, Rye, East Sussex TN31 7ND
Tel/Fax 01797 225139
Website www.Alandickinson.com

CHARTERED BUILDING SURVEYOR, HISTORIC BUILDINGS CONSULTANT: Sussex and Kent based sole practitioner specialising in surveys, restoration and alterations to period buildings backed by an understanding of historical development. Measured surveys, archival studies and archaeological analysis are also offered for planning and research purposes.

▶ DRIVERS JONAS

6 Grosvenor Street, London W1K 4DJ
Tel 020 7896 8339 Fax 020 7896 8021

CHARTERED SURVEYORS: Drivers Jonas is a multi-disciplined firm covering all aspects of property consultancy. Their strong Construction Unit includes specialists in conservation, led by Peter Bickerstaff BSc DipBldgCons FRICS (RICS Accredited). Major projects include the refurbishment of North Mymms Park, work and surveys at Hampton Court Palace and the re-presentation of John Wesley's House as a museum. They specialise in planned maintenance programming for large complexes of historic buildings such as Winchester and Atlantic Colleges, Westminster Cathedral; and in fire insurance assessments of historic buildings including Leeds Castle, Heveningham Hall and Tonbridge School. Other services include: dilapidations, party walls and rights of light, planning consultancy and property management.

▶ GIBBON, LAWSON, MCKEE LIMITED

41A Thistle Street Lane South West, Edinburgh EH2 1EW
Tel 0131 225 4235 Fax 0131 220 0499

BUILDING SURVEYORS: As building surveyors GLM are often asked to undertake detailed condition surveys of historic buildings and to report with budget costs setting out the priorities for maintenance and repair expenditure. They also provide a cost effective design and project management service with in-house architectural support. *See also: profile entries in Architects section, page 29 and Fire Protection Consultants section, page 182.*

▶ THE HARTLEY CONSERVATION PARTNERSHIP

Argyll House 12, Gentle Street, Frome, Somerset BA11 1JA
Tel 01373 466618 Fax 01373 466428
E-mail hartleyconservation@btinternet.com
Website www.heritageconservation.co.uk

HISTORIC BUILDINGS CONSERVATION CONSULTANTS: An independent practice with RICS accreditation in building conservation, providing a comprehensive specialist service to organisations and individuals owning buildings of special architectural and historic interest. Contact Phillip Hartley MA(Arch Hist) Dipl Bldg Cons MRICS or Tracey Hartley BSc Hons IHBC MRICS Grad Dipl Cons AA.

▶ HILL BEILD ASSOCIATES

Post Office House, Stockton, Warminster, Wiltshire BA12 0SE
Tel 01985 850067 Fax 01985 850082
E-mail hillbeild@ndirect.co.uk

CHARTERED BUILDING SURVEYORS AND BUILDING HISTORIANS: Multi-discipline consultancy working across England and Wales for commercial, private and public sector clients. They specialise in historic building conservation and restoration, together with interpretative surveys, both above and below ground archaeological investigations, measured and record surveys, analysis and background historical research. New complementary designs for all types of work from internal refurbishments to new extensions or conversions. Recent projects have included fire and structural damage restoration, historic building conversions, archaeological recording and building dismantling, pre design stage integrated standing structure analysis for other professionals, town centre and group building surveys and historic environmental impact studies.

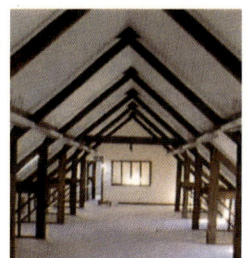

▶ HODKINSON MALLINSON PARTNERSHIP

1 Derby Place, Hoole, Chester CH2 3NP
Tel 01244 329505 Fax 01244 312403
E-mail surveyors@hodmal.co.uk

CHARTERED BUILDING SURVEYORS: The practice has expertise in the conservation and restoration of historic buildings including designs for contemporary structures and for appropriate forms of repair with minimal intervention. All aspects are dealt with, from the preparation of detailed condition surveys to specification, site supervision and project management.

▶ JUBB & JUBB LTD

2 Rodford Cottages, Rodford, South Gloucestershire BS37 8QG
Tel 01454 316313

CHARTERED BUILDING SURVEYORS

SURVEYORS

▶ OXLEY CONSERVATION

Orchard House, Cherry Tree Close, Stoke Row, Henley on Thames RG9 5RD
Tel 01491 682288 Fax 01491 682204
Contact Richard Oxley BSc DipBldgCons ARICS

HISTORIC BUILDINGS CONSULTANCY: The appropriate and sympathetic repair of old buildings, using traditional materials and methods, is strongly advocated by Oxley Conservation. The services provided include: condition surveys; the specification and supervision of conservation repairs; independent damp and timber reports based upon environmental control of problems; surveys of historic timber framed buildings; quinquennial inspections; environmental assessments – including improving energy efficiency; listed building consent and planning applications. Contact Richard Oxley, who is RICS Accredited in Building Conservation, to obtain a personal service. *See also: The Need For Roofs To Breathe, page 79.*

▶ RICKARDS CONSERVATION

105 St John's Hill, Sevenoaks, Kent TN13 3PE
Tel 01732 741677 Fax 01732 740149

HISTORIC BUILDINGS CONSERVATION CONSULTANCY: Rickards Conservation is an independent professional building conservation practice providing a comprehensive range of services to care for historic and listed buildings, including churches and ruins. Specialists in structural and condition surveys, defects analysis and repair. Stephen Rickards GradDiplCons(AA) FRICS IHBC ARPS is a Chartered Building Surveyor, Accredited by the RICS for conservation work, and holds the Architectural Association postgraduate Diploma in Building Conservation. He was responsible for conservation work praised by the Civic Trust in Chatham Historic Dockyard. Clients include the SPAB, National Trust and PCC. Conservation Award winner in 1999.

▶ RIDGE AND PARTNERS

Midland House, West Way, Botley, Oxford OX2 0PJ
Tel 01865 794777 Fax 01865 268725

QUANTITY AND BUILDING SURVEYORS: Qualified and experienced surveyors who understand the sensitive requirements of conservation and restoration of historic and listed buildings. Projects include Blenheim Palace, Stroud Museum in the Park, Wallace Collection, Royal Agricultural College, Oxford University and Old LMS Station. Members of SPAB and the Georgian Group.

▶ SHEPPARD & CO

7 Brentgovel Street, Bury St Edmunds, Suffolk IP33 1EB
Tel 01284 767521 Fax 01284 724550
▸ 52 Burleigh Street, Cambridge CB1 1DJ
Tel 01223 460258 Fax 01223 368746

CHARTERED SURVEYORS: Sheppard & Co provide sympathetic advice for the conservation of historic buildings. A complete service is offered including the initial building survey report, detailed drawings, advice on specialist contractors and project management. Current schemes include the restoration and conversion of The Old Snuff Factory, Bury St Edmunds (Listed Grade II) and Manor Farm, Thriplow, a Grade II* 15th century hall house near Cambridge. The company are also project managers for Bury St Edmunds Town Trust. Visit Sheppard & Co's website at www.sheppardand.co.uk.

▶ TFT CULTURAL HERITAGE

211 Piccadilly, London W1V 9LD
Tel 020 7917 9590 Fax 020 7917 9591
E-mail tftheritage@cs.com
Website www.tftsurveyors.co.uk

HERITAGE CONSULTANCY: TFT Cultural Heritage is a specialist division of Tuffin Ferraby & Taylor, working throughout the UK. Its services include: condition surveys, sensitive repair, strategic management, conservation planning, and development advice and master planning in historic areas. Its canvas is broad and diverse: city streets, towns and villages; World Heritage Sites and conservation areas; individual buildings, archaeological sites and landed estates; parkland and landscapes of outstanding interest. Such spaces and places are our legacy from the past; yet, they must work to survive. The practice promotes solutions that will respect their value and significance, whilst enriching their contribution to society.

▶ W R DUNN & CO

27 Front Street, Acomb, York YO24 3BW
Tel 01904 784421 Fax 01904 784679

CHARTERED BUILDING SURVEYORS, ARCHITECTS AND HISTORIC BUILDING CONSULTANTS: Established in 1986, W R Dunn & Co undertake commissions throughout the UK on behalf of English Heritage, the Home Office, the MoD, Harewood House Trust Ltd, several local authorities and councils and many private and commercial clients. The practice principal is an RICS Accredited Surveyor in Building Conservation and is included on the GHBAU Consultants Register. Past schemes include re-roofing a Grade I listed stately home, repairs/alterations to a Grade I listed stable block and repairs to castles, lodges, halls, monuments, walled gardens, dwellings and other structures. The practice also provides quantity surveying, project management services plus health and safety advice.

▶ WAKEMANS

11/12 Highfield Road, Edgbaston, Birmingham B15 3EB
Tel 0121 454 4581 Fax 0121 454 5206 E-mail c.deacon@wakemans.com
Contact Christopher Deacon BSc(Hons) MRICS Dip Bldg Cons

BUILDING CONSERVATION, BUILDING SURVEYING, PROJECT MANAGEMENT, QUANTITY SURVEYING, PLANNING SUPERVISION: Wakemans is a nation-wide operation with building conservation experience. The firm can survey, provide cost and technical advice, liaise with other professionals and provide project management services for repair, refurbishment and alteration work to historic buildings. Christopher Deacon is an approved quinquennial surveyor to the Diocese of Birmingham. Advice on grants and funding is also available.

▶ WARD & DALE SMITH, CHARTERED BUILDING SURVEYORS

The Walker Hall, Market Square, Evesham, Worcs WR11 4RW
Tel 01386 446623 Fax 01386 48215 E-mail wds@ricsonline.org

HISTORIC BUILDING CONSULTANCY INCLUDING CONDITION SURVEYS

▶ WATKINSON + COSGRAVE

Fleet House, 62 Highgate Road, London NW5 1PA
Tel 020 7485 6016 Fax 020 7284 4058
E-mail watcos@cwcom.net

CHARTERED BUILDING SURVEYORS AND STRUCTURAL ENGINEERS: Watkinson + Cosgrave has 40 years experience of repairing and restoring buildings, including initial appraisals, structural surveys, feasibility studies, design, specification, supervision of works and contract management.

▶ WATSON BERTRAM & FELL

5 Gay Street, Bath, Somerset BA1 2PH
Tel 01225 337273 Fax 01225 448537
E-mail wbfbath@compuserve.com

ARCHITECTS AND SURVEYORS: *See also: profile entry in Architects section, page 36.*

NON-DESTRUCTIVE INVESTIGATIONS

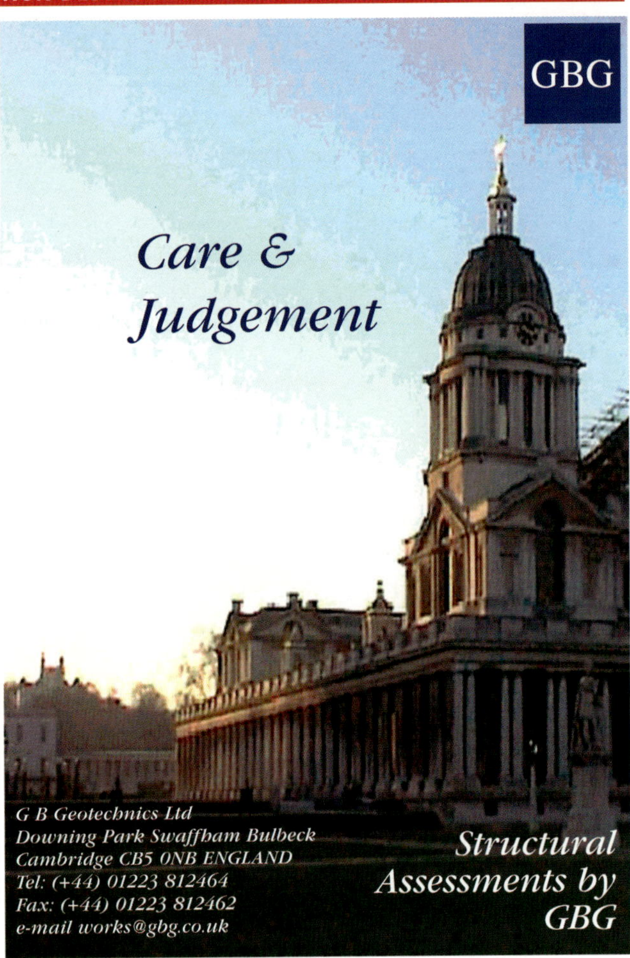

▶ DEMAUS BUILDING DIAGNOSTICS LTD

Stagbatch Farm, Leominster, Herefordshire HR6 9DA
Tel 01568 615662
E-mail info@demaus.co.uk
Website www.demaus.co.uk

INFRARED THERMOGRAPHY, ULTRASOUND, MICRODRILLING AND ENDOSCOPY: Demaus Building Diagnostics Ltd specialises in the non-destructive investigation and assessment of historic buildings using a wide range of advanced techniques. Thermographic imaging, using the latest most sensitive equipment provides a very wide range of information quickly and economically. Services also include the location and accurate measurement of decay in concealed structural timber, the assessment of fire damage, the identification of concealed structural alterations and failures, and advice on repair and conservation. Clients range from English Heritage, The Royal Household, the National Trust and Department of National Heritage to private individuals.

▶ HUTTON+ROSTRON ENVIRONMENTAL INVESTIGATIONS LIMITED

Netley House, Gomshall, Surrey GU5 9QA
Tel 01483 203221 Fax 01483 202911
E-mail ei@handr.co.uk Website www.handr.co.uk
Contact Tim Hutton MA MSc MRCVS

CONSULTANTS ON BUILDING FAILURES AND ENVIRONMENTS: *See also: profile entry in Damp & Timber Decay section, page 176.*

MEASURING TOOLS

▶ LEICA GEOSYSTEMS LTD

Davy Avenue, Knowlhill, Milton Keynes MK5 8LB
Tel 01908 246286 Fax 01908 246259
Website www.disto.com

LASER MEASURING DEVICES: *See also: display entry on page 49.*

MEASURED SURVEYS

▶ ARCHIVE

Rose Cottage, Lower Buckenhill, Fownhope, Hereford HR1 4PX
Tel/Fax 01432 860673

SPECIALIST PHOTOGRAPHIC SERVICES: Archive provides comprehensive photographic services to conservation bodies and architects, supplying rectified photographic surveys and building condition archives in colour and monochrome as required. Archival processing of monochrome negatives and prints is available, and database indexes, contact reference sheets and scaled photographic prints assembled into elevations are supplied to clients' requirements. Archive is a contractor to Cadw, supplier to English Heritage, many archaeological units and conservation architectural practices. Estimates for coverage at a wide range of levels are supplied subsequent to site visits. Recent and current contracts include Wigmore Castle, Bath Spa Project, Tintern Abbey, Plas Mawr (Conwy) and Blaenavon.

▶ KINGSLAND SURVEYORS LIMITED

53a High Street, Leatherhead, Surrey KT22 8AG
Tel 01372 362059 Fax 01372 363059
Website surking.co.uk

SPECIALIST IN ARCHITECTURAL AND PHOTOGRAMMETRY SURVEY WORKS: Producing 2D and 3D digital floor plans and detailed elevations, providing a comprehensive and accurate record. Kings – architectural, land and utility surveyors, formerly known as McDowells – have over the years undertaken a wide range of survey works on major contracts, both in the United Kingdom and overseas since the company's foundation in 1963. UK clients include National Westminster Bank, NHS Trust hospitals, Metropolitan Police and numerous architects.

▶ NATIONAL MONUMENTS RECORD

Kemble Drive, Swindon, Wiltshire SN2 2GZ
Tel 01793 414802 Fax 01793 414924

MEASURED SURVEYS OF SITES AND LANDSCAPES: *See also: display entry in Archives section, page 14.*

▶ PLOWMAN CRAVEN & ASSOCIATES

141 Lower Luton Road, Harpenden, Herts AL5 5EQ
Tel 01582 765566 Fax 01582 765370

ARCHITECTURAL PHOTOGRAMMETRY, 3D LASER SCANNING, BUILDING SURVEYS AND ALL ASPECTS OF MEASURED, PHOTOGRAPHIC AND DIGITAL SURVEY: With over 30 years experience as surveyors, PCA are one of the leading firms of professional 'measured' surveyors. Clients include English Heritage, Cadw, Historic Scotland, Royal Household, Palace of Westminster, Church Commissioners and many others. Projects have included Windsor (before and after fire), Hampton Court and many smaller restoration tasks. Data supplied in 2D, 3D digital form.

▶ STRUCTURAL PERSPECTIVES

48 Holdsworth Road, Holmfield, Halifax HX2 9SZ
Tel/Fax 01422 240789

SPECIALIST RECORDING OF HISTORIC STRUCTURES: Structural Perspectives provides comprehensive recording and analysis of historic structures and sites, including measured drawings (high quality hand-drawing and judicious use of EDM) and surveying services. *See also: display entry and profile entry in Archaeologists section, page 16.*

PHOTO-REAL VISUALISATIONS

ROBERT DEMAUS

Figure 1

Figure 2

Figure 3

Figure 4

Presented with an empty site, and a set of plans and elevations, most people find it very difficult to visualise how the actual finished building will look. Elevation drawings, however carefully prepared, almost always fail to portray accurately the scale and proportion of a building, and can give little or no sense of the interplay of different materials and colours, wherein small changes can have a substantial effect. At a basic level, when elevations are drawn to scale, the roofs always appear disproportionate to the walls.

To help bridge the gap between the proposed and the actual, architectural illustrators have been commissioned to produce perspectives and views that go some way towards achieving a realistic representation, but they usually fall well short of true realism. Another significant disadvantage of traditionally produced artists' impressions (apart from a tendency to be more impressionistic than realistic) is that they are usually commissioned only at the end of the design process, so that they do not inform and influence the decision-making. Furthermore, once produced, they cannot be readily altered to assess the effect that any changes in the proposals might have.

Photo-real visualisations, often using a combination of manipulated photographic computer-generated images, have been used for some years in new build projects by architects and developers and in the reconstructions of long-lost structures so beloved by the current plethora of archaeological programmes on television. Whilst the most dramatic images are often created by 3-dimensional 'fly-throughs', these are often more entertaining than informative, relying on movement to disguise a lack of detail and accuracy. The computing power and man hours (and consequent cost) required to create such fly-throughs is considerable, and tends to put the process outside the reach of most projects. Besides, we do not really need to know what the building might look like through the eye of a bird or worm. The production of photo-realistic stills is, however, economically viable, even on small projects, and can offer considerable benefits to all involved. At present, and for the foreseeable future, still images remain the best method of visualising buildings with sufficient accurate detail.

So what relevance do these altered images have in the field of building conservation? After all, the buildings with which we are involved generally exist, and can be viewed 'in reality'.

In a great majority of cases, the repair or rehabilitation of a building involves physical changes to the building that will alter its appearance, and the effect of these changes on the building and its surroundings can be of concern, not only to the owner, but also to any number of interest groups, both locally and sometimes nationally. Very often these changes can be better understood using visualisations rather than scale drawings. This ability to assess readily the effect of even minor variations, such as the proportions of glazing bars, as well as major changes, is of great value at the consultative stage of a project. It is not only clients, the general public and members of local authority planning committees (who often have difficulty reading and understanding conventional scale drawings) that benefit from these visualisations. Professionals, who regularly work with drawings, also find that accurate visualisations allow them a far better appreciation of the various options available.

The Staffordshire farmhouse shown here *(Figure 1)* has had a hard cement render applied some time in the mid 20th century. It was known to contain some timber framing but the extent and condition of the surviving frame was unknown. A sensitive thermal imaging camera was able to reveal the pattern and extent of the framing beneath the render *(Figure 2)*. Once the exact positions of structural timbers were known, their condition could be assessed using 1mm diameter micro-drills through the render, enabling a complete assessment of the frame to be made without any removal of the render. Using visualisations based on thermographic images, it was then possible to create accurate images of how the building would look if the render was removed, and furthermore to assess the effect of different finishes on the concealed structure *(Figures 3 and 4)*. In this way, all the important decisions about the proposed work could be made before any contract was let.

Figure 5

Figure 6

Figure 7

Figure 8

Figure 9

In another case, the Georgian house illustrated *(Figure 5)* had suffered a number of unfortunate changes, the most obvious being the conservatory appended to one side. The owner was understandably keen to improve the elevations. In addition to replacing the conservatory, the effects of relatively minor changes such as the replacement of the front door, removal of paint from stone lintels and the rebuilding of the chimney tops could be considered *(Figure 6)*. By using visualisations, the effect of the proposals and their interaction with the building could be assessed very much more readily. The visualisations were then used and accepted as a listed building consent application.

In many cases, it can be difficult to persuade the public and/or those in positions of authority that the regeneration of a dilapidated building is worthwhile, and will produce benefits to the wider community beyond the building itself. The Victorian Pump Rooms *(Figure 7)* were considered by many to be an unworthy eyesore: the visualisation *(Figure 8)* showed very clearly how the building would look if restored. The completed building *(Figure 9)* demonstrates how accurate such a visualisation can be. This ability to raise awareness of the potential benefit of regeneration schemes for individual buildings or for larger areas is of particular value, where the need to raise funds, and public awareness and support, at an early stage can be an essential primer for a project.

It is becoming increasingly apparent that many stone buildings are degrading more rapidly and/or allowing far greater water penetration, because lime render or limewash shelter coats have been removed, or simply eroded away, and also because so many have been re-pointed in cement. Many people have become accustomed to the look of the exposed stone and are wary of the visual impact that re-applying such shelter coats would have. Whilst photo-real images will not necessarily persuade them of the benefits, they often can be a useful aid. The stone of the house in *Figure 10* has degraded very rapidly, due partly to cement re-pointing and partly to the friable nature of the stone itself. There is no surviving evidence that the building was ever rendered, but there is little doubt that removal of the pointing and the application of a shelter coat of lime render would be of benefit. *Figure 11* shows the visual effect such a scheme would have. It also shows, as always, that the return to more appropriate windows can be of great benefit. *Figure 12* takes the process a step further to assess the impact of introducing a veranda, which is thought to have existed at the end of the 19th century.

The images used to illustrate this article have been created using a combination of manipulated photographic images of existing buildings and components, and elements created entirely within the computer.

There is a wide variety of software available for image manipulation. The industry standard is Adobe Photoshop, which is a very sophisticated and complex program that can be used to create images from scratch, as well as manipulate imported photographs. Although the basic features of the program can be learnt relatively quickly, training and practice is required to master the more powerful features. If the visualisations are to be built up from measured survey drawings, these can be imported from any standard CAD package or scanned from printed drawings.

Manipulating images often involves very large computer files of more than 100MB, with up to 40 or more layers, depending on the complexity of the image. The computer used for such work needs to have a high specification with a fast processor and at least 256MB of RAM. In addition a large hard disc is necessary for file storage, or removable data storage can be used. The higher the specification, the faster the big files can be handled, but the cost of hardware and software to achieve reasonably high quality manipulations need not be more than £2,000–£3,000. However, the training and experience necessary to achieve good results can involve a considerably greater investment in time and money. A comprehensive photographic library of buildings, architectural features and details is also necessary.

It is of course essential that anyone intending to create realistic visualisations of old structures has a deep understanding of the way buildings were constructed. The more accurate and realistic the image created, the more apparent any errors become, even if they are quite small.

Increasingly, conservation officers and planners are accepting visualisations as part of listed building consent applications, and in many cases express a preference for them over conventional scale drawings.

ROBERT DEMAUS is a director of Demaus Building Diagnostics Ltd which specialises in the non-destructive investigation and assessment of historic buildings. For further information see page 50.

Figure 10

Figure 11

Figure 12

CALDEY ISLAND, PEMBROKESHIR

The conservation of the Grade II* listed complex of monastery buildings is being assisted by the Heritage Lottery Fun
The buildings were largely constructed between 1910–13 and are considered the best example of large scale Arts and Crafts work in Wale

Photograph © David Ward, Heritage Lottery Fun

Chapter 2

BUILDING CONTRACTORS

BUILDING CONTRACTORS

Contractor	Type	Page
SCOTLAND		
Carpenters Oak & Woodland Co Ltd	tf	72
Heritage Engineering	ml	76
Stoneguard	bu	66
NORTH OF ENGLAND		
Ceiling Access Platforms	sl	193
Wildwood Joinery	tf	142
YORKSHIRE		
Anelays	bu tf	58
Burrows Davies Limited	bu	58
C J Ellmore & Co Ltd	bu	59
Kirby Peel	tf	73
Stoneguard	bu	66
Weaver Construction Ltd	bu	69
NORTH-WEST		
Altham Hardwood Centre	tf	72
Bernard A Shepherd Ltd	bu tf	73
Burleigh Stone Cleaning & Restoration	bu	58
Cameron (UK) PLC	bu	60
Chester Masonry Group	bu	59
Dorothea Restorations Ltd	ml	76
Maysand	bu tf	62
S & J Whitehead Ltd	bu	66
Stoneguard	bu	66
Wm Langshaw & Sons	bu	68
WALES		
Belfield Timber Co	tf	73
Bricknell Construction	bu	58
Taylor Dalton, Heritage Building Contractors	bu	67
WEST MIDLANDS		
Capps & Capps Ltd	bu tf	60
Linford-Bridgeman Limited	bu	61
Russcott Conservation	bu	111
Sandy & Co (Contractors) Ltd	bu	65
Stoneguard	bu	66
T J Crump Oakwrights Limited	bu tf	74
Tamworth Scaffolding Company	sl	83
Timothy Williams (Builders) Ltd	bu	66
Treasure & Son Ltd	bu	67
William Sapcote & Sons Ltd	bu	68
EAST MIDLANDS		
C R Crane & Son Ltd	bu tf	60
Church Conservation Limited	bu	84
Hirst Conservation	bu	62
Historic Buildings Conservation Limited	bu	39
Marsh Brothers Engineering Services Ltd	ml	151
SOUTH-EAST – NORTH OF THE THAMES		
Arthur E Woodward	bu	58
Between Time	bu tf	58
Boshers (Cholsey) Ltd	bu	58
Calder Traditional Building Services	bu	60
The Chiltern Partnership	ml	76
Dunne & Co Ltd	bu	60
Hall Construction	bu	60
The Hertfordshire Roofing & Renovation Company	bu	59
I J P Building Conservation	bu ml tf	61
J G Matthews Ltd	bu tf	62
Knowles & Son (Oxford) Ltd	bu	62
Mathias Builders	bu	98
McCurdy & Co Ltd	bu tf	73
Mowlem Rattee & Kett	bu	62
Period Projects Limited	tf	162
Sindall Ltd	bu	64
Splitlath Limited	bu tf	64
Stoneguard	bu	66
T J Evers Ltd	bu	66
Tain Crafts (UK) Ltd	bu tf	67
LONDON		
Charles Adams Limited	bu	60
David Ball Restoration (London) Limited	bu	60
Holloway White Allom	bu	62
Sindall Ltd	bu	64
Taywood Engineering Limited	bu	68
Timothy Williams (Builders) Ltd	bu	66
Wallis	bu	55
Wates Construction Limited	bu	69
SOUTH-EAST – SOUTH OF THE THAMES		
Anthony Hicks	tf	72
Antique Buildings Ltd	tf	72
Busby's Builders	bu tf	59
Chalk Down Lime Ltd	bu tf	60
Hampshire Oak Carpenters Ltd	tf	73
Haslemere Builders Limited	bu tf	60
Heritage Oak Buildings	tf	72
I W Payne & Company Ltd	bu tf	61
Jameson Joinery	tf	73
Longley	bu	62
Millway Builders Ltd	bu	61
Mott Graves Projects Ltd	bu	63
Oakwrights Limited	tf	74
R W Armstrong & Sons Limited	bu	64
Simmonds of Wrotham	bu	65
Stonewest	bu	64
W T Specialist Contracts	bu	69
Wallis	bu	55
SOUTH-WEST		
Bosence & Co	bu	45
Carpenters Oak & Woodland Co Ltd	tf	72
Carrek Limited	bu	60
Dorothea Restorations Ltd	ml	76
Hill Beild Associates	bu	46
Melcombe Regis Construction	bu	61
St Blaise Ltd	bu tf	65
Stoneguard	bu	66

KEY
- bu building contractors
- ml millwrights
- sl scaffolding
- tf timber frame builders

Regional designation is according to office location. Many firms operate nationally.

THE LANDFILL TAX ENVIRONMENTAL CREDIT SCHEME

FRANCIS GOLDEN

Since 1996 landfill operators have had to charge Landfill Tax on waste deposited on their sites. To the great surprise of Customs and Excise, who had to implement the new tax, the landfill operators were allowed by the Government to make regular donations to a variety of approved 'environmental' bodies for certain types of projects in lieu of handing the tax over to Customs and Excise. As well as supporting the protection of the natural environment, projects were allowed to benefit where they supported the *"protection, maintenance, repair, restoration of a building or other structure which is a place for religious worship or of historical or architectural interest, open to the public and situated in the vicinity of a landfill site".*

Miniscule requirements for eligibility opened the door for a tidal wave of applications which have benefited churches and cathedrals throughout the UK, the sums realised being either one-off payments of a few hundred to a hundred thousand pounds, or a continuous series of payments as work progressed.

In the year 2000 the total tax raised by the landfill operators was about £84 million, of which approximately £7.1 million went towards the costs of restoring and repairing churches, chapels and buildings of architectural or historic interest.

THE MATCH FUNDING REQUIREMENT

To qualify for funding under the Landfill Tax Credit Scheme, there has to be a degree of match funding by a third party. The requirement is a little complex. The landfill operator can credit the Landfill Tax due to Customs and Excise with 90 per cent of a donation to an enrolled 'environmental' body (such as a church), but that credit has to be a maximum of 20 per cent of the tax due. For example:

Landfill Tax due to Customs	£10,000
Credit claimed @ 20% by landfill operator	-2,000
Net paid to Customs by landfill operator	8,000
As only 90% of the landfill operator's donation can come from the credit, the actual donation made is £2,222 (credit x 100/90):	
Donation by landfill operator to good cause	£2,222
Tax credit	-2,000
Shortfall	**£222**

The shortfall (in this case £222) may be carried by the landfill operator, but all landfill operators (in the experience of the author) expect to be recompensed for the amount. Under the rules this element of the funding has to come from a third party. It cannot be paid by the beneficiary.

OTHER KEY POINTS

- Landfill operators are under no obligation to give any donation.
- Although there is no requirement for funding to come from the landfill operator of the site which is nearby, many operators will only donate to buildings in the vicinity of their own site.
- Where third party funding is to be obtained, applications made to different organisations must address the specific funding criteria of each.
- Organisations should find a donor to make a commitment before registering in case they are unable to gain funding from the landfill tax operators, wasting their £100 registration fee.

At present 31 per cent of tax credits goes to recycling waste. In the 2001 budget, the Chancellor expressed a wish that this should be increased. Since then an 'indicative target' of 65 per cent of landfill tax credits has been allocated to sustainable waste management projects, with a third of tax credits within this category allocated specifically to recycling projects (House of Commons Hansard Written Answers for 10 May 2001). No reduction in the amount available for historic buildings and churches has been suggested, but we will need to keep an eye on future Budget reports.

FURTHER INFORMATION

To enrol as an environmental body contact:
Entrust Tel 0161 972 0044

For a list of landfill operators in the UK contact:
Customs and Excise Tel 0845 0109000

Addresses are given under Useful Contacts page 226.

FJ GOLDEN is a former Customs and Excise Senior Executive Officer. He advises on VAT on works to listed buildings and on applications under the Landfill Tax Credit Scheme. For further details see Highmead, page 38.

The Windsor Castle fire took 15 hours and a million and a half gallons of water to put out the blaze. The next five years would be spent restoring the Castle to its former glory.

INSURING YOUR HISTORIC BUILDING DURING REPAIRS AND ALTERATIONS

As the fire at Windsor Castle showed, historic buildings are more at risk during building works than at any other time. **MATTHEW MULLEE** of La Playa examines insurance issues for repair, alteration or extension work.

Whether or not you're considering restoration or alteration work, make sure you set yourself up with watertight insurance from the outset. If you own an historic building, whether it is a stately home or an ancient cottage, a standard policy from a general insurance provider could fall well short of expectations if you need to make a claim. If your building is also listed, it is protected by law, limiting repair options and making some higher costs unavoidable: it is the responsibility of the owner to insure it properly.

CHOOSING A SUITABLE INSURANCE POLICY

It does pay to use a specialist insurance broker who can provide advice about the choice of insurer, negotiate better premium rates and cover, and will usually arrange a 'risk management' appraisal. They will also think ahead to issues which will come up if the property needs to be repaired, altered or extended:

Building value – the difference between 'market' value and 'reinstatement' value is crucial for historic buildings. You need to determine how much it would cost to rebuild the entire building using like materials and methods of construction. A cursory note of the exterior construction is really not sufficient to calculate the rebuilding cost: from the outside, a building may appear to date from the 18th century, but the inner timber structure may be 15th century. Special features could be overlooked, such as fireplace mantles, plaster mouldings, carved timber panelling. Insurers like Independent, Chubb and Hiscox provide specialist appraisers.

Cover for repairs – repair contracts of substantial historic buildings can be long and complicated, giving rise to many extra costs. The standard 'cost per square metre' tables used by non-specialist insurers and mortgage company surveyors to calculate rebuilding costs are simply not up to the job as far as historic buildings are concerned, especially if authentic materials are required. The Association of British Insurers advises that these tables are suitable only for houses of 'average quality finish' and that if a pre-1920 house needs be reinstated to its original style, a professional valuation is required.

A CASE IN POINT

A Georgian home in the Midlands was converted into seven units and insured under a standard policy for £7 million. Moving to a specialist insurer who conducted a risk management appraisal, the residents discovered that the cost of rebuilding the property would have been over £15 million. In the event of a fire, the residents could have suffered a financial loss of over £8m, which would NOT BE PAID by the insurer. The revised policy offered double the cover for only 25 per cent extra premium.

WHEN YOU COME TO REPAIRS AND ALTERATIONS

Having decided on the new wing to be added to the property, or renovation project to be undertaken, talk to your insurer about extending your cover for the works.

Some insurers reserve the right to refuse cover for works over £75,000, or where a JCT Minor Works 6.3b contract (the standard contract between builder and client) has not been signed. Others will happily re-insure if the paperwork is in place. For example, Chubb Insurance will ask you to complete a questionnaire providing details such as the name of the contractor, their insurers and the scope of their insurance policy (the amount of their public liability cover, and the limit of indemnity), as well as a full description of works to be carried out. They will also need to know the contract value, contract period, details of security at the site, fire protection during the works and whether a JCT Clause is applicable.

If the property is uninhabited during the works, extra theft cover may need to be put in place. If the property is open to the public, the owner's public liability cover will need reviewing, too.

WHO DOES WHAT?

The parties involved in the building works will usually be co-ordinated by an architect or surveyor who will oversee the project. They will tender the work out to trusted building contractors and craftsmen. The choice of contractors is important as specialist materials and skills may be required to achieve a proper repair, and a poor job could affect the market value of the property. If the work is part of an insurance claim, most non-specialist insurers will insist that a building contractor from their own panel is used, but insurers such as Independent, Chubb and Hiscox allow their clients to use the most appropriate craftsmen for the repair work.

Who is liable for what?

RISK	RESPONSIBILITY
Existing property	Owner
Works in progress	Owner
Negligent damage	Contractor's public liability

The architect/surveyor will probably use the JCT Minor Works 6.3b contract conditions to form the basis of the contract between you and the builder. This usually places the onus on the owner as the employer to insure against material damage to the project – hence the need to brief the insurance company fully. You need an insurer and broker who understand these contract conditions.

This onus on the property owner is the subject of some debate among insurers, and may change in the future so that the contractor becomes responsible for the works' insurance.

The main contractor has a responsibility to ensure that no damage is caused to the property and must maintain adequate public liability insurance. The indemnity limit should reflect the maximum potential loss. For smaller homes an indemnity limit of £2 million will be adequate; for larger properties, the sum insured will need to be increased accordingly. Any sub-contractors who are employed must also carry the same indemnity limit and their insurance details should be verified before they start work.

Many claims arise from 'hot works'. Blowtorches and welding equipment for example, if left smouldering, may start a blaze and all naked flames have been banned from some historic properties. Where works are permissible all reputable contractors should have a hot works permit that requires them to monitor an area worked on for at least an hour after the work has been completed. Check that the contractor or any sub-contractor does not have a heat exclusion in their liability policy, as this will affect your insurer's recovery rights.

If the building being worked on is close to other people's property, the owner should consider extending the contractor's liability insurance to cover non-negligent damage that may be caused during the works; the JCT 21 2 1 contract.

CROMWELL'S ESCAPE ROUTE

If the property is listed due to an association with historical events or figures, legislation may require an archaeological excavation of the debris left following damage – potentially a costly process. In the case of one Oxfordshire property, the tunnels leading from the cellars to the surrounding fields were used by Oliver Cromwell as an escape route. Due to their historical significance, the specialist insurance appraiser carefully included a nominal sum in the rebuilding cost to excavate the entire length of the tunnels.

EXPOSURES

Repair and alteration works will make your property particularly susceptible to damage:

Fire – risks may be increased due to temporary wiring, exposed electrical wiring, electrical shorts brought about by pulling wiring, or damage to wiring caused by sloppy demolition of surrounding walls and ceilings. Flammable welding gases and paint removal torches are also dangerous.

Water – damaged pipes and exposure to the elements add to the risks.

Mechanical systems – such as ventilation, heating, air conditioning and plumbing systems, need special attention. Natural gas piping is a primary concern and often gets damaged during renovation works.

Roof – renovations pose increased risks from both water and fire damage. Water damage claims are more common, but roof fires started by tar kettles, welding and other hot works are not uncommon.

Theft/vandalism – a work site may be exposed to theft, particularly of architectural items such as fireplaces.

Public safety – if your property is open to the public, holes, walkways and other obstructions may be a hazard.

MINIMISING THE RISK

Good risk management is vital. A specialist insurance policy will make allowance for the costs of temporary shoring to prevent further collapse of the building, and for the cost of protecting the property from the elements and theft. Here are a few ways to reduce risk exposure:

Housekeeping – renovation projects often generate a substantial amount of debris. This debris contributes significantly to the flammable load contained within a building, so keep it cleared with all debris removed from the site regularly.

Electrics – faulty electrical wiring and appliances cause most fires. It is essential that all wiring and older electrical appliances are certified safe by a qualified electrician.

Fire protection – extend into the site such fire protection measures as fire blankets and fire extinguishers as well as panic buttons and smoke detectors linked to a monitoring station. For larger homes, a visit from the local fire officer should be arranged so that the brigade is familiar with the access to your property, the layout and the nearest water supply. A smoking ban should be mandatory on the site. If repairs are being undertaken on a large scale, it may be worth installing a temporary sprinkler system.

Water – ensure the property is inspected for weatherproofing at the end of each working day. You could also install leak sensors on pipes, leak trays under pipes in high risk areas, and/or automatic isolator valves.

Theft/vandalism – scaffolding may make the building more accessible to intruders, so it might be necessary to consider employing security staff.

Security measures should be carefully designed so as not to compromise or damage the historic fabric or integrity of the building.

THE COST OF INSURING YOUR REPAIR WORKS

You will need to allow for an extra charge for the extended cover during the works. Also, once the works are completed (and your property's reinstatement value has increased with a brand new wing or a magnificent face-lift) the sum insured, and consequently the premium, will increase.

WORK WITH SPECIALISTS

Just as it is necessary to employ specialist craftsmen to restore a moulding, it really is worth choosing a specialist broker who can advise on cover and deal with claims settlements and disaster recovery quickly and easily. This will protect both the fabric and the value of your home – not to mention your sanity – when it comes to the crunch.

Proper planning today by owners and insurance companies will ensure the survival of historic buildings tomorrow.

USEFUL CONTACTS:

La Playa (specialist insurance broker):
www.laplaya.co.uk Tel 01223 522411

Chubb Insurance
www.chubb.com

Independent Insurance
www.independent-insurance.co.uk

MATTHEW MULLEE heads La Playa's Private Homes division, and has a wealth of specialist knowledge in arranging insurance for period and listed buildings, as well as in household contents including fine art, jewellery and antiques. Tel 01223 578152 (direct) E-mail matt.mullee@laplaya.co.uk

BUILDING CONTRACTORS

▶ ANELAYS

William Anelay Limited, Murton Way, Osbaldwick, York YO1 5UW
Tel 01904 412624 Fax 01904 413535
E-mail office@williamanelay.co.uk Website www.williamanelay.co.uk

BUILDING AND RESTORATION CONTRACTORS: The company is a family concern, founded in 1747 in Doncaster. Now based in York, William Anelay continues to prosper specialising in restoration and repairs to historic buildings, employing highly skilled craftsmen in masonry, joinery, carpentry, leadwork and general building work. Their hand craftsmanship is only supplemented by the latest technology. Anelays offers a personal service, taking pride in their work, giving attention to detail and achieving very high standards. Examples of recent work can be seen throughout the United Kingdom including: Harewood House; Temple Newsam House, Leeds; Middleham Castle; The Old Wellington, Manchester; and St Magnus Cathedral, Orkney.

▶ ARTHUR E WOODWARD

1 Hall Mews, Leigh Hall Road, Leigh-on-Sea, Essex SS9 1QZ
Tel/Fax 01702 476345

SPECIALISTS IN GOOD QUALITY CARPENTRY AND BUILDING WORK: Arthur E Woodward has been established since 1963 undertaking refurbishment work in period buildings, including Georgian and Tudor to a high quality. Also building contracting specialising in structural work and shoring up of buildings. Traditional on-site joinery, whether you need a window made for a stone turret or a door repaired, this can be carried out by a Woodward craftsman. Mouldings, skirtings, architraves, handrails, can all be matched to your requirements. Member of the British Woodworking Federation.

▶ BERNARD A SHEPHERD LTD

33A Cumberland Street, Macclesfield, Cheshire SK10 1DD
Tel 01625 432477 Fax 01625 432488

CONSTRUCTION AND CONSERVATION CONTRACTORS: This highly experienced company operates primarily in the North West of England. *See also: profile entry in Timber Frame Builders section, page 73.*

▶ BOSHERS (CHOLSEY) LTD

6 Reading Road, Cholsey, Wallingford, Oxon OX10 9HN
Tel 01491 651242 Fax 01491 651800 Contact CWL Bosher

BUILDING CONTRACTORS: This family company has been working on historic buildings in Berkshire, Oxfordshire and neighbouring counties for over 175 years and employs many of its own tradesmen and apprentices. Boshers actively seek to be involved in work of a high quality and of a wide variety, especially to listed and other buildings. Various works have won awards. They have their own extensive joinery and timber milling facilities and are also able to provide specialist plasterwork, decoration and leadwork to the highest standards.

▶ BURLEIGH STONE CLEANING & RESTORATION CO LTD

The Old Stables, 56 Balliol Road, Bootle, Merseyside L20 7EJ
Tel 0151 922 3366 Fax 0151 922 3377
E-mail info@burleighstone.co.uk
Website www.burleighstone.co.uk

BUILDING CONTRACTORS: A comprehensive high quality service in historical and listed buildings, archaeological contracting and advice. *See also: display entry in Masonry Cleaning section, page 172.*

▶ BURROWS DAVIES LIMITED

The Stoneyard, West End, Strensall, York YO32 5WH
Tel 01904 491849 Fax 01904 491910

MASONRY AND RESTORATION SPECIALISTS: *See also: profile entry in Stone section, page 105.*

BUILDING CONTRACTORS

▶ BUSBY'S BUILDERS

Buzwood, Basingstoke Road, Old Alresford, Hants SO24 9DL
Tel/Fax 01962 732076

SPECIALIST BUILDING CONTRACTORS: A small family firm with experienced craftsman used to working on older properties and listed buildings including churches. Recent contracts include the complete renovation and refurbishment of a thatched timber barn at Armsworth; extension and renovations to an 1867 rendered lodge; re-instatement of three main period rooms at Tichborne House following dry rot eradication; a new cottage in the style of, and together with, the re-building of barns and garages at Cheriton. Busby's work within a 25 mile radius of Old Alresford.

▶ C J ELLMORE & CO LTD

Henshaw Works, Henshaw Lane, Yeadon, Leeds LS19 7RZ
Tel 0113 250 2881 Fax 0113 239 1227
E-mail mail@ellmore.co.uk
Website www.ellmore.co.uk

BUILDING CONTRACTORS: Ellmore Construction is a family run business, established in 1972. They have a well equipped joiners shop capable of handling large and specialist joinery contracts supported by a fully skilled site workforce, including stone masons and carpenters. The company undertakes refurbishment and restoration of historic buildings including churches, cathedrals, town halls, monuments, retail and financial premises in addition to new-build and maintenance work throughout the North of England. The company is interested in contracts up to the value of £1.5 million. For more information please contact Andrew Ellmore.

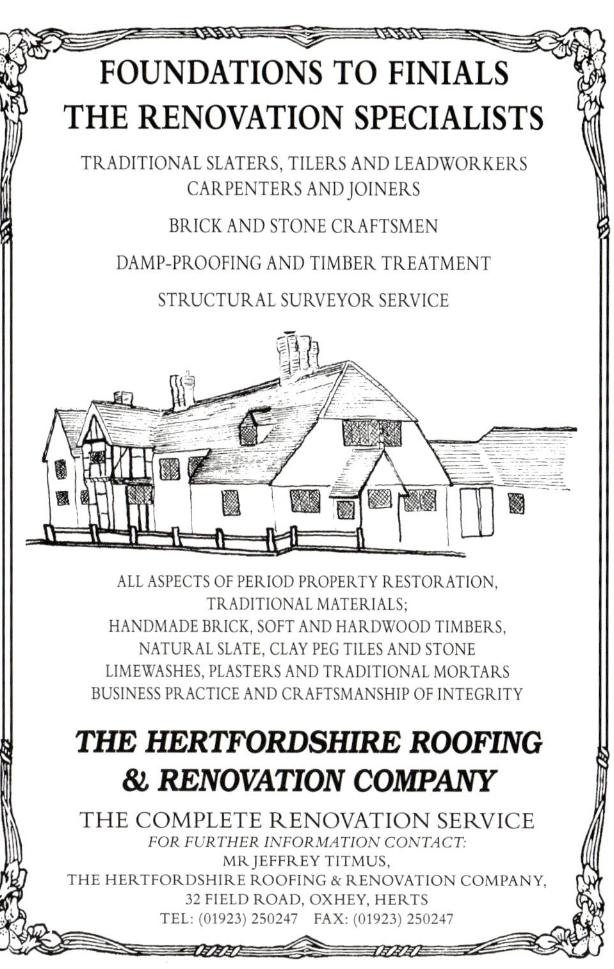

BUILDING CONTRACTORS

▶ C R CRANE & SON LTD

Manor Farm, Main Road, Nether Broughton, Leics LE14 3HB
Tel 01664 823366 Fax 01664 823534
Website www.crcrane.co.uk

SPECIALIST BUILDING AND JOINERY CONTRACTORS: Chartered Builders established 1910 by the current Managing Director's grandfather. Winner of conservation awards, specialising in traditional repair works using SPAB and English Heritage methods to churches, cathedrals, barns, follies - they undertake contracts up to £2 million. Their apprentice trained craftsmen are experienced in timber framing, leadwork, masonry, ironworks, brickwork, lime mortars and plasters, backed up by CIOB qualified staff. Consultancy service for historical research, structural surveys, planning, feasibility studies and cost/value analysis. Their in-house joinery works produces traditional windows, doors, panelling, staircases and ecclesiastical joinery.

▶ CALDER TRADITIONAL BUILDING SERVICES

'Woodhurst' Cattlegate Road, Crews Hill, Enfield, Middx EN2 8AU
Tel 01707 876515 Fax 01707 872413

SYMPATHETIC RENOVATIONS: Restoration and structural alterations and repairs to all period properties. Lime mortars, plastering, rendering and limewash. Wrought and cast ironwork. Working exclusively in London north of the Thames up to Hertford town. Established 1966, this family business portfolio of projects includes many important historic buildings, mostly in residential but also in commercial present day uses. Calder's work is fully insured and guaranteed. *See also: Calder Group display entry in this section, page 59.*

▶ CAMERON (UK) PLC

Cockshades, Wybunbury, Nantwich, Cheshire CW5 7HA
Tel 01270 841122 Fax 01270 841520
E-mail cameron@dial.pipex.com

BUILDING RESTORATION AND CONSERVATION: Having in-house design services and their own masonry yard, Cameron offer a nation-wide, comprehensive service, from initial survey, through design and supply to fixing and restoration of natural and cast stone, brick, terracotta and faience. *See also: display entry in Stone section, page 107.*

▶ CAPPS & CAPPS LIMITED

The Sawmill, Sarnesfield, Herefordshire HR4 8RH
Tel 01544 318877 Fax 01544 318399

HISTORIC BUILDING CONTRACTOR: Long established specialist building contractor Capps & Capps Limited provides a full range of skills and expertise for the conservation and repair of churches, castles, cathedrals, houses and other historic buildings. Based in Herefordshire the company provides quality oak for many of its projects. Capps & Capps has a well-equipped joiners shop capable of handling a range of specialist joinery projects which is supported by a fully skilled site workforce. The company employs a team of highly skilled stonemasons, many of whom are working presently at Hereford Cathedral. Clients include the National Trust, English Heritage, The Dean and Chapter of Hereford, Merthyr County Borough Council and private clients.

▶ CARREK LIMITED

Mason's Yard, Wells Cathedral, Wells, Somerset BA5 2PA
Tel 01749 689000 Fax 01749 689089
Website www.carrek.co.uk

HISTORIC BUILDING REPAIR COMPANY: Carrek provides a full range of skills and expertise for the conservation and repair of historic buildings, monuments and sculpture. Specialist trades include; stone and plaster conservation, sensitive cleaning including, JOS and DOFF systems, lime rendering, plain and decorative plastering, stonemasonry and carving, carpentry and traditional joinery and leadwork. Carrek will assist with the preparation of reports and specifications and act as consultants for the analysis of historic mortars, plasters and paints. Clients include The Churches Conservation Trust, the National Trust, English Heritage and numerous private individuals. Recent or current contracts include: Wells Cathedral Chain Gate; Temple Church, Bristol; Muchelney Abbey; Oare Church; Smeatons Tower; Blackfriars, Gloucester; Godolphin House, Cornwall.

▶ CHALK DOWN LIME LTD

102 Fairlight Road, Hastings, East Sussex TN35 5EL
Tel 01424 443301 Fax 01580 830096 Mobile 0771 873 8708
E-mail chalkdownlime@supanet.com

MAINTENANCE AND REPAIR OF HISTORIC BUILDINGS: *See also: profile entry in the Mortars & Renders section, page 168.*

▶ CHARLES ADAMS LIMITED

27 Palace Gate, Kensington, London W8 5LS
Tel 020 7584 9106 Fax 020 7590 7585
E-mail charlesadams27@hotmail.com

SPECIALISTS IN PROPERTY RESTORATION AND RENOVATION: Over 120 years ago Charles Adams Daw founded C A Daw & Son. The firm constructed some 200 substantial properties in Kensington, Bayswater and Mayfair. Later the company built the Queen Anne style houses and flats in Palace Gate, together with purpose-built offices, which are still in use by the company today. Charles Adams Limited, named after the founder, specialises in high quality renovation and restoration projects. The company's team all have competent track records working with top architects and interior designers for many important clients, including the National Trust and English Heritage.

▶ DAVID BALL RESTORATION (LONDON) LIMITED

104A Consort Road, London SE15 2PR
Tel 020 7277 7775 Fax 020 7635 0556
E-mail mail@dbr.uk.com

SPECIALIST BUILDING CONTRACTORS: *See also: display entry and profile entry in Stone section, page 109.*

▶ DUNNE AND CO LTD

Ashbrooke, Chalkhouse Green, near Reading, Berks RG4 9AN
Tel 0118 972 2364 Fax 0118 972 1120
E-mail dunneandco@aol.com

BUILDING AND RESTORATION: A long established company employing a large team of specialist crafts people from masons, joiners and roofers to specialist decorators in the restoration and renovation of period and listed buildings using traditional materials. Mainly working in Berkshire and Oxfordshire, Dunne and Co has carried out projects in France and The Channel Islands. Excellent references available.

▶ HALL CONSTRUCTION

Unit 6, Tannery Yard, Witney Street, Burford, Oxfordshire OX18 4DQ
Tel 01993 822110 Fax 01993 823880

CHARTERED BUILDING COMPANY: Phillip Hall is a corporate member of the Chartered Institute of Building. He founded the company in 1983 and has gathered around him a team of artisans skilled in traditional methods. Hall Construction's local reputation and success is reflected in the ongoing and phased developments which have spanned several years. In following the highest standards of workmanship and professionalism Hall Construction has been recognised in quality award schemes since 1989, at both regional and national level. Recent projects have included restoration, refurbishment and also new building work.

▶ HASLEMERE BUILDERS LIMITED

Cylinders Lane, Fisher Street, Northchapel, West Sussex GU28 9EL
Tel 01428 707282 Fax 01428 708040

BUILDING CONTRACTORS: A small family run business carrying out a variety of work from small repairs to major projects. Relying on traditional methods and high standards of craftsmanship they are able to provide a comprehensive service covering all aspects and trades. Working mainly in West Sussex, Surrey and North East Hampshire they have been involved in the extension, repair, restoration and refurbishment of period, listed and historic buildings. Haslemere Builders has been a member of the CIOB Chartered Building Company Scheme since 1992 and was the main contractor on an 18th century Mill restoration project which was awarded a commendation in the Civic Trust Awards.

BUILDING CONTRACTORS

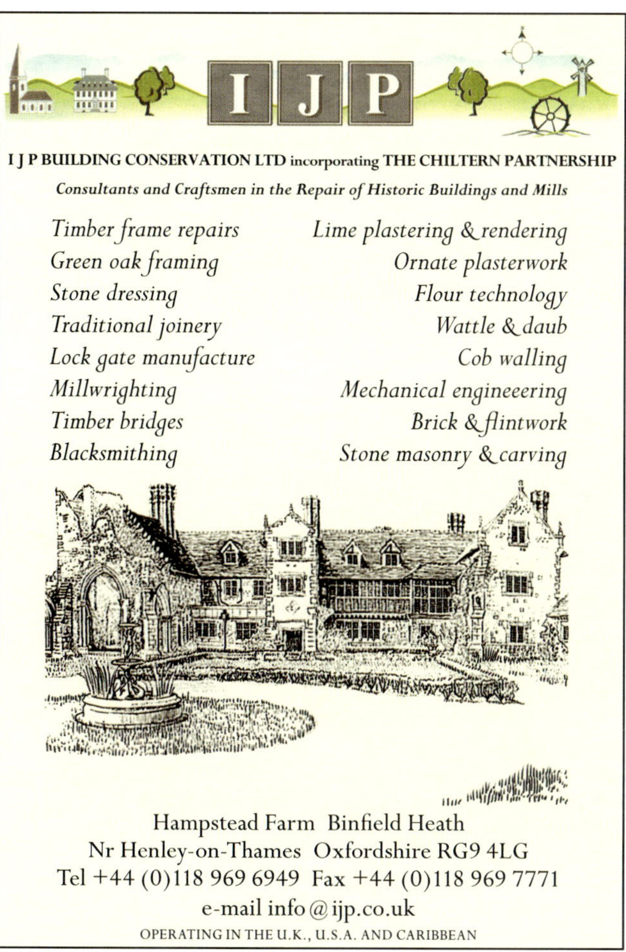

I J P
I J P BUILDING CONSERVATION LTD incorporating THE CHILTERN PARTNERSHIP
Consultants and Craftsmen in the Repair of Historic Buildings and Mills
Timber frame repairs
Green oak framing
Stone dressing
Traditional joinery
Lock gate manufacture
Millwrighting
Timber bridges
Blacksmithing
Lime plastering & rendering
Ornate plasterwork
Flour technology
Wattle & daub
Cob walling
Mechanical engineeering
Brick & flintwork
Stone masonry & carving
Hampstead Farm Binfield Heath
Nr Henley-on-Thames Oxfordshire RG9 4LG
Tel +44 (0)118 969 6949 Fax +44 (0)118 969 7771
e-mail info@ijp.co.uk
OPERATING IN THE U.K., U.S.A. AND CARIBBEAN

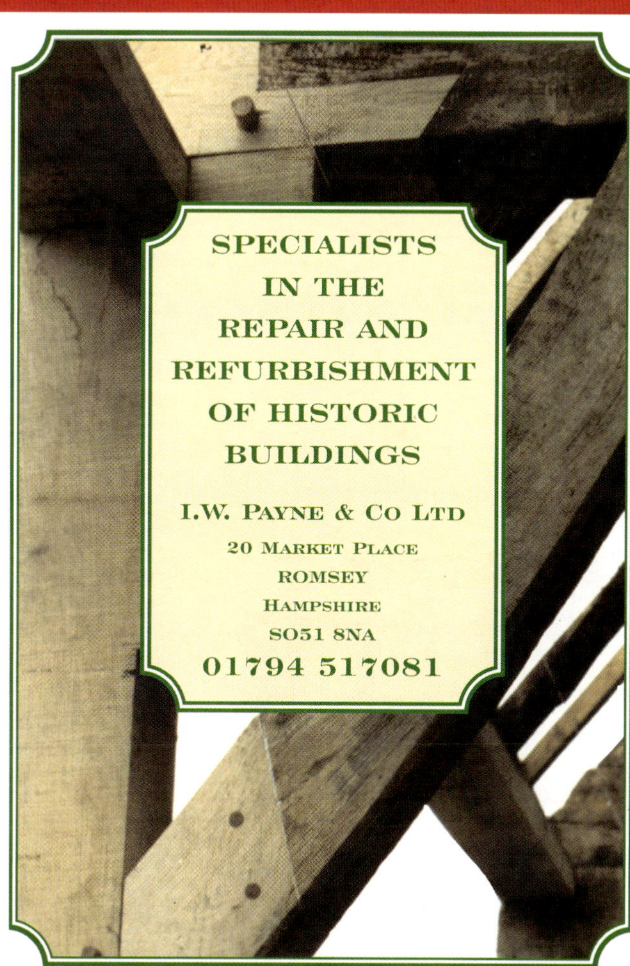

For 120 years Linford - Bridgeman has been entrusted with the restoration *and* conservation *of* some *of* Europe's finest historic buildings *and* monuments.

Our particular strength is in working as *a* main *or* management contractor backed up by *a* broad range *of* our own specialist single trade skills including:

Stonework
Leadwork
Joinery
Fine Crafts

We also offer *a* comprehensive technical service *for* structural inspection *and* reports with the fee discounted against any work required.

Linford - Bridgeman Limited
Quonians
Lichfield
Staffordshire WS13 7LB

Tel: 01543 414234 Fax: 01543 410065

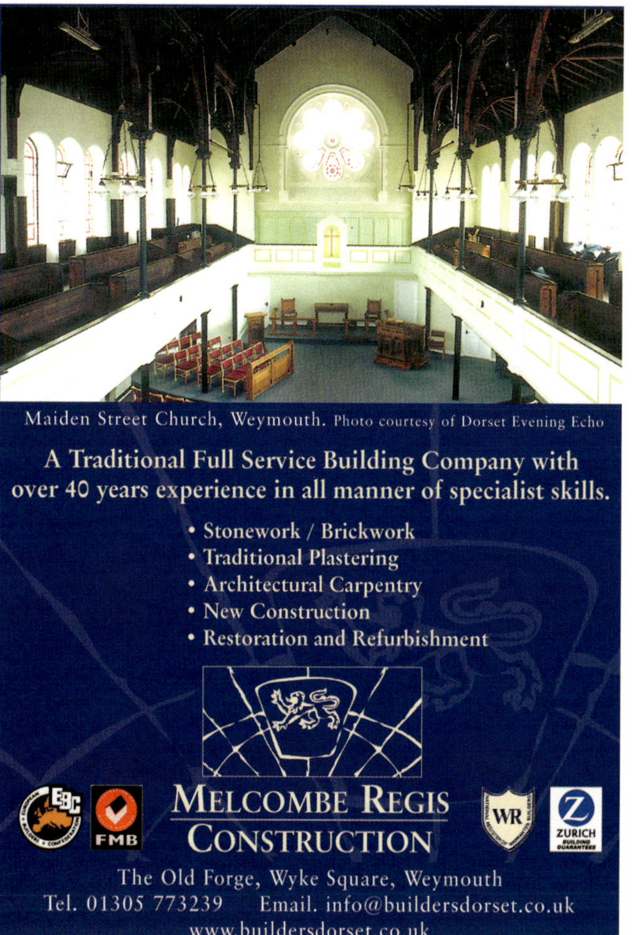
Maiden Street Church, Weymouth. Photo courtesy of Dorset Evening Echo

A Traditional Full Service Building Company with over 40 years experience in all manner of specialist skills.

- Stonework / Brickwork
- Traditional Plastering
- Architectural Carpentry
- New Construction
- Restoration and Refurbishment

MELCOMBE REGIS CONSTRUCTION
The Old Forge, Wyke Square, Weymouth
Tel. 01305 773239 Email. info@buildersdorset.co.uk
www.buildersdorset.co.uk

▶ HIRST CONSERVATION

Laughton, Sleaford, Lincolnshire NG34 0HE
Tel 01529 497449 Fax 01529 497518
E-mail hirst@hirst-conservation.com
Website www.hirst-conservation.com

SPECIALIST BUILDING AND ART CONSERVATORS: Consultancy and conservation work to painted and applied decoration on plaster, stone, canvas, wood and metal substrates. Restoration and recreation of historic decorative schemes. Also specialist building works including joinery, sculpture, marble, stonework, stone cleaning, stucco, pargetting, wall and floor plasters. Surveys, specifications and analysis services available. Hirst's policy is to produce work of the highest quality, the approach to which is fully justified with regard to contemporary conservation ethics, both during the period of work on site and in documentation which follows the completion of projects. Their team of conservators represents many different skills and disciplines and their combined knowledge and experience is used to develop comprehensive conservation practices. *See also: display entry on the inside front cover.*

▶ HOLLOWAY WHITE ALLOM

43 South Audley Street, Grosvenor Square, London W1K 2PU
Tel 020 7499 3962 Fax 020 7629 1571 E-mail hwa@laing.com

BUILDING CONTRACTORS AND HIGH QUALITY REFURBISHMENT: Based in Mayfair since 1902, its 1999 turnover was £34 million. With a directly employed staff of 100, 107 skilled craftsmen and 10 apprentices in traditional trades and crafts, it is particularly well known for refurbishment and restoration, fine decorative finishes and attention to detail on listed historic buildings. Its current work is being carried out in historic monuments, country estates and single residences in the heart of London. Holloway White Allom is a wholly owned subsidiary of the Laing Group which gives it full access to all their resources.

▶ J G MATTHEWS LIMITED

7 Currie Street, Hertford SG13 7DA
Tel 01992 550173 Fax 01992 584998
Website info@buildingconservation.co.uk

BUILDING CONTRACTORS: J G Matthews, managed by a chartered surveyor, have over 25 years experience with old buildings. Their qualified and experienced workforce carries out repairs and maintenance to listed buildings and timber frames and artfully crafts barn conversions. The company works with lime render and skilfully carries out brick repairs using lime mortars. J G Matthews also specialises in purpose made joinery including doors, windows and sash repairs. Members of Hertfordshire Building Preservation Trust and Society for the Protection of Ancient Buildings. Clients include Gascoyne Cecil Estates (Hatfield House), Diocese of St Albans, Hertfordshire BPT.

▶ KIRBY PEEL

37 Slack Lane, Crofton, Wakefield, W Yorks WF4 1HH
Tel/Fax 01924 862713 (Timber Framing) Tel 01132 813452 (Joinery)

TIMBER FRAME CONSERVATION AND ARCHITECTURAL JOINERY: *See also: profile entry in Timber Frame Builders section, page 73.*

▶ KNOWLES & SON (OXFORD) LTD

Holywell House, Osney Mead, Oxford OX2 0EA
Tel 01865 249681 Fax 01865 790601 E-mail build@knowle.co.uk

BUILDING CONTRACTOR: Seventh generation Knowles & Son, established in 1797, currently turn over £12 million and has 140 directly employed staff. Building on a reputation of craftsmanship to the highest quality, one of the company's specialities is the conservation and repair of historic and ecclesiastical buildings. Continually working for the Oxford Colleges and other bodies such as the National Trust has ensured this traditional craftsmanship has been maintained. Knowles & Son has its own specialist joinery division, and operates within a 60 mile radius of Oxford on minor works up to major contracts to a value of £5 million.

▶ LONGLEY A division of Kier Regional Ltd

East Park, Crawley, West Sussex RH10 8EU
Tel 01273 561212 Fax 01273 564333

BUILDING RESTORATION AND CONSERVATION IN THE CENTRAL HOME COUNTIES: Longley has over 130 years experience working to maintain the nation's heritage buildings. Appropriate skills and technical expertise have been used on projects such as the Hampton Court Palace restoration following the tragic fire, the re-roofing of the National Trust's Abbey at Romsey, Newhaven Fort restoration in conjunction with English Heritage, and the rebuilding of Tonbridge School Chapel after the fire which destroyed the listed structure. Contracts range from £100,000 to £5 million and can benefit from Wallis joinery and stonework as Longley is also part of the Wallis division within the Kier Group. For further information contact Graham Todd, General Manager. *See also: Wallis display entry in Building Contractors section, page 55.*

▶ MAYSAND LIMITED

109-111 Windsor Road, Oldham, Lancs OL8 1RH
Tel 0161 628 8888 Fax 0161 627 0996 E-mail sales@maysand.co.uk

MASONRY AND TIMBER CONSERVATION CONTRACTORS: *See also: display entry in Stone section, page 110.*

▶ MOWLEM RATTEE & KETT

Digital Park, Station Road, Longstanton, Cambridge CB4 5FB
Tel 01954 262600 Fax 01954 262601

SPECIALIST BUILDING AND RESTORATION CONTRACTORS: Building on 160 years of experience Mowlem Rattee & Kett has been responsible for building and repairing some of the country's finest historic buildings. Management skills have developed alongside the traditional craft based activities to provide a sensitive and commercial approach to restoration and refurbishment contracts. Operating as a specialist division of the Mowlem Group, few companies can demonstrate the same depth of project management experience and comprehensive track record. Key projects include work on the Houses of Parliament, Westminster Abbey, Ely Cathedral and the £9 million restoration of the Albert Memorial.

BUILDING CONTRACTORS

BUILDING CONTRACTORS

▶ R W ARMSTRONG & SONS LIMITED

Aldermaston Road, Sherborne St John, Basingstoke, Hampshire RG24 9JZ
Tel 01256 850177 Fax 01256 851089 Website www.rwarmstrong.co.uk

HIGH QUALITY BUILDING AND DESIGN: R W Armstrong & Sons is a family company spanning three generations which provides architects and interior designers a comprehensive service for substantial works to large country homes. From the experienced management team through long serving directly employed site personnel and traditionally trained apprentices making up a directly employed workforce of over 100 craftsmen, the firm is one of the few with the resources to cover all aspects of building work. Armstrong craftsmen include those trained and experienced in paint techniques, decorative plaster cornice and mouldings, traditional lime putty in brickwork, flintwork and plastering. Also, the company runs a large joinery shop in which it reproduces period doors, windows, stairs and mouldings. An approved contractor for English Heritage, SPAB, National Trust and local authority conservation departments. *See also: display entry in this section, page 64.*

▶ S & J WHITEHEAD Est 1872

Sawmills and Stoneyard, Derker Street, Lower Moor, Oldham OL1 4EE
Tel 0161 624 4395 Fax 0161 627 2952

CONSERVATION AND RESTORATION CONTRACTORS STONEMASONRY AND LEADWORK SPECIALISTS: For over 125 years S & J Whitehead have been proud of their contribution to the North West skyline with fine examples of their work and craftsmanship to be seen on cathedrals, churches, historic and public buildings throughout the region. Clients include the National Trust, Manchester, Blackburn and Bradford Cathedrals, Capesthorne Hall, Whalley Abbey, Skipton Castle, St Georges Hall, Liverpool. 1994 Calder Leadwork Award Winner. Suppliers of fine masonry and leadwork products. *See also: display entry in this section, page 66.*

▶ SANDY & CO (CONTRACTORS) LIMITED

Grey Friars Place, Stafford ST16 2SD
Tel 01785 258164 Fax 01785 256526
E-mail info@sandy.co.uk Website www.sandy.co.uk

HISTORIC BUILDING CONTRACTORS: Established in 1903, Sandy & Co is a well known firm of high quality building contractors which specialises in work to historic buildings and churches. Conservation, restoration and repair services are provided throughout the Midlands and across the United Kingdom for large and interesting projects. Sandy & Co works with top architects, local government and private individuals. Recent projects have included 17/19 High Street, Kinver for which they won the Carpenters' Award, and works to Windsor Castle and the Tower of London. Please ring Peter Godwin to discuss your project requirements.

▶ SIMMONDS OF WROTHAM

The Square, Wrotham, Sevenoaks, Kent TN15 7AH
Tel 01732 883079 Fax 01732 884055

RESTORATION BUILDING CONTRACTORS AND DEVELOPERS: Simmonds employ a long standing team of craftsmen, experienced in matching traditional methods of construction with current practice. Barn and oast conversions are a speciality. Recent projects include work to a Grade I listed church and several other churches, the complete renovation of a listed 16th century farmhouse, with its listed outbuildings of a thatched barn, oast house and stables. Operating in London, Kent and surrounding areas.

▶ SINDALL LTD

Conservation Office, 7 Avro Court, Ermine Business Park, Huntingdon, Cambs PE29 6WF
Tel 01480 414500 Fax 01480 414514

CONSTRUCTION SPECIALIST – CONSERVATION, RESTORATION, RENOVATION: Since 1895 Sindall has maintained its reputation for sympathetic work in the repair, reinstatement, conservation and refurbishment of historic and listed buildings, and for new-build in classical styles, in an area encompassing London, East Anglia and the Northern Home Counties within the contract value range of £10,000 to £10 million. Clients include the National Trust; English Heritage; Crown Estate Commissioners; Historic Royal Palaces; The Cadogan Estate; church PCCs and dioceses; local authorities and county councils; universities; developers and private owners. Directly employed labour skills and the building services company give Sindall the flexibility and expertise to adapt to the particular demands of each project. *See also: display entry on page 64.*

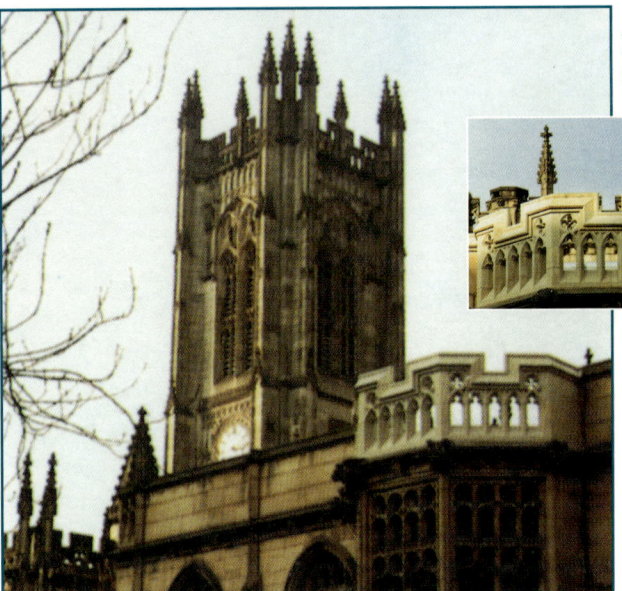

▶ ST BLAISE LTD

Westhill Barn, Evershot, Dorchester, Dorset DT2 0LD
Tel 01935 83662 Fax 01935 83017
E-mail stblaise@compuserve.com

MAIN AND MANAGEMENT CONTRACTORS IN HISTORIC BUILDING REPAIR: A leading conservation contractor committed to long term employment of skilled craftspeople in the traditional trades. Competitive nationally with bases in London, Kent, Dorset and Edinburgh, and especially competitive on larger or more demanding projects. A wealth of experience gained by working for the country's finest architects on its best buildings. Many churches, castles, follies, grottoes and towers. Consultancy for English Heritage, Historic Scotland, Cadw, Historic Royal Palaces, National Trust regions and many professional practices. There are significant cost and programme benefits by appointing a main contractor who maintains the best skills in-house. *See also: display entry in this section, page 65.*

▶ STONEGUARD

St Martins House, High Street, Ruislip, Middlesex HA4 7AU
Tel 01895 675577 Fax 01895 679125
- **Bath, Tel 01225 754025**
- **Birmingham, Tel 0121 452 5053**
- **Bradford, Tel 01274 656463**
- **Manchester, Tel 0161 773 3398**
- **Stirling, Tel 01786 831144**

E-mail sales@stoneguard.co.uk
Website www.stoneguard.co.uk

BUILDING RESTORATION AND CONSERVATION: Stoneguard is one of the country's leading refurbishment contractors, operating nationwide from offices in London, Leeds, Manchester, Bath, Birmingham and Stirling with over 120 operatives and skilled craftsmen. Stoneguard provides a complete masonry contracting service covering all aspects of work from a simple face-lift involving masonry repair and sensitive cleaning, to a complete structural refurbishment.

▶ T J EVERS LTD

New Road, Tiptree, Colchester, Essex CO5 0HQ
Tel 01621 815787 Fax 01621 818085
E-mail office@tjevers.co.uk

BUILDING RESTORATION AND CONSERVATION, SPECIALIST JOINERY MANUFACTURERS: Established in 1918, a traditional contractor offering a quality service in the conservation and restoration of historic and ancient buildings. The high calibre of their craft skills, traditional and modern construction techniques, coupled with quality management skills, they believe enables them to offer a service that is second to none. T J Evers' in-house specialist joinery division produces period and bespoke joinery to the highest standard of craftsmanship for many clients, including English Heritage. Small and major works departments enable them to carry out contracts both large and small.

▶ TIMOTHY WILLIAMS (BUILDERS) LIMITED

Howman House, The Square, Stow on the Wold, Nr Cheltenham, Glos GL54 1AF
Tel/Fax 01451 832554
E-mail twbuild@aol.com
Contact Tim Williams BSc (Hons) ACIOB
- **27 Palace Gate, Kensington, London W8 5LS**

Tel 020 7590 7588
Contact Neil Garrett ACIOB

SPECIALIST BUILDING AND JOINERY CONTRACTORS: Fourth generation family business specialising in high quality building renovation, refurbishment and the installation of purpose made joinery to domestic period and listed property in the Cotswolds and Central London. Adept at managing all trades in the construction process from lime plasterwork to stone slate roofs. Client list and project portfolio available on request.

▶ TREASURE & SON LTD

Temeside, Ludlow, Shropshire SY8 1JW Tel 01584 872161 Fax 01584 874876

CONSERVATION AND RESTORATION CONTRACTORS: Established in 1747, Treasure & Son is a family firm specialising in all aspects of building, from work on scheduled monuments to the restoration of Georgian and half-timbered houses. The company employs 50 time-served craftsmen and has worked recently at: Warwick Castle; Church of St. Bartholomew; Richards Castle; Ludlow Castle; Knowle Lime Kilns; and many private houses in the West Midlands and border country areas. In 1997, as main contractor, Treasure & Son Ltd won the Building of the Year Award for its work at the Mappa Mundi and Chained Library in Hereford. The company has its own in-house joinery shop which received The Carpenters Award in 1997 for high quality joinery at the Mappa Mundi building.

▶ WALLIS

47 Homesdale Road, Bromley, Kent BR2 9TN Tel 020 8464 3377 Fax 020 8464 5847

BUILDING CONTRACTORS: Over many years Wallis, with its associated companies Wallis Joinery and Broadmead Cast Stone and now incorporating Longley, has earned an exceptional reputation for the quality and excellence of their restoration and refurbishment work. In recent years Wallis has carried out work in the Foreign and Commonwealth Office (Old Public Offices Whitehall), Marlborough House, Windsor Castle, The National Portrait Gallery, The Natural History Museum, The British Museum, London Oratory, The Guildhall, Somerset House, Ightham Mote, Canada House, Eltham Palace and Hampton Court Palace. In 1991 the Old Public Offices, Whitehall, won the coveted Europa Nostra Award "For the magnificent and meticulous restoration to the original design of one of the finest examples of Victorian architecture in the United Kingdom". As part of the Kier Group, Wallis combines the advantages of a substantial and secure organisation with the skills and resources capable of undertaking the most demanding projects large or small in their chosen field. Wallis undertakes contracts throughout London and the South East of England and has offices in London and Maidstone. *See also: display entry on page 55.*

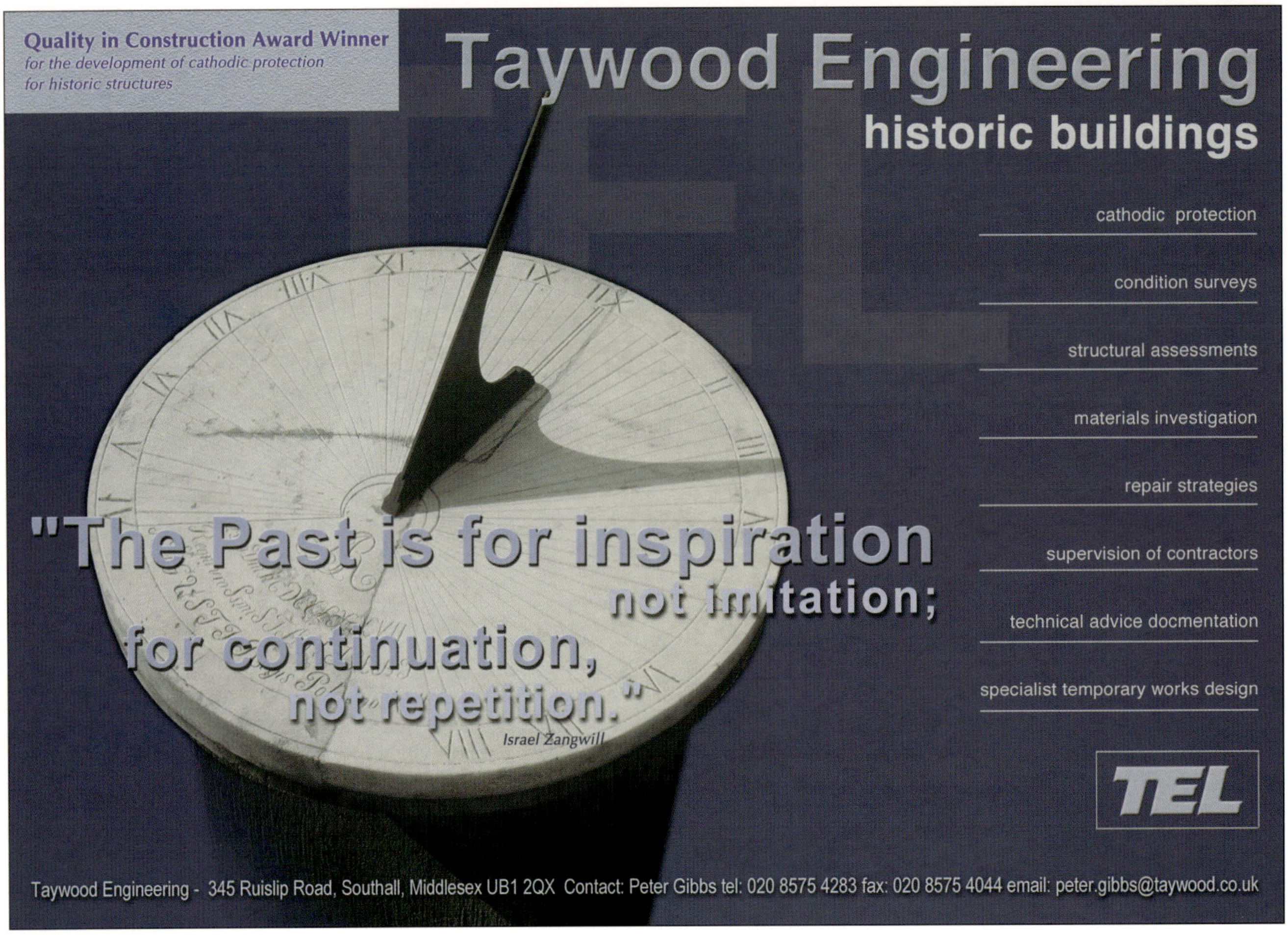

BUILDING CONTRACTORS

▶ WATES CONSTRUCTION

1260 London Road, Norbury, London SW16 4EG
Tel 020 8764 5000 Fax 020 8679 5570
Website www.wates.co.uk

RESTORATION AND REFURBISHMENT OF LISTED AND HISTORIC BUILDINGS: *See also: display entry above.*

▶ WEAVER CONSTRUCTION LIMITED

Harlington House, Harlington Road, Mexborough, South Yorkshire S64 0LE
Tel 01709 586201 Fax 01709 583329

BUILDING CONTRACTORS: Weaver Construction Limited, established in 1919, has a reputation based on its flexible response to the requirements of clients in both the public and private sectors, combined with the skills and experience to undertake refurbishment works of a demanding yet aesthetically pleasing nature, as well as sympathetic new-build in historic settings.

▶ WT SPECIALIST CONTRACTS LTD

21 Phoenix Place Industrial Estate, Lewes, East Sussex BN7 2QJ
Tel 01273 479764 Fax 01273 479765
E-mail info@wtgroup.co.uk
Website www.wtgroup.co.uk

STRUCTURAL STABILISATION: WT Specialist Contracts Ltd provides a complete structural repair and stabilisation service for residential, commercial, historic and listed buildings of all ages, styles and construction materials. With particular expertise in the sympathetic, non-disruptive restoration of period housing, churches and castles, the company has undertaken many successful projects for English Heritage and Historic Scotland. Using innovative, specialist repair products and techniques, WT's experienced engineers devise and implement concealed, cost-effective and reliable solutions for all situations where cracked or unstable masonry requires skilled drilling, tying, stitching, anchoring, reinforcing, pointing or grouting to overcome all common forms of structural failure.

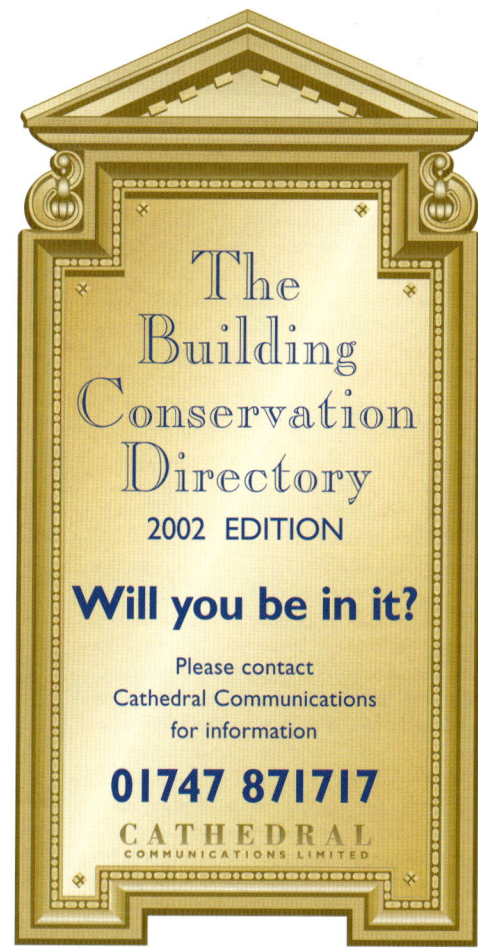

WATTLE & DAUB

IAN PRITCHETT

A timber framed building with wattle and daub infill, limewashed in the medieval manner

For many people the wonderful irregularities of wattle and daub walls and the undulations of a distorted roofline form part of the attraction of a medieval timber framed building. The walls gain their character from the timber frame which forms the load bearing structure of the building, leaving open areas between that have to be infilled to keep the weather out. The type of infilling varies according to the function and status of the building, its location within the country and the locally available materials. It is probably fair to assume that if a material was readily available and could be adapted for use it would have been used as an infill to a timber framed building at sometime, somewhere.

New woven hazel ready for daubing

Wattle and daub is one of the most common infills, easily recognisable by the appearance of irregular and often bulging panels that are normally plastered and painted. It is an arrangement of small timbers (wattle) that form a matrix to support a mud-based daub. The timbers normally fall into two groups, the primary timbers or staves, which are held fast within the frame and the secondary timbers or withies which are nailed or tied to, or woven around the staves. Arrangement and sizes of panels vary from area to area as does the orientation of the staves. The daub was applied simultaneously from both sides in 'cats' (damp, workable balls) pressed into and around the wattle in order to form a homogeneous mass. As the daub dried it was often keyed by scratching or 'pecking'. Once the daub had hardened, the surface was dampened to receive a lime plaster covering. The surface plaster was usually made of lime and sand or other aggregates reinforced with animal hair or plant fibre. The plaster was finished flush, or in some cases, it would continue across the panels and timbers alike. This would allow less important timbers to be concealed and only principal members to be shown. The plaster may be smooth trowelled, rough cast or even parged (incised and/or built up with a pattern or design).

PERFORMANCE

Wattle and daub may not be the most rigid material, but therein lies its strength. It is able to accommodate even the most severe structural movement; it is usually well sprung into the timber frame and offers support to weakening timbers that other forms of infill might not. Wattle and daub is not lightweight or flimsy. Its weight is not dissimilar to bricks, however its insulation is better and from a security point of view it can be far more difficult to break through than brick. Although wattle and daub is porous and moisture is absorbed when it rains, moisture levels are kept low because the daub acts like blotting paper to disperse the moisture and because of the high rate of evaporation from its surface.

In moderate, sheltered conditions and if well maintained, a wattle and daub panel should last indefinitely. Examples of 700 years old are known to exist.

Traditional infill panels in timber framed buildings can perform extremely well if properly constructed and maintained. Although in some areas of the country it was normal for infill panels to have protective plaster coatings which extended over the timber frame, it has become fashionable to remove plaster to expose timbers. This is likely to compromise the performance of the building and accelerate the decay of the previously protected structure. It is unreasonable to expect to have a timber frame exposed on both sides and not have draughts and/or some water penetration whether the infill panel is traditional or modern.

Where timber framing was not plastered over it was normal practice to limewash it each spring. Although this was partly for hygienic reasons (being slightly caustic, fresh limewash acts as a mild biocide and disinfectant), it had the tremendous benefit of filling minor cracks caused by seasonal movement. Medieval buildings would have looked quite different from the more recent black and white interpretation that we see so often today.

In some cases weatherboarding or tile hanging may have been added over the infill panels, particularly on exposed gables, to protect them from the weather. Removing the protective covering can lead to the recurrence of old problems all over again. It would be wise to learn from our forebear's experience and consider alterations only after careful thought and for good reasons, not purely on aesthetic grounds.

Decay is often caused by the introduction of hard cement in new renders and repairs, and by the use of modern impervious paints. This is because cement based renders are brittle and often crack, especially at the junction with the timber frame. When it rains, water runs down the face of the panels because both the cement and the modern paints are impervious, soaking right into the wall behind wherever a crack is found. Thus the daub will get wetter and wetter over time, leading to the decay of the timber frame and wattles as well as soggy, unstable daub. Only soft, porous and flexible finishes such as haired lime plaster and lime wash should ever be applied to daub.

REPAIR CONSIDERATIONS

Through the passage of time buildings may become neglected and some damage is inevitable. Knowing whether a damaged panel should be repaired or replaced, even with experience, requires careful consideration, weighing up many factors such as age, importance, rarity, position and function within the building, condition and cost.

Although cost has deliberately been put at the end of this list it will, in many cases, be the deciding factor. Age, importance and rarity can be difficult to define without research, however, bear in mind that all elements of ancient fabric are important and that the loss of any eats away at our heritage.

REPAIR TO WATTLES

Repair to daub can normally be implemented, even in the most extreme cases, providing the wattles are still in good condition or repairable. Whereas repair to a panel where the wattles have been totally consumed by fungal decay or insect attack can be very difficult even where much of the daub/plaster survives. Deterioration may be found in the wattles if they have been damp, particularly if they are not oak or contain sapwood, and hazel seems to be particularly prone to decay by woodworm (common furniture beetle). Wattle panels with insect attack *may* need some localised treatment, but are often strong enough to carry the daub. Introducing additional support can increase their strength. This can take the form of new staves or withies or timber battens or stainless steel mesh fixed across weakened areas. Each repair will be different, depending on the circumstances. In general finding the right solution is a matter of ingenuity based on the defects and conditions found.

Repair to a wattle panel may not be too difficult if the daub has already fallen away. The wattle behind does not need to be absolutely rigid, but should be strong enough to carry the new daub. It may be necessary to hold the wattle firm whilst applying the new daub.

Where daub is still in place the repair of a wattle panel can be much more challenging. In some cases it may be possible to re-support or re-fix loose daub by using non-ferrous wire ties or screws and washers. In some cases it will be necessary to hold a panel carefully in position, or even totally remove it in one piece, while repairs are carried out to the timber frame, and then put it back. In this case specialist advice is essential if a disaster is to be avoided.

REPAIR TO DAUB

Some shrinkage is normal even in the most successful of historic daubs, and gaps around the edge of the panel are usually caused by a combination of shrinkage within the daub and the timber frame seasoning. These gaps allow the panel to move, so to keep it weathertight they should be filled. They can easily be filled with daub or lime mortar. If problems are experienced with excessive shrinkage it is either because there is too much suction in the existing daub, or the repair mix is unsuitable, but it is always easier to control the shrinkage of a whole panel with the same moisture content. When areas of daub have failed or become detached, they can be repaired by applying new daub to fill the missing areas (after careful preparation and pre-wetting).

Problems can sometimes be overcome by additional wetting of the existing daub or by modifying the repair mix. The ingredients used in an original daub mix were normally used because they were locally available and cheap, they may not have been ideal. Nevertheless the first recommendation for a compatible material would always be to use the original material. Old daub salvaged from damaged panels can be broken up and mixed with a little water to make it useable again. It may be necessary to add additional material to bulk it out, or modify its performance. However, the required performance of a repair mix may be different from the requirement for a whole panel. A useful tip is to mix one part daub with one part of a good coarse lime mortar to achieve a better-behaved material.

NEW DAUB MIXES

Daub is generally made up of a combination of ingredients shown in the table below.

Binders	Aggregates	Reinforcement	Others
Clay	Earth	Straw	Dung
Lime	Sand	Hair	Blood
Chalk dust	Crushed chalk	Flax	Urine
Lime stone dust	Crushed stone	Hay or grass	Dung

The binder holds the mix together, the aggregates give it bulk and dimensional stability, the reinforcement helps hold it all together, control shrinkage and provide long term flexibility. Some locally available materials may contain more than one of the aggregates and other ingredients – for example, sub soil may contain clay, sand and earth. There is some debate over whether dung was deliberately added to daub mixes. It is probably reasonable to assume that the presence of dung in daub mixes was due to using old straw from animal sheds (why use fresh straw when it is valuable for animal bedding?) and using animals to do the hard work of treading the daub.

Historically, daub was a cheap material and lime was relatively expensive, so it is unlikely that lime was included in daub except under special circumstances. It is far more likely that the expensive lime would have been reserved for the plaster and limewash, where it would be necessary.

There are probably as many daub mixes as there are daub buildings. Try experimenting with locally available materials. Remember to only add enough water to make the mix workable, not so much as to cause excessive shrinkage. Another tip is to mix the ingredients (without hair or straw) in advance and leave the mix to 'temper'. It can then be re-mixed when required and the reinforcement added. This will allow any dry ingredients to soak up water and for the whole mix to have an even moisture content.

REPAIR TO SURFACE PLASTER

Where the surface plaster has failed but the daub behind is still sound it is normally possible to repair the plaster. It may be that the whole topcoat to the panel has failed or been removed in the past in which case it will be necessary to replace the whole area. Detached plaster can sometimes be re-secured to the daub behind by means of small stainless steel screws and washers, or re-adhered to the daub surface with a lime mix.

If you are faced with having to repair or replace areas of lime plaster and carry out minor repairs to the daub behind, it may be sensible to consider using lime plaster for the daub repairs as well as the plastering. This is often a sensible approach since it means only having to deal with one type of material and can minimise the shrinkage problems that may occur with small daub repairs.

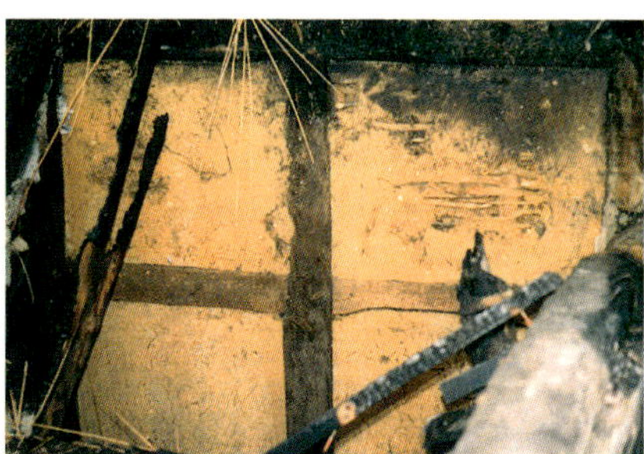

The durability of wattle and daub is illustrated by this wall, still standing after fire burnt the roof off.

A wattle and daub panel in need of repair

The eroded surface of a daub panel revealing its hair and straw binder

REPLACEMENT PANELS

If a wattle and daub panel is beyond repair or missing altogether then a replacement panel will be required. Before removing any panels of a listed building consult your local conservation officer. Listed building consent will normally be necessary and you may be required to carry out recording of the existing panel/panels before proceeding. Some buildings constructed before the 18th century were decorated with wall paintings to brighten the home. These important works of early art vary from simple patterns of repeated motifs to fine works of art and *trompe l'oeil* architectural elements. Whenever considering the removal of a panel it is essential to be aware that original wall paintings or patterns could be hidden beneath the layers of limewash, plaster or panelling.

Before deciding upon a design for your new panels it is necessary to understand why the old ones have failed and to address these reasons. For example, there is clearly no point in replacing a panel damaged by a leaking gutter if the leak is still there.

Wattle and daub is the natural choice for a replacement panel. The evidence of the previous panel will normally dictate the species and pattern of the wattles. Most properly constructed and maintained wattle and daub panels will out-live their builder.

Combining as it does our understanding of traditional performance and the needs of old buildings, wattle and daub has proved itself over time. Properly maintained, the infill panels not only keep the weather out but also create an environment where the structural timber frame is not at risk. Not only is wattle and daub the sound choice from a constructional viewpoint it is also the most environmentally friendly approach. The materials are renewable, from sustainable resources, and minimal energy is consumed in their production.

FURTHER READING

SPAB technical pamphlet No 11, *Infill Panels for Timber Framed Buildings*

Ashurst, John *Practical Building Conservation Volume 3 Mortars, Plasters & Renders*

Wright, Adela *Craft Techniques for Traditional Buildings*

Paper by John McCann published in *Transactions of the Ancient Monuments Society* Volume 31, 1987

It is likely that repairs to infill panels will take place alongside repairs to the structural timber frame, so see also: SPAB Technical Pamphlet No 12 *The Repair of Timber Frames and Roofs*

IAN PRITCHETT is managing director of IJP Building Conservation, the company he founded in 1986. He lectures widely on building conservation issues including lime in particular, and is currently revising the SPAB's technical pamphlet No 11, *Panel Infilling to Timber Framed Buildings*. In addition to repairing all kinds of historic buildings, IJP offers a series of *Practical Courses in Building Conservation*, and supplies traditional and ecological building materials through a subsidiary company, the Old House Store. For company details see pages 61 and 158.

► ALTHAM HARDWOOD CENTRE LTD

Altham Corn Mill, Burnley Road, Altham, Accrington, Lancs BB5 5UP
Tel 01282 771618 Fax 01282 777932
Website www.oak-beams.co.uk

HAND CUT OAK SPECIALISTS: The co-operative's main product is hand cut and dressed oak beams, the controlled irregularity of which is completely in keeping with any restoration – see their website to see how impressive that is. Derek Goffin, their manager, has trained at post-graduate level, PG Dip Building Conservation (timber), at the Weald & Downland Open Air Museum. The co-operative has worked for the National Trust and other well known clients and has made, for example: large roof trusses – a 30′ long naturally arched oak footbridge; a cruck spire folly; and one-piece 11′ long tapered elm columns. In fact, any traditional or imaginative project in green oak or home-grown timber is within Altham Hardwood's capability.

► ANELAYS

William Anelay Limited, Murton Way, Osbaldwick, York YO1 5UW
Tel 01904 412624 Fax 01904 413535
E-mail office@williamanelay.co.uk
Website www.williamanelay.co.uk

BUILDING AND RESTORATION CONTRACTORS: *See also: profile entry in Building Contractors section, page 58.*

► ANTIQUE BUILDINGS LIMITED

Hunterswood Farm, Dunsfold, Surrey GU8 4NP
Tel 01483 200477 Fax 01483 200752

OAK BEAMS, BARN FRAMES, HANDMADE PEG TILES AND BRICKS: *See also: display entry in Reconstructed Buildings section, page 162.*

TIMBER FRAME BUILDERS

▶ BELFIELD TIMBER CO

Pen-Y-Bryn, Glascoed Road, St Asaph, Denbighshire, North Wales LL17 0LH
Tel 01745 585929 Fax 01745 583984
E-mail belfieldtimber@cwcom.net

TIMBER FRAMERS AND OAKWRIGHTS: Specialising in the vernacular building of traditional oak framed homes, extensions, conservatories, garages and restoration projects. Belfield Timber supplies individual trusses, beams and joists, and has a constant selection of interesting inglenook beams, carefully selected from character Welsh oak. Belfield's timber mill in North Wales also manufactures quality hardwood flooring in a variety of grain and grades.

▶ BERNARD A SHEPHERD LTD

33A Cumberland Street, Macclesfield, Cheshire SK10 1DD
Tel 01625 432477 Fax 01625 432488

SPECIALIST TIMBER FRAME BUILDERS: Building contractors for construction and conservation projects. The company is directed by Bernard Shepherd who has over 30 years experience in construction having completed numerous timber frame restoration projects, complex stonework schemes, and restoration of historic buildings alongside new building work such as motor showrooms, schools and residential developments. Their clients include The Church Conservation Trust, English Heritage, the National Trust, Lloyds Bank and many others. The company operates primarily in the North West of England with an experienced workforce and supporting administration staff.

▶ HAMPSHIRE OAK CARPENTERS LTD

4 Broadwater Road, Townhill Park, Southampton SO18 2EB
Tel/Fax 01329 835551

TIMBER FRAME BUILDERS: Hampshire Oak Carpenters Ltd are specialists in all aspects of timber framing repairs and the construction of new timber framed buildings. They have a large workshop in Hampshire and welcome visitors who would like to see what they do.

▶ I J P BUILDING CONSERVATION LTD

Hampstead Farm, Binfield Heath, Nr Henley-on-Thames, Oxfordshire RG9 4LG
Tel 0118 969 6949 Fax 0118 969 7771
E-mail info@ijp.co.uk Website www.ijp.co.uk

CONSTRUCTION AND REPAIR OF TIMBER FRAME BUILDINGS: *See also: display entry in Building Contractors section, page 61.*

▶ I W PAYNE AND COMPANY LIMITED

20 Market Place, Romsey, Hampshire
Tel 01794 517081

SPECIALISTS IN HISTORIC BUILDINGS: I W Payne & Co Ltd are specialists in the engineering, architecture and repair of historic buildings, particularly in timber framed and masonry construction, where they provide consultancy advice as well as the practical, traditional and engineering skills required to carry out the work. They employ carpenters, masons, plasterers and leadworkers. They have carried out work for English Heritage, the National Trust and many local authorities, as well as private and business clients. Most has been in complicated repair projects where realignment of timber framework and masonry walls has resulted in economic repair solutions. *See also: display entry in Building Contractors section, page 61.*

▶ KIRBY PEEL

37, Slack Lane, Crofton, Wakefield, W Yorks WF4 1HH
Tel/Fax 01924 862713 (Timber Framing) Tel 01132 813452 (Joinery)

TIMBER FRAME CONSERVATION AND ARCHITECTURAL JOINERY: Specialists in the conservation of timber framed buildings and the production and repair of architectural joinery. For the past 15 years Kirby Peel has worked primarily on prestigious conservation and restoration projects. In-house skills and an in-depth understanding of historic timber buildings and their fixtures and fittings have found Kirby Peel providing a specialist service to both private clients and professional organisations alike. Recent contracts include: Margaret Clitheroe Shrine, The Shambles, York; Harewood House near Leeds; Ackworth Hall near Pontefract; Little Tallboys, Keevil, Wiltshire; and most recently, timber frame repairs at Hopton Hall, Mirfield.

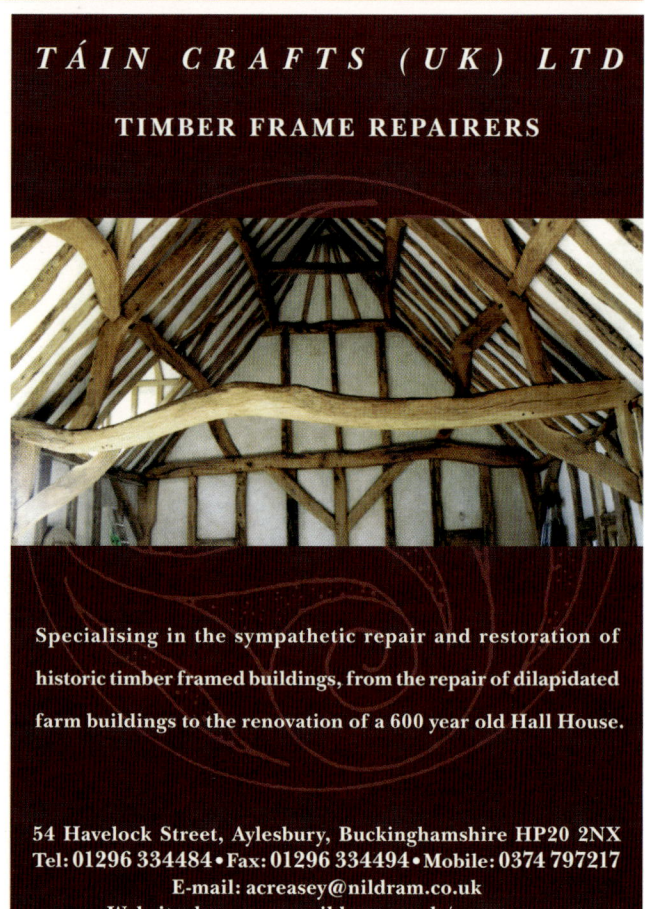

▶ OAKWRIGHTS LTD

West End Lane, Frensham, Farnham, Surrey GU10 3EP
Tel 01252 794325 Fax 01252 795947
E-mail info@ oakwrights.com
Website www.oakwrights.com

TIMBER FRAME BUILDERS: Oakwrights is a team of craftsmen and a civil engineer specialising in design, construction and restoration of timber framed buildings. With combined in-house professional and technical skills we are able to provide cost-effective solutions avoiding the need for independent architects and engineers. Complex conservation and new-build projects are undertaken from planning through to completion. (Oakwrights also provides services in support of a main contractor or architect). Work carried out includes design and construction of new houses, extensions and outbuildings and repair of historic houses and barns.

▶ SPLITLATH LTD

Forest Lane, Craswall, Herefordshire HR2 0PL
Tel 01981 510611 Fax 01981 510342
E-mail forestlawn@compuserve.com

OAK FRAME REPAIRS: *See also: display entry in Building Contractors section, page 64.*

▶ ST BLAISE LTD

Westhill Barn, Evershot, Dorchester, Dorset DT2 0LD
Tel 01935 83662 Fax 01935 83017
E-mail stblaise@compuserve.com

HEAVY OAK CARPENTRY, TIMBER FRAME REPAIR, STEEL AND RESIN REPAIR, ALLIED WET TRADES: *See also: entries in Building Contractors section, pages 65 and 66.*

▶ TAIN CRAFTS (UK) LTD

54 Havelock Street, Aylesbury, Bucks HP20 2NX
Tel 01296 334484 Fax 01296 334494

STRUCTURAL CONSULTANCY, DESIGN AND REPAIRS: Tain Crafts are specialists in the repair, restoration and alteration of historic timber framed buildings. Their work has ranged from the repair of dilapidated farm buildings to the renovation of 600 year old Hall House, reinstating many original details. They also carry out a complete design and construction service, including planning permission, specification and working buildings. They have successfully secured contracts from City Centre Restaurants for Cafe Uno projects in Marlow, Farnham and Henley. In Old Amersham they rebuilt the upper floors of Lloyds Chemists, restoring it to its former timber framed glory. *See also: display entries on this page and in Building Contractors section, page 67.*

▶ WEALD AND DOWNLAND OPEN AIR MUSEUM

Singleton, Chichester, West Sussex PO18 0EU
Tel 01243 811363 Fax 01243 811475
E-mail wealddown@mistral.co.uk
Website www.wealddown.co.uk

CONSERVATION SUPPLIES AND SERVICES: The Museum offers various supplies for use in timber frame building and traditional building methods. *See also: profile entry in Courses & Training section, page 221.*

A delightful example of 17th century pargeting on the Ancient House, Ipswich

Different pargeted textures on The Salisbury Arms Hotel, Hertford with a plastered vine frieze at the jetting

PARGETING

TIM BUXBAUM

Pargeting is the ornamentation of plastered and rendered building facades that would otherwise be smooth, lined-out or roughcast. The term was once also used to include internal decoration. Pargeting ranges from simple geometric surface patterning to exuberant sculptural relief of figures, flowers and sea monsters, but it is only skin deep, applied onto masonry or a lathed, timber-framed wall.

THE HISTORY OF PARGETING

English plasterwork became increasingly elaborate in the 16th century and the dramatic external decoration of Henry VIII's Nonsuch Palace (1538) was contemporary with early plaster friezes in the great houses. Some of the most opulent pargeting was produced over the next 150 years with a high point around 1660 (for example, Ancient House, Ipswich, and the Sun Inn, Saffron Walden), then the technique began to fall out of fashion.

In the last decades of the 19th century architects like Norman Shaw became interested in the 'arts and crafts' skills of an earlier age and there was a revival of interest in pargeting. But it was more controlled, precise and scholarly, better suited to more regular buildings. Irrespective of fashion, calm almost dateless pargeting flourished in many country districts. Beware – dating pargeting can be risky – what you see may have been remade several times in an approximation of the original pattern!

ITS REGIONAL DISTRIBUTION

East Anglia is the traditional home of pargeting, but it can also be seen in Kent and is documented as far away as York and the West Country. There was plenty of pargeting in London before the 1666 Fire. Neglect, redevelopment, fire and changing taste are the main enemies of pargeting which may simply survive in East Anglia because of a slower rate of change and less industrialisation.

STYLES AND TECHNIQUES

The simplest pargeting takes the form of lines scratched with a stick across wet plaster to create, for example, a lattice within a border. More complexity comes from using fingers and combs or moulded templates, incising or impressing chevrons, scallops, herringbones, guilloches, fantails, rope patterns and interchanging squares. It is important that the individual strokes are sublimated to overall effect and precision is often less important than correct texture, itself affected greatly by orientation to sunlight and ability to weather. The rhythm of the chosen pattern, its scale, weight and proportion directly relate to the type of building being embellished, so that subtlety and understatement can be more effective than over-enthusiasm.

Precast decorations can also be used, with friezes and three dimensionality added, given suitable structure. There is no reason why contemporary designs cannot be devised to suit new buildings.

MATERIALS USED

The original raw material is parge, a mixture of sand and lime with a binder like hair, traditionally used for parging flues and underlining roof tiles to reduce drafts. Many additional ingredients are recorded, including stable urine, loam, soot, tallow, road scrapings, cheese, dung, blood and salt, the aim being to produce a viscous material slowly curing to something leather hard. If it cured too quickly it would be difficult to work up a complex pattern; if it cured too slowly the frost might catch it. Traditional mixes might be applied in two or three layers finished with a limewash sheltercoat (repeated layers may obscure the design); later pargeting often contains cement, sometimes in sufficiently high proportions for the ornamentation to appear harsh, with a greater risk of cracking.

Detail of a frieze type typical of Yoxford, Suffolk with an impressed dotted pattern

RECOMMENDED READING

Ashurst, John; *Mortars Plaster and Renders in Conservation.* Gower, 1988

Essex County Council; *Traditional Building Materials in Essex: Pargetting.* Essex County Council, 1982

Millar, William; *Plastering Plain and Decorative.* 1897 reprinted 1998, Donhead, Shaftesbury.

Penoyre, John and Jane; *Decorative Plasterwork in the Houses of Somerset 1500-1700.* Somerset County Council, 1994

Sandon, Eric; *Suffolk Houses – a Study of Domestic Architecture.*

Schofield, John; *Medieval London Houses.* Yale University Press, New Haven & London, 1994

TIM BUXBAUM, a former SPAB scholar, runs a small architectural practice in Suffolk. He has written Shire volumes including *Pargeting and Icehouses* and has a particular interest in old buildings and construction techniques. He can be contacted by e-mail at TimBuxbaum@aol.com.

MILLWRIGHTS

▶ DOROTHEA RESTORATIONS LTD

INCORPORATING ERNEST HOLE (ENGINEERS) LTD
Northern office & works: New Road, Whaley Bridge, via Stockport, Cheshire SK23 7JG
Tel 01663 733544 Fax 01663 734521
▶ Southern office & works: Riverside Business Park, St Annes Road, St Annes Park, Bristol BS4 4ED
Tel 0117 971 5337 Fax 0117 977 1677

RESTORATION OF ARCHITECTURAL METALWORK, TRADITIONAL MACHINERY AND MILLS: *See also: display entry in Architectural Metalwork section, page 147.*

▶ HERITAGE ENGINEERING

22-24 Carmyle Avenue, Foxley, Glasgow G32 8HJ
Tel 0141 763 0007 Fax 0141 763 0583
E-mail lndherco@aol.com

HERITAGE CONSULTANTS AND RESTORERS: Heritage Engineering is a leading UK multi-disciplined heritage organisation engaged in a wide range of engineering restoration projects, working in both metal and timber. They have extensive experience of restoration of mills and mill machinery, waterwheels and waterway systems. Other specialisations include restoration and new-build of architectural metalwork including bandstands and conservatories, restoration of bells and bell mechanisms, timber engineering including lockgates and bridges and heritage consultancy. Clients include English Heritage, Historic Scotland, the National Trust, The National Museum of Scotland, British Waterways, architects, local authorities and construction companies.

▶ I J P BUILDING CONSERVATION LTD

Incorporating The Chiltern Partnership
Hampstead Farm, Binfield Heath, Nr Henley-on-Thames, Oxfordshire RG9 4LG
Tel 0118 969 6949 Fax 0118 969 7771
E-mail info@ijp.co.uk
Website www.ijp.co.uk

APPLIED CONSERVATION TO HISTORIC BUILDINGS AND STRUCTURES: Leading restoration millwrights and engineers in restoring historic wind and water mills, with over 130 mills repaired or restored. All aspects are undertaken, from design, surveys, and feasibility and cost studies to completion of all elements of project work such as masonry and brickwork using lime mortars, wrought iron work, steel fabrication, engineering, bearings, castings, hydraulics and stone dressing. They also have an extensive joinery workshop. Working with leading architects, English Heritage, SPAB and national authorities, often for long term maintenance and monitoring, mills and all vernacular structures are covered ie colleges, castles, manor houses, bridges. Each project is carefully researched and executed; exceptional, unusual or difficult projects are normal in the UK or abroad. *See also: display entry on this page.*

BLACKSMITH, KEVIN DOLAN
preparing a bar to repair 18th century railings in Hillsborough
Environment and Heritage Service, Crown Copyright DOE

ROOFING

ROOFING CONTRACTORS

		Page
SCOTLAND		
Adam Cooper, Master Thatcher	th	85
YORKSHIRE		
Aire Valley Roofing	rc	82
Anelays	me rc	87
Bradford Roofing Contractors Ltd	rc	83
C J Ellmore & Co Ltd	rc	59
Weaver Construction Limited	rc	69
NORTH-WEST		
Barry Milne Thatcher	th	85
Northwest Lead	me	86
S & J Whitehead Ltd	me rc	66
Wm Langshaw & Sons	rc	68
WALES		
Welsh Heritage Thatching	th	85
WEST MIDLANDS		
A V Brown	me	87
Linford-Bridgeman Limited	rc	61
Treasure & Son Ltd	rc	67
EAST MIDLANDS		
C R Crane & Son Ltd	me rc	60
Church Conservation Limited	me rc sj	84
H & W Sellors Ltd	rc	108
Norman & Underwood	me	86
SOUTH-EAST – NORTH OF THE THAMES		
B & S Fowler Master Thatchers	th	85
Bardsley & Brown	th	85
Between Time	me rc	58
Boshers (Cholsey) Ltd	me rc	58
C E L Architectural Metal Roofing	me rc	86
Calder Traditional Roofing Services	me rc	82
E G Swingler & Sons	me rc	82
Hall Construction	rc	60
The Hertfordshire Roofing & Renovation Company	rc	82
J G Matthews Limited	rc	62
Master Thatchers of Oxford	th	85
Mowlem Rattee & Kett	rc	62
Andrew Rees	th	85
Russell & Buckingham	th	85
Sindall Ltd	rc	64
LONDON		
Holloway White Allom	rc	62
Sindall Ltd	rc	64
SOUTH-EAST – SOUTH OF THE THAMES		
Haslemere Builders Limited	rc	60
I W Payne & Company Ltd	rc	61
John Williams & Company Ltd	me rc	82
Karl Terry Roofing Contractors Ltd	me rc	83
R W Armstrong & Sons Limited	rc	64
Redpath Buchanan	sj	84
Simmonds of Wrotham	rc	65
Wallis	rc	55
SOUTH-WEST		
A C Wallbridge & Co Ltd	sj	84
Bosence & Co	rc	45
Busby's Builders	me rc	59
Carrek Limited	me rc	60
Heritage Structural Ventilation Ltd	rc	90
Ian Rose Thatchers	th	85
M J Read	th	85
St Blaise Ltd	me	65
Wessex Thatchers	th	85

KEY
me metal sheet roofing
rc roofing contractors
sj steeplejacks
th thatch
Regional designation is according to office location. Many firms operate nationally.

ROOFING PRODUCTS AND MATERIALS

		Page
Ace Demolition – The Reclamation Centre	pt rs rt	159
Aire Valley Roofing	sr	82
Ajeer Limited	pt rs rt	159
Aldershaw Handmade Clay Tiles Ltd	rt	91
Allison Stone (Wigan) Limited	sr	106
Alumasc Building Products Ltd	rd	88
Antique Buildings Limited	pt rs rt	162
Artisan Oak Buildings Ltd	pt rs rt	142
Babylon Tile Works	bt rt	91
Bradford Roofing Contractors Ltd	rs rt sr	83
The Bulmer Brick and Tile Co Ltd	rt	99
Bursledon Brickworks Conservation Centre	bt	167
C E L Architectural Metal Roofing	ld rd rf rs	86
Calder Traditional Roofing Services	rd	82
Carpenter Oak & Woodland Co Ltd	bt os	76
The Cast Iron Company	rd rl	152
Castaway Cast Products and Woodware	rd	146
Cathedral Works Organisation (Chichester) Limited	bt	107
Chalk Down Lime Ltd	bt	168
Clement Windows Group Ltd	rl	89
Cobar Services	rs rt	160
Conplex	rd	88
Conservation Building Products Limited	pt rs rt	160
Daniel Platt	rt	100
Dreadnought Tiles	rf rt	91
Drummonds Architectural Antiques Limited	pt rs rt	160
Eurocom	rd	86
Fine Iron	wv	147
G & N Marshman	bt	91
I J P Building Conservation	bt	61
J & J W Longbottom Limited	rd	87
Karl Terry Roofing Contractors Ltd	rf rs rt	83
Keyline	bt	115
Keymer Tiles Ltd	rt	91
The Metal Window Co Ltd	rl	90
Mike Wye & Associates	bt	168
Norman & Underwood	rd	86
Northwest Lead	rd	86
Osiris Lead Limited	ld	87
Richards Brackets	rd	88
Robus Architectural Ceramics	pt rf rt	97
Saint-Gobain Pipelines Plc	rd	89
Salmon (Plumbing) Ltd	ld	87
Sandtoft Roof Tiles	rt	91
Smithbrook Building Products Ltd	rt	91
Solopark Plc	pt rs rt	161
South West Reclamation	pt rs rt	162
Symonds Salvage	pt rs rt	162
The Traditional Lime Co	bt	170
Timothy Williams (Builders) Limited	sr	66
Ty-Mawr Lime Ltd	bt	170
V & A Traditional Lead Castings Ltd	rd	88
W J Haysom & Son	sr	114
Walcot Reclamation Ltd	pt rs rt	162
Weald & Downland Open Air Museum	bt	85
West Meon Pottery	pt	91
Whippletree Hardwoods	bt	144

KEY
bt battens, laths and tile pegs
ld lead sheet
pt chimney pots
os oak shingles
rd roof drainage
rf roof features
rl roof lights and lantern lights
rs roofing slates
rt clay roof tiles
sr stone roofing slates

THE NEED FOR ROOFS TO BREATHE

RICHARD OXLEY

Dampness and mould growth on a tie beam caused by the use of impervious roofing felt

Internal valley gutters are a perennial concern as their drainage is vulnerable to blockage by leaves in the autumn and snow in the winter in particular

It is now commonly accepted, at least within conservation circles, that it is important not to restrict the ability of traditional building materials and structures to 'breathe'. However, attention has tended to focus on the damage caused by the use of impervious modern paint systems and cement-rich mortars and renders, and the one part of an old building where the assessment of performance and attention to detail is often neglected is the roof. Yet this is one of the principal areas where ventilation could readily take place, particularly in traditionally detailed tiled or slated roofs.

Today, the performance of many historic roofs has been impaired by the introduction of roofing felts and insulation. Although many roofing felts are now marketed as being vapour permeable, until recently almost all felts were impervious. Roofing felt was introduced primarily to act as a secondary barrier against wind-driven snow and rain, but its use also causes a reduction in air movement within the roof space, particularly if the roofing felt is impervious, and this effect is often compounded by the introduction of insulation. Fibreglass quilt or resin fibre materials, for example, are often laid over eaves and applied to the underside of the roof, in contact with the roofing felt.

In addition, most historic buildings in active use will be subjected to an increase in humidity caused by modern lifestyles. Particular concerns include the production of water vapour due to:

- increased use of the building by the occupants (we now spend more time inside than we used to)
- bathing, particularly showers
- cooking
- water tanks within the roof that are not provided with fitted lids.

Of equal concern is the reduction in natural ventilation caused by:

- the installation of double or secondary glazing
- the reduced use of open fires
- blocking-up disused fireplaces and flues.

BACKGROUND

Traditional variations of a physical secondary barrier against wind-driven snow and rain include reeds laid between the tiles and the battens, and a coating of mortar known as 'torching' to the underside of the tiles or slates. Torching is most commonly encountered to the underside of old stone slate roofs. Both techniques allow roofs to breathe.

Roofing felt was first introduced on a regular basis in the 1930s, when it generally comprised of thin building paper. After the Second World War heavy duty bitumen and plastic felts were commonly used, with increasingly impervious materials becoming more common as time went on, until relatively recently when more vapour permeable felts started being introduced *(see page 80)*.

Insulation started being introduced on a massive scale after the 1970s oil crisis due to the need to make buildings, in particular domestic dwellings, more energy efficient. Insulating our homes is also seen as one of the most effective ways of reducing the demand for fossil fuels, cutting pollution and global warming caused by carbon dioxide.

The lessons that are being learnt from experience of the problems caused by earlier improvements to the energy efficiency of older buildings now need to be heeded, particularly in view of the proposed revision of Part L of the Building Regulations which will increase insulation requirements yet again.

ASSESSING THE PERFORMANCE OF A ROOF

It is important that a holistic approach is adopted for the performance of a building to be understood, as the roof and walls cannot be taken in isolation: they are an integral part of the building and have an active and continuing relationship with the rest of the building, its environment and its occupants.

The two most commonly encountered ways in which the performance of the roof of an old building has been dramatically and detrimentally altered are by the introduction of insulation and impervious roofing felt.

TYPICAL EFFECTS OF INTRODUCING INSULATION AND IMPERVIOUS ROOFING FELT

- The introduction of insulation over a ceiling creates a 'cold roof' *(see diagram)*.
- Roofing felt significantly reduces the air movement in the roof space.
- Moist air from the accommodation readily finds its way into the roof space through the ceiling and holes in ill-fitting hatches.
- The amount of evaporation that can take place within the roof is considerably reduced by the introduction of the roofing felt.
- Increased amounts of dampness and moist air are now present within the roof space.
- The timbers in the roof space are therefore increasingly subjected to the conditions conducive for active fungal decay and wood boring insect infestation.
- Any drop in the air temperature provides the atmospheric conditions for the condensation of the moist air to take place.
- The impervious roofing felt provides a high level of resistance to the passage of water vapour and a cold contact surface

upon which warm moist air can condense. In these circumstances the rafters in contact with the felt may remain damp most of the time, causing the surface of the rafter to become stained and, in the worst case, rotten – as illustrated below.

As can be seen, the introduction of both a roofing felt and insulation has provided an environment susceptible to condensation, which in turn increases the risk of dampness and associated timber defects.

TYPICAL INSULATION DEFECTS

- Insulation laid so that it covers the eaves, significantly reduces ventilation to the roof.
- Insulation is often laid in contact with the roofing felt. Where the felt is impervious any contact condensation will run down the felt and make the insulation damp.
- Many modern insulation quilts such as fibreglass, in comparison with alternatives that are now readily available, retain moisture. Where an impervious roofing felt has been used, this type of insulation may not dry out readily. In the worst case the insulation becomes a soggy mass at the bottom, causing the feet of the rafters and the ends of the joists to decay.

This brief overview of some of the problems that can be encountered where the roof of an old building has been provided with a secondary barrier and insulation, illustrates that there is a need to evaluate the influence that any changes in the traditional 'breathing' performance of the roofs of old buildings is having.

HOW TO IMPROVE WEATHER-TIGHTNESS AND INSULATION WITHOUT JEOPARDISING THE TRADITIONAL 'BREATHING' PERFORMANCE OF THE ROOF

The options available for improving thermal performance will be largely dictated by the following:

- If the building is listed, consent will be required for almost all alteration work, including work to the roof, inside or out, and some restrictions may also apply to external alterations if the building is in a conservation area.
- Building Regulations apply if the work involves alterations to an existing building rather than repair. These give some well-defined guidance and measures which may conflict with the needs, requirements and performance of a historic building, particularly after the proposed revision of Part L of the Building Regulations is introduced.
- The financial circumstances of the owner may restrict the work which can be carried out. For example, it is far easier to improve the weather-tightness and insulation of the roof of an old building once the existing roof coverings have been carefully removed, but this may not be possible within the budget.

Before making any improvements:

- Ensure that the intervention or loss of historic fabric is kept to an absolute minimum.
- Make sure that the structural performance of the roof will not be adversely affected.
- Be confident that the traditional 'breathing' performance of the roof is maintained, or reinstated.
- Take time to carefully select the materials and methods to be used, to ensure that they are compatible with traditional performance requirements - this usually means that the materials and 'systems' need to be vapour permeable.

THE MATERIALS AND METHODS AVAILABLE

Examples of some of the materials that could be used to improve the weather-tightness and insulation of the roof of an old building without jeopardising the traditional 'breathing' performance include vapour permeable roofing felt and sheep's wool insulation. In addition to these two materials, attention should also be given to extracting water vapour at source – particularly in the kitchen and bathrooms. Providing improved ventilation or installing mechanical extractors will significantly reduce the risk of condensation.

'COLD ROOF': simplified cross section of a building provided with insulation and roofing felt

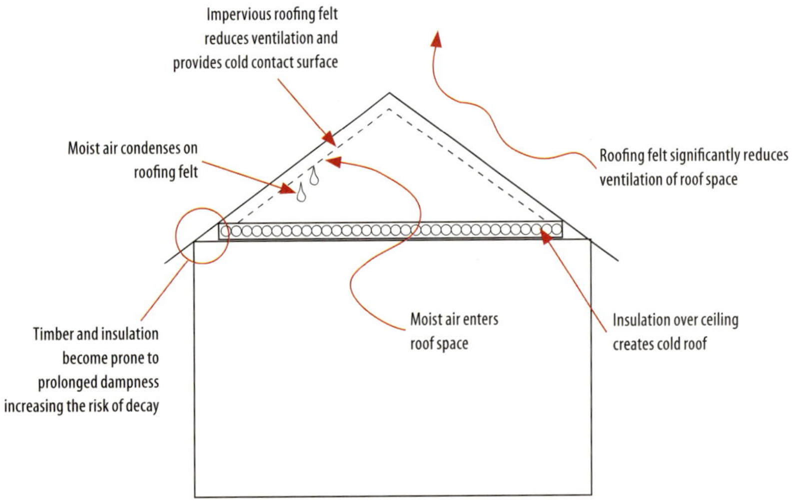

Impervious roofing felt reduces ventilation and provides cold contact surface

Moist air condenses on roofing felt

Roofing felt significantly reduces ventilation of roof space

Timber and insulation become prone to prolonged dampness increasing the risk of decay

Moist air enters roof space

Insulation over ceiling creates cold roof

VAPOUR PERMEABLE ROOFING FELT

One technical advance made in recent times is the production of a roofing felt which allows some movement of water vapour through it. This is a vast improvement on some of the impervious felts previously used.

However, it is important to appreciate that these new roofing felts have not been tried and tested over any significant period of time. Although designed and promoted as being vapour permeable, a roofing felt is a compromise that both allows the movement of water vapour and, by providing a secondary barrier, significantly reduces the risk of water penetration (which is most important when regular inspection and maintenance is not standard practice).

The provision of a properly detailed roofing felt requires the roof coverings to be stripped, so it may need to be delayed until the condition of the roof justifies stripping it, or the opportunity arises for this work to be carried out.

As can be seen, great care is needed when introducing any barriers to historic buildings to prevent any restriction in the movement and evaporation of water vapour. It is essential that the positioning and detailing of any vapour barrier is correct, and on the warm side of

BATS AND BIRDS

It is not uncommon to find that bats use a roof space. Bats do not pose a significant threat to the building fabric or the health of the occupants and, under Section 9 of the *Wildlife and Countryside Act (1981)*, it is an offence to intentionally damage, destroy or obstruct access to any place used by bats, even when bats are apparently absent, or to disturb bats while roosting, just as it is an offence to intentionally kill, injure or take a bat.

Where bats are present or there is evidence that bats have used, or are using a roof, English Nature* or a local bat group should be contacted for informed advice and guidance *before* any roofing works are programmed and initiated.

If there is an active bat roost, works will need to be programmed to cause the minimal amount of disturbance *and* measures provided to allow bats to continue to use the roof space upon completion of the roofing works.

It also needs to be appreciated that *The Wildlife and Countryside Act 1981* gives protection, with certain exceptions, to all birds, their nests and eggs. It is an offence to:

- kill, injure or take any wild bird
- take, damage or destroy the nest of any wild bird while it is in use or being built.

Consequently, roofing works need to be specifically programmed where nesting birds are present so that any disturbance is minimised to reduce the risk of the birds deserting their nests or young nestlings.

* English Nature can be contacted on 01733 455000.

CASE STUDY

The roof before works implemented and, on the right, nearing completion

The question of how best to insulate and ventilate a historic roof cannot be answered with a single solution; different situations require different solutions. Each building should be considered individually and holistically. This case study shows how one particular set of problems was addressed.

In this case the roof coverings, which consisted of stone slate to one slope and concrete tiles to the other, were to be left intact. The brief was to provide a finish suitable to accommodate the conversion of an unused upper floor 'loft' space and improve the thermal insulation.

The factors to be taken into consideration included that there was no roofing felt, and the building was in an exposed situation where there was a strong risk of wind-blown snow and rain penetrating the roof space. There was a risk that water penetration could occur in sufficient quantities to cause staining and discolouring of the ceilings or even timber decay.

In this case lime mortar torching to the underside of the roof coverings was used in the vernacular tradition to reduce the risk of wind-blown snow or rain. The subsequent risk of water penetration caused by loose or dislodged slates or tiles causing the torching to break down was mitigated by the use of a lightweight vapour permeable building paper to the underside of the rafters. The building paper also acts as a means of reducing draughts, which improved the effectiveness of the insulation. The disadvantage of this detail is that any water that penetrates the roof coverings could run down the paper and collect at the wallplate.

For this reason the client was made aware of the need for regular inspections and maintenance. Insulation was provided between the rafters with natural wool insulation and a lightweight mineral wool insulation board was used to provide a face upon which to apply two coats of lime plaster, which was subsequently limewashed. (An alternative to the lightweight mineral wool insulation board is the use of reed board, which is natural water reed held together with wire.)

The benefit of using a natural insulation material in this manner is that the permeability of the building's envelope is maximised whilst allowing the upper floor accommodation to be used to its full potential.

Although some of the materials were relatively new, they were already commonly used and readily available *(see stockists in this Directory, page 158)*. These and other similar materials provide the best compromise where the thermal insulation of a building needs to be improved whilst maximising permeability and thereby reducing the risks of condensation and any associated problems. Not only is this in the best interests of the building, the improvement of the thermal performance is also in the best interests of the environment – as long as some thought has gone into the environmental impact and sustainability of the materials selected.

A PROPRIETARY VENTILATOR NOT VISIBLE EXTERNALLY

One alternative solution to the problems of roof ventilation is the lapVent. Made from recycled materials, this simple ventilator is designed to maintain a constant passage of air between overlapping sheets of underlay. Unlike most modern roof ventilators, this one does not protrude above the roof-line (see page 90).

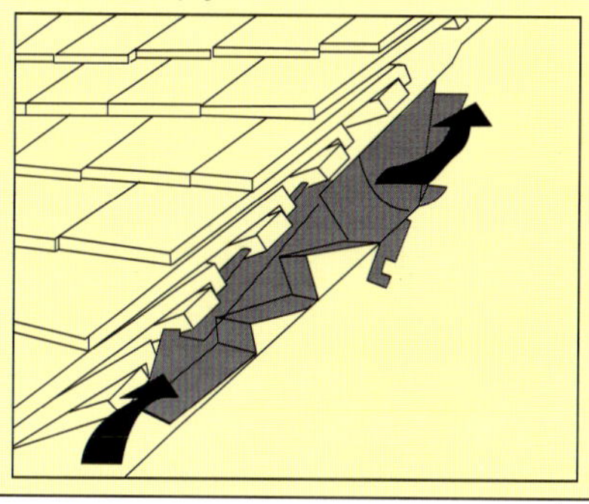

the insulation, otherwise water vapour may be trapped against timbers, ceilings or insulation and causing problems for the building.

The reluctance to carry out regular inspection and maintenance has lead to an increased reliance on the use of vapour barriers and roofing felts. It is preferable to carry out regular maintenance wherever possible, rather than change the performance of an historic building, bearing in mind that the building has performed well and survived for, in some cases, centuries without these barriers. Where the decision is made that a roofing felt is to be introduced it needs to be ensured that it is of an appropriate type and correctly detailed.

SHEEP'S WOOL INSULATION

The insulation material that has probably received the most publicity recently is sheep's wool. The benefits of this material are:

- it is mainly natural (although resin fibres have to be added for bonding and where some rigidity is required)
- it is a 'breathable' material as it is porous and hygroscopic – yet it still retains its thermal insulation capabilities when damp
- it is not an irritant to those who use the material.

Sheep's wool is derived from renewable resources and has a relatively low embodied energy. It therefore has environmental advantages over many other conventional alternatives. These factors, together with 'life cycle costings', should be an important consideration when assessing what materials to use in the repair and improvement of all buildings, not just historic buildings.

Richard Oxley is an RICS accredited surveyor in building conservation with his own independent historic buildings consultancy practice near Henley on Thames, Oxfordshire. See Oxley Conservation, page 47.

▶ AIRE VALLEY ROOFING

22 Grange Park Drive, Cottingly, Bingley BD16 1NR
Tel 01274 568878/01274 560840 Fax 01274 560840

SPECIALIST IN YORKSHIRE STONE SLATE: Aire Valley Roofing have been specialising in all aspects of traditional Yorkshire Stone random slating for almost 20 years, mainly to buildings in the Yorkshire Dales and surrounding districts. All works are carried out using their large stocks of the best reclaimed materials by an experienced team of slaters using timber or steel pegs. Included along with their stone slate works are random Westmoreland, Burlington and Welsh slate. Contracts undertaken predominantly in the £5,000 to £150,000 price range. Aire Valley Roofing are members of the Federation of Master Builders and are Warranted Builders. All enquiries welcomed.

▶ C E L ARCHITECTURAL METAL ROOFING LIMITED

Progress House, 256 Station Road, Whittlesey, Peterborough PE7 2HA
Tel 01733 206633 Fax 01733 206644

TRADITIONAL METAL ROOFING: Manufacturers of traditional sand cast lead sheet and flashings also reproduction lead rainwater goods, decorative mouldings, plaques and much more. Skilled craftsmen from the company's contracting division use authentic metal construction materials including lead, copper, zinc and stainless steel for restoration or replacement of roofs and other areas for all types of ecclesiastical and heritage buildings. *See also: display entry in Metal Sheet Roofing section, page 86.*

▶ CALDER TRADITIONAL ROOFING SERVICES

'Woodhurst' Cattlegate Road, Crews Hill, Enfield, Middx EN2 8AU
Tel 01707 876515 Fax 01707 872413

TRADITIONAL SLATE, TILE, LEAD COPPER AND ZINC WORK, ROOF AND LANTERN LIGHTS AND CLASSIC RAINWATER PRODUCTS: Working exclusively in London north of the Thames up to Hertford town, Calder is a family business who's craftsmen sympathetically install or repair virtually every type of traditional roof. Established 1966, their portfolio of successful projects includes many important historic buildings, mostly in residential but also in commercial present-day uses. Calder's work is fully insured and guaranteed. Please contact the company to discuss your requirements. *See also: Calder Group display entry in Building Contractors section, page 59.*

▶ E G SWINGLER & SONS

27/29 Weedon Road, Northampton NN5 5BE
Tel 01604 755055 Fax 01604 587506

SPECIALIST ROOFING CONTRACTORS: E G Swingler & Sons, established in 1964, is a family business which for the past 15 years has worked almost exclusively on historic and listed buildings. Within a 50 mile range of Northampton, they offer the full range of traditional building services with special emphases on traditional slate, tile and lead roofing commissions, in the £10,000 to £100,000 range. There are currently nine in-house craftsmen who work under the close supervision of Robert Swingler. The craftsmen's work is supported by a ready supply of the best reclaimed and new materials. Recently the company has completed the re-roofing of listed buildings in Norfolk, Leicestershire and Warwickshire. *See also: display entry on this page.*

▶ THE HERTFORDSHIRE ROOFING AND RENOVATION COMPANY

32 Field Road, Oxhey, Herts WD1 4DR Tel/Fax 01923 250247

ROOFING AND CONSERVATION SPECIALIST CONTRACTORS: *See also: display entry in Building Contractors section, page 59.*

▶ JOHN WILLIAMS & COMPANY LTD

Stone Street, Lympne, Hythe, Kent CT21 4LD
Tel 01303 265198 Fax 01303 261513

ROOFING CONTRACTORS: Established 1870, John Williams & Company Ltd's craftsmen specialise in the fixing of traditional roofing materials using skills and techniques passed down through generations. The company is proud to have been involved in the refurbishment of many important buildings including the re-slating of Tower Bridge, The Royal Chelsea Hospital and St James' Palace, random slating to Rochester Cathedral and Portchester Castle, Kent peg tiling to Westenhanger Barns and lead roofing to Saltwood Castle. Members of the National Federation of Roofing Contractors and The Lead Contractors Association, John Williams & Company Ltd offer comprehensive contracting throughout Kent, East Sussex and South East London.

ROOFING CONTRACTORS

SCAFFOLDING

TSC

TAMWORTH SCAFFOLDING COMPANY

Contract Scaffolding Specialists

SPECIALISTS IN SCAFFOLDING AND WEATHER PROTECTION TO HISTORIC, PERIOD AND ECCLESIASTICAL BUILDINGS

10 Wembury, Amington Fields, Tamworth, Staffordshire B77 3RE

Tel 01827 56990

STEEPLEJACKS & LIGHTNING PROTECTION

▶ A C WALLBRIDGE & CO LTD

Windsor Road, Salisbury, Wiltshire SP2 7DX
Tel 01722 322750 Fax 01722 328593

STEEPLEJACKS, LIGHTNING CONDUCTOR ENGINEERS: Wallbridge has been providing comprehensive and professional lightning conductor installation and maintenance services for the past 20 years. Its work is carried out by the company's own staff of trade certified engineers. All of the work carried out by Wallbridge complies with the current Health and Safety Regulations and to the BS6651 standard. Wallbridge are full members of the National Federation of Lightning Conductor Engineers and are party to and work within the industries agreed methods of safety and quality control. This ensures workmanship to the correct BS standards.

▶ KAIZEN STONE CLEANING

Highland House, West End Road, Kempston, Beds MK43 8RU
Tel 01234 840301 Fax 01234 840302
E-mail sales@kaizen-cleaning.co.uk
Website www.kaizen-cleaning.co.uk

MASONRY CLEANING AT HEIGHT: *See also: display entry in Masonry Cleaning section, page 172.*

ROOFING SLATE

▶ STATS CONSULTANCY

(STATS Limited, founded 1974)
Porterswood House, Porters Wood, St Albans, Herts AL3 6PQ
Tel 01727 833261 Fax 01727 835682
E-mail ian.sims@stats.co.uk
Contact Dr Ian Sims

SPECIALIST ENGINEERING, MATERIALS AND ENVIRONMENTAL CONSULTANTS: *See also: profile entry in Materials Analysis section, page 45.*

www.buildingconservation.com
ARTICLES
This article is reproduced from The Building Conservation Directory 1997

Articles · Books · Courses · Directory · Events

AUTHOR
TIM DONLON BSc, MCIOB, works for Church Conservatoin Limited of Nottingham as Technical Contracts Manager. The Company, which is a member company of the National Federation of Master Steeplejacks and Lightning Conductor Engineers, specialises in the care and conservation of historic properties.

FURTHER INFORMATION

Related Articles

Strategic Fire Protection for Historic Buildings - Richard Forrest

Repairs to Lead Sheet Roofings and Flashings - F C Coote

Products & Services Directory

Lightning protection

Roofing contractors

Lightning Protection for Historic Buildings

Tim Donlon

Lightning, an intensely bright spark or streak of light through the air to ground, has terrified and excited mankind for centuries. Once considered to be in the realm of the gods, it was only in the past 200 years that the theory of lightning has been transferred to the scientific realm, initially through the endeavours of Benjamin Franklin (1751).

There are different types of lightning: cloud to ground, cloud to cloud and within a single cloud, not to mention such rare variations as ribbon and ball lightning. On average a cloud to ground strike would be in the order of 20,000 amperes with a duration of 0.2 seconds, and at its peak, the power released can be 100 megawatts per metre.

Although relatively few deaths occur (10 per year in the UK) (Smith, Keith *Principles of Applied Climatology*. McGraw Hill 1977), the possibility of a lightning strike to the structure of a building such as a small church is much greater, at around 1:500 (British Standard 6651:1992) per year in the UK. Mechanical effects and damage are primarily caused by the explosive expansion of air heated to around 30,000ºC, by the ignition of dust, and by flying debris. Electrical circuits may also be damaged by the electo-magnetic field generated.

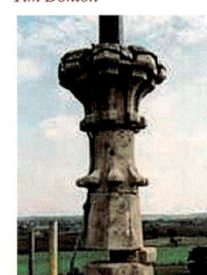

Figure 1: Eastern maudit Church, Northamptonshire: An example of the damage caused below a spire-top air terminal by a side flash. Passing through masonry from an internal vane rod, the lightning caused the displacement of solid stonework.

In a temperate climate lightning is caused by 'frontal' storms which usually occur as a cold front meets warm air, wedging it upwards thousands of feet into the atmosphere. Thunderclouds (cumulonimbus) develop as the moisture condenses and then freezes with altitude. Their development gives rise to charges building up; in general, ice particles in the upper part of the clouds are positively charged, while water droplets lower down contain a negative charge. At the base of the cloud, this negative charge induces a positive charge at 'ground' level. As the cloud continues to grow, the charge increases until the voltage difference between the ground and cloud is so great that the resistance of the air between the two is broken down and a lightning discharge occurs. Recent photographs taken from the space shuttle also show discharges from the top of the thundercloud to the stratosphere; previously such a

INTERESTED? see www.buildingconservation.com

CATHEDRAL COMMUNICATIONS LIMITED

THATCHERS

❶ B & S FOWLER MASTER THATCHERS
54 Westland Road, Faringdon, Oxon SN7 7EY
Tel 01367 242185 Mobile 07770 593681
Website www.roofthatching.co.uk
Complete rethatches, ridges and patching, insurance work undertaken, free estimates. Established for 25 years. Member of the Oxfordshire, Bucks and Berks Master Thatchers Association.

❷ BARDSLEY & BROWN
1 Marlston Cottages, Marlston, Berkshire RG18 9UN
Tel 01635 201546/01635 255149
High quality craftsmanship from a well established business. All thatching work quoted for. Best thatched house (OBBMTA) 2000 winners

❸ ADAM COOPER, MASTER THATCHER
The Grieves Cottage, Pert Farm, Northwaterbridge, Laurencekirk AB30 1QP
Tel/Fax 01674 840 538
New thatch, rethatch and repairs. Free estimates and advice.

❹ IAN ROSE THATCHERS
The Coach House, Ashleigh Manor, Atherington, Umberleigh, Devon EX37 9HW
Tel 01769 560031 Fax 01769 560100
Quality workmanship from an established family business using wheat reed and water reed. Listed buildings, insurance and small building works undertaken. Member of Devon & Cornwall Master Thatchers Ass'n.

❺ MASTER THATCHERS OF OXFORD
17 The Green, Steventon, Abingdon, Oxon OX13 6RR
Tel/Fax 01235 832286 Mobile 07966 229418
E-mail bodthedog@aol.com
Specialists in both straw and water reed. Only quality nitrate tested materials used. Fire retardant treatments. Warranty underwritten by the NFU Mutual. Trained and experienced staff. Conservation advice.

❻ BARRY MILNE THATCHER
15 Fleetwood Crescent, Banks, Southport, Lancs PR9 8HF
Tel 01704 231510
Specialist in combed wheat reed and water reed. Established 25 years. Secretary and Member of North of England and Scotland Master Thatchers Association.

❼ M J READ
3 Ridge Cottages, Chilmark, Salisbury, Wiltshire SP3 5BS
Tel 01722 716631
Third generation thatcher in combed wheat reed, long straw and Norfolk reed. Timber repairs undertaken. All work guaranteed. Guild member.

❽ ANDREW REES
20 Rowntree Way, Saffron Walden, Essex CB11 4DG
Tel 01799 513743 Mobile 07712 840596
A highly experienced thatcher, using long straw and water reed. Local council approved. Please ring for a free estimate locally.

❾ RUSSELL & BUCKINGHAM
37 Hanborough Close, Eynsham, Witney, Oxford OX29 4NR
Tel 01865 883818 Mobile 0860 919257
Superior work carried out by traditionally trained craftsmen with over 30 years experience. Only quality selected materials used.

❿ WEALD AND DOWNLAND OPEN AIR MUSEUM
Singleton, Chichester, West Sussex PO18 0EU
Tel 01243 811363 Fax 01243 811475
E-mail wealddown@mistral.co.uk Website www.wealddown.co.uk
Specially grown traditional varieties of longstraw and wheat reed supplied, also hazel spars produced on site by our sparmaker in residence.

⓫ WELSH HERITAGE THATCHING
8 Brynbach Road, Brynbamman, Ammanford, Carmarthenshire SA18 1BH
Tel/Fax 01269 824585 Mobile 07703 379917
E-mail craftsman@thatching-uk.com Website www.thatching-uk.com
Thatch renovations and repairs, thatched new build, museum thatch maintenance, authentic thatch for film sets, thatched garden buildings. Free advice and estimates.

⓬ WESSEX THATCHERS
2 Trotts Lane, Eling, Totton, Hampshire SO40 4UE
Tel 02380 667637 Mobile 07771 534261/0790 1597814
Complete thatching service. Council approved. Water reed. Wheat reed. Longstraw. Proven quality and reliability. Superior work by experienced craftsmen.

METAL SHEET ROOFING

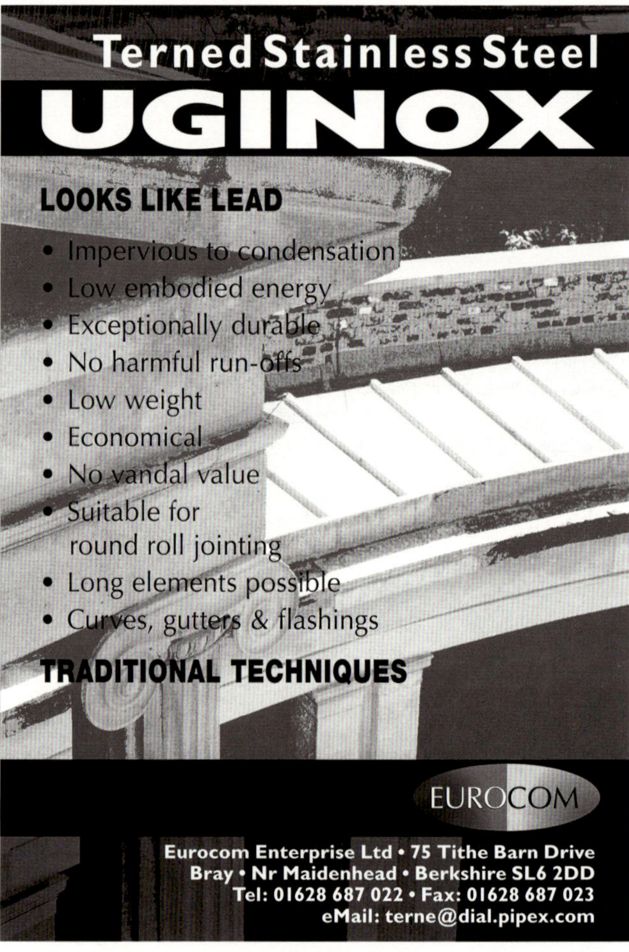

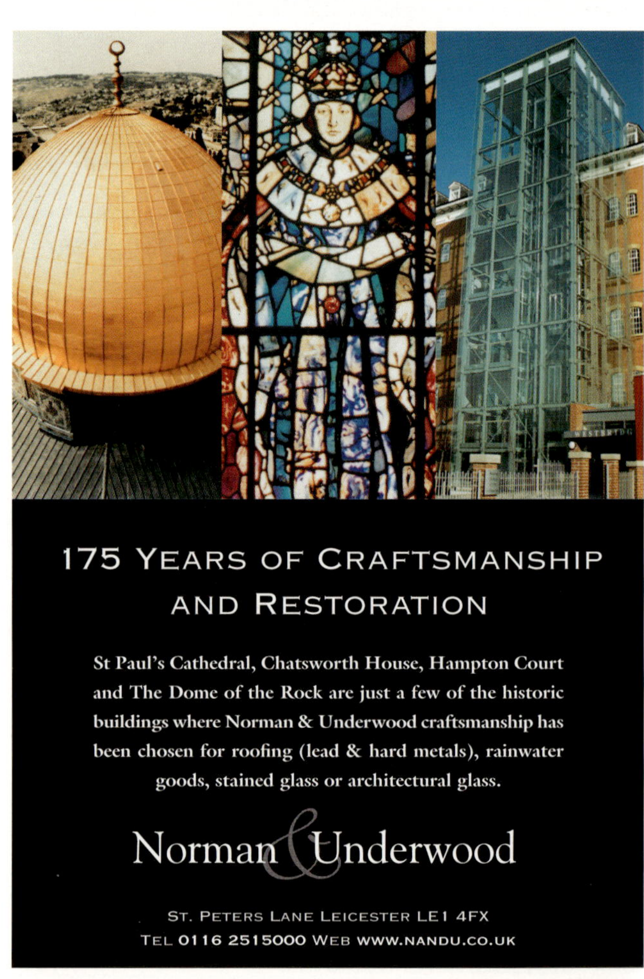

METAL SHEET ROOFING

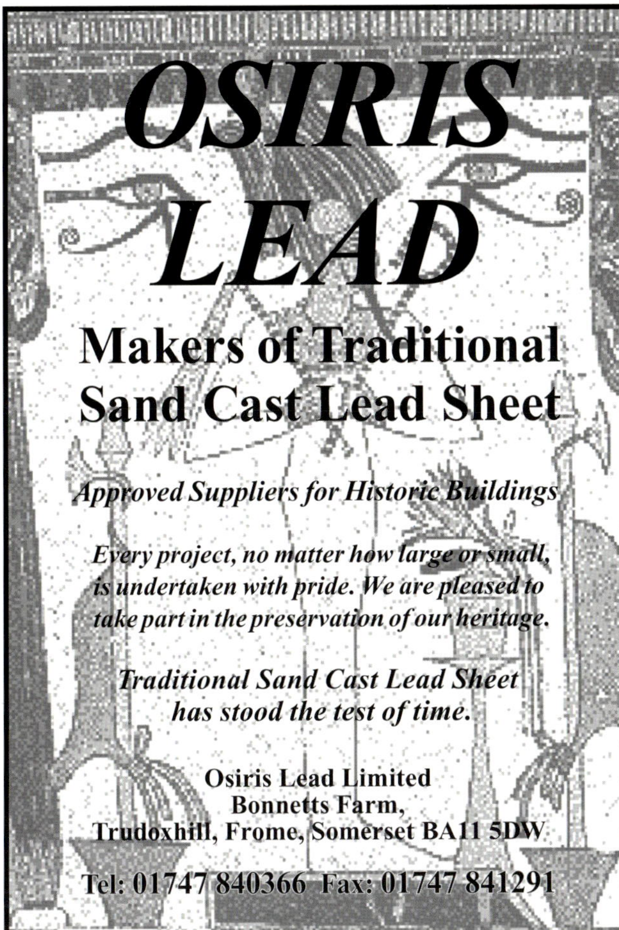

▶ A V BROWN

82, Poolbrook Road, Malvern, Worcestershire WR14 3JD
Tel/Fax 01684 562969 Mobile 0860 625447

LEAD ROOFING CONTRACTOR: Established in 1976, this family firm has undertaken lead and copper roofing for over 20 years. Their workforce of craftsmen has worked on churches, castles and all types of houses throughout their area. Clients include local authorities, architects, builders, water authorities and owners of listed buildings. All types of work are undertaken including cast milled lead and copper sheet roofing.

▶ ANELAYS

William Anelay Limited, Murton Way, Osbaldwick, York YO1 5UW
Tel 01904 412624 Fax 01904 413535
E-mail office@williamanelay.co.uk
Website www.williamanelay.co.uk

BUILDING AND RESTORATION CONTRACTORS: *See also: profile entry in Building Contractors section, page 58.*

▶ CALDER TRADITIONAL ROOFING SERVICES

'Woodhurst' Cattlegate Road, Crews Hill, Enfield, Middx EN2 8AU
Tel 01707 876515 Fax 01707 872413

LEAD, COPPER, ZINC AND CLASSIC RAINWATER PRODUCTS: *See also: profile entry in Roofing Contractors section, page 59.*

ROOF DRAINAGE

▶ J & J W LONGBOTTOM LIMITED

(inc Sloan & Davidson Ltd)
Bridge Foundry, Holmfirth, Huddersfield HD7 1AW
Tel 01484 682141 Fax 01484 681513

CAST IRON RAINWATER AND SOIL: A traditional foundry, producing a comprehensive range, comprising pipes, gutters, and all the necessary fittings for rainwater, soil (BS 416) and smoke. An extremely extensive pattern range of moulded gutters (half round, ogee, box, valley) (including curves) and of ornamental rainwater heads is produced. Substantial stocks of all products are continually maintained and prompt delivery throughout the UK can be effected. Their comprehensive catalogue is available on request.

▶ SAINT-GOBAIN PIPELINES

Sinclair Works, PO Box 3, Ketley, Telford, Shropshire TF1 5AD
Tel 01952 262500 Fax 01952 262555 Literature Hotline 0800 328 7458
Website www.saint-gobain-pipelines.co.uk (soil and drain)

CAST IRON RAINWATER SYSTEMS: The classical range of traditional cast iron rainwater and gutter systems has been manufactured at Sinclair Works for over 100 years. The range includes a choice of standard half round, beaded and deep half round, OG, Notts OG, moulded no 46, and box gutter systems, with downpipes available in round and rectangular profiles. Most of the profiles are offered in a series of diameters, and all offer a full range of fittings. A new gutter jointing kit has recently been introduced to the half round gutter profile, speeding up installation on site. The Sinclair Classical range is manufactured under a BS EN ISO9002:1994 Quality Assurance Scheme and in accordance with BS460 and is the only system with BBA Approval (Cert.97/3434). *See also: display entry in this section, page 89.*

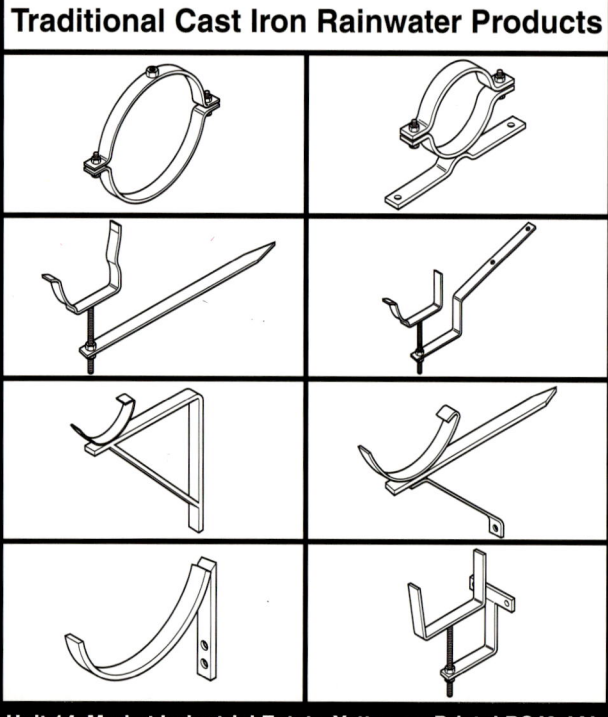

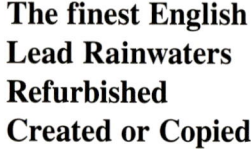

ROOF DRAINAGE

ROOF LIGHTS

ROOF LIGHTS

IN THE DESIGN STUDIO

I LIKE THE WORK YOU'RE DOING ON THE MEDWAY PROJECT, IAN, BUT......

MAY I SUGGEST THAT THE ROOFLIGHTS YOU SPECIFY WOULD LOOK MUCH BETTER IF THEY WERE **THE CONSERVATION ROOFLIGHT** FROM THE METAL WINDOW COMPANY....

...THEY ARE FINELY DETAILED AND LOW IN PROFILE...

....WHICH WOULD MAKE THEM SIT INCONSPICUOUSLY INTO THE ROOF AND NOT DOMINATE IT... AS WELL AS BEING LIKED BY PLANNERS

I DO LIKE THE LOOK OF THE CONSERVATION ROOFLIGHT BUT WONDER IF ITS PERFORMANCE IS UP TO MODERN STANDARDS....

YOU NEEDN'T WORRY ON THAT SCORE....

....WITH THEIR 'THERMOLINER' DECOUPLED LINING WHICH ELIMINATES CONDENSATION PROBLEMS....

....THE BRE TEST RESULTS THAT PROVE THE DESIGN TO BE WEATHERTIGHT IN THE SEVEREST WEATHER

....AND THEIR USE OF 'K' AND SAFETY GLASS AS STANDARD...

THEY COMBINE UP-TO-DATE PERFORMANCE AND SAFETY WITH THOSE FINE LINES AND LOW PROFILE

YES I CAN SEE THEY HAVE A BETTER PERFORMANCE SPEC THAN MOST OTHER ROOFLIGHTS I'VE COME ACROSS..... ON THEY GO!

The Metal Window Company, Phone: 01993 830613 Fax: 01993 831066 www.metalwindow.co.uk

ROOF LIGHTS

▶ **CLEMENT WINDOWS GROUP LTD**

Clement House, Haslemere, Surrey GU27 1HR
Tel 01428 643393
Website www.clementwg.co.uk

ROOF LIGHTS, METAL WINDOWS AND LANTERN LIGHTS: *See also: display entry on page 89.*

▶ **THE METAL WINDOW CO LTD**

Unit 8 Wychwood Business Centre, Milton Road, Shipton-under-Wychwood, Oxon OX7 6XU
Tel 01993 830613 Fax 01993 831066
E-mail info@metalwindow.co.uk

THE CONSERVATION ROOFLIGHT: *See also: display entry above.*

ROOF VENTS

BUILDINGS MUST BREATHE TO SURVIVE

THE AWARD WINNING CHOICE

"PROVEN LOGIC – ORIGINAL TECHNOLOGY"

- 'Completely unexposed' and non-load bearing
- Prevents structural alteration/material damage
- The roof's inbuilt defenses remain un-breached
- Unique safety and storm protection features
- BRE tested – 20,000mm² filtered free-vent area
- Universally adaptable + mech. extraction facility
- Made from best high impact 'recycled' material
- Cost effective input with a 'lifetime' guarantee
- Retrofit *inside* at anytime or during construction

IN SUPPORT OF THE BUILT ENVIRONMENT

Heritage Structural Ventilation Ltd
33 Queens Road – Bournemouth – Dorset BH2 6BN
Fax/Phone: 01202 769 255 – Email: info@lapvent.co.uk
Website: www.lapvent.co.uk

CLAY TILES & ROOF FEATURES

▶ ALDERSHAW HANDMADE CLAY TILES LTD

Pokehold Wood, Kent Street, Sedlescombe, East Sussex TN33 0SD
Tel 01424 756777 Fax 01424 756888
E-mail sussexterracotta@wealden.freeserve.co.uk
Website www.sussexterracotta.co.uk www.aldershaw.co.uk

HANDMADE CLAY TILES: Truly handmade clay tiles and fittings in the widest range of colours are manufactured from Aldershaw's own Wadhurst clays in the heart of the Sussex countryside. Aldershaw's roofs have a patina normally only associated with a bygone age. The company specialises in restoration, in Kent peg tiles, mathematical tiles, fireplace briquettes and Gault clay tiles for the Cambridge and Ely areas. Special sizes and shapes are no problem. Sussex Terracotta produces handmade floor tiles in a vast range of colours shapes and sizes. The tiles are made to look mature from the day they are laid. Recent work includes 1,000 yards in the Crypt of St Paul's Cathedral.

▶ BABYLON TILE WORKS

Babylon Lane, Hawkenbury, nr Staplehurst, Tonbridge, Kent TN12 0EG
Tel 01622 843018 Fax 01622 843398
Website www.babylontileworks.co.uk

KENT PEG TILES AND FITTINGS: Babylon Tile Works produces their unique Kent peg tiles using traditional moulding methods with modern quality control to ensure the highest tile quality and performance. The tiles are made from Babylon's own Kent clay which is available on site. Two natural colours are available – Terracotta and Antique – giving you the warm rich colours of the Weald. Please contact Melvin Gash for further information and friendly specification advice.

▶ G & N MARSHMAN

1 Nell Ball, Plaistow, Billingshurst, West Sussex RH14 0QB
Tel 01798 342427

MANUFACTURERS OF RIVEN OAK, CHESTNUT PLASTERERS' LATHS AND TILE AND STONE SLATE BATTENS: *See also: profile entry in Plasterwork section, page 214.*

▶ SANDTOFT ROOF TILES

Sandtoft, Doncaster, South Yorkshire DN8 5SY
Tel 01427 871200 Fax 01427 871220 Contact Nick Oldridge

ROOF TILES: Sandtoft Roof Tiles has a history of excellence as rich and respected as the tiles it produces. Founded in 1904, not only does Sandtoft boast the largest product portfolio, but it is the largest independent tile producer in the UK, employing nearly 400 people at five manufacturing plants. Sandtoft's Heritage Service has provided replacement tiles for the finest historic roofs in England, including the Victoria and Albert Museum, Cambridge University, Eton College and the Verulamium Museum in St Albans. Combining traditional techniques with modern technology, its dedicated craftsmen recreate even the most difficult of historic tiles, from Roman under-and-overs to bespoke handmade hips, valleys and ridges.

▶ SMITHBROOK BUILDINGS PRODUCTS LIMITED

Pollingfold Farm Barn, Ellen's Green, Rudgwick, Horsham RH12 3AS
Tel 01403 824170 Fax 01403 824171

GLAZED CLAY PANTILES, GLAZED BRICKS, ENCAUSTIC TILES: These are some of the speciality clay products offered by Smithbrook Building Products. Of particular interest to conservation officers and architects are the glazed pantiles to match colours used predominantly on 1930s buildings, and non-interlocking black glazed pantiles used extensively in East Anglia. Unobtrusive modern ventilation systems are also available.

▶ WEST MEON POTTERY

Church Lane, West Meon, Petersfield, Hampshire GU32 1JW
Tel 01730 829434

ARCHITECTURAL CERAMICS: *See also: profile entry in Architectural Terracotta section, page 98.*

MASONRY SERVICES

SCOTLAND		Page
Abbey Heritage Ltd	sm ss tc	104
Balmoral Stone	sm tc	106
Cameron (UK) Plc	br sm tc	107
Graciela Ainsworth	sm ss	114
Heritage Engineering	ss	115
Holden Conservation Ltd	sm ss tc	115
Nicholas Boyes Stone Conservation	sm ss	114
Stoneguard	cr sm tc	113
YORKSHIRE		
A D Calvert Architectural Stone Supplies Ltd	sm ss	104
Abbey Heritage Ltd	sm ss tc	104
Anelays	br sm	104
Burrows Davies Limited	sm	105
C J Ellmore & Co Ltd	br sm	59
Hanna Conservation	ss	114
Stoneguard	cr sm tc	113
Weaver Construction Limited	br sm	69
Yorkshire Decorative Plasterers	mm	216
NORTH-WEST		
Allison Stone (Wigan) Limited	sm ss	106
Bernard A Shepherd Ltd	br sm	73
Burleigh Stone Cleaning & Restoration Co Ltd	br sm	105
Burnaby Cleaning Co Ltd	sm	171
Cameron (UK) Plc	br sm tc	107
Maysand Limited	br sm tc	110
S & J Whitehead Ltd	sm	66
Stone Central	sm	112
Stoneguard	cr sm tc	113
Wm Langshaw & Sons	br sm	68
WALES		
Abbey Masonry and Restoration	sm ss	104
The Stone Doctors	sm ss	112
Taylor Dalton, Heritage Building Contractors	sm	67
William Taylor Stonemasons	sm	112
WEST MIDLANDS		
Capps & Capps Limited	sm	60
Centreline Architectural Sculpture	sm ss	107
Eura Conservation Ltd	ss	115
Linford-Bridgeman Limited	br sm	61
Russcott Conservation and Masonry	sm	111
Stonecraft	dw	113
Stoneguard	cr sm tc	113
Timothy Williams (Builders) Ltd	sm	66
Treasure & Son Ltd	br sm	67
Wells Masonry Services Ltd	sm	114
William Sapcote & Sons Ltd	sm	68
EAST MIDLANDS		
Boden & Ward Stonemasons Ltd	sm ss	106
C R Crane & Son Ltd	br sm	60
Church Conservation Limited	sm	84
H & W Sellors Ltd	sm	108
Hirst Conservation	ea	115
Leander Architectural	ss	115
R M H Eaton Stonemasonry	sm	111
Weldon Stone Enterprises Ltd	sm	114
SOUTH-EAST – NORTH OF THE THAMES		
A F Jones (Stonemasons)	sm ss	105
Abbey Heritage Ltd	sm ss tc	104
Ashby Stone Masonry Limited	dw sm ss	105
Beckwith Tuckpointing	br sm	99
Between Time	br sm	58
Boshers (Cholsey) Ltd	br sm	58
Calder Traditional Building Services	br sm	60
Cliveden Conservation Workshop Ltd	sm ss	108
Hall Construction	dw br sm	60
The Hertfordshire Roofing & Renovation Company	br sm	59
I J P Building Conservation	br ea sm ss	61
J G Matthews Ltd	br sm	62
J Joslin (Contractors) Ltd	sm	108
Gerard C J Lynch	br	98
Mathias Builders	br	98
Mosaic Marble and Granite Ltd	sm	96
Mowlem Rattee & Kett	br sm ss	62
Richard Noviss Sculptor & Stonemason	sm ss	110
Protovale Oxford Ltd	cr	179
Sindall Ltd	br sm	64
Stoneguard	cr sm tc	113
LONDON		
John Bysouth - Stone Consultant	sm	106
Cameron Taylor Bedford	cr	40
Carthy Conservation Ltd	sm ss tc	107
David Ball Restoration (London) Limited	br sm tc	109
Holden Conservation Ltd	sm ss tc	115
Holloway White Allom	br sm	62
Ken Negus Ltd	br sm tc	109
Kim Meredew	sm	109
O'Reilly Period Cornice Restoration & Cleaning	mm	215
PAYE Stonework & Restoration	sm	111
Plowden & Smith	ss	115
Priest Restoration	sm	111
Sindall Ltd	br sm	64
Spencer & Richman	ss	190
Stonewest Ltd	br sm	98
Taylor Pearce Restoration Services Limited	ss	115
Timothy Williams (Builders) Ltd	sm	66
SOUTH-EAST – SOUTH OF THE THAMES		
C Ginn Building Restoration	sm	106
Cathedral Works Organisation (Chichester) Limited	sm ss	107
Chalk Down Lime Ltd	sm	168
Haslemere Builders Limited	br sm	60
I W Payne & Company Ltd	br sm	61
Mott Graves Projects Ltd	sm ss	110
R W Armstrong & Sons Limited	br	64
Robus Architectural Ceramics	ss tc	97
Simmonds of Wrotham	br sm	65
Timothy J Shepherd Historic Brickwork Specialist	br	99
Wallis	br sm	55
SOUTH-WEST		
The Bath Stone Group	sm	105
Busby's Builders	br sm	59
Carrek Ltd	br sm	107
Cranborne Stone	mm	116
Nicholas Durnan	sm ss	114
Jonathan Rhind. Architects	ea	31
McMarmilloyd Limited	ss	96
Mike Wye & Associates	ea	168
Nimbus Conservation Ltd	sm ss	110
Somerset Stone Masons Ltd	sm	112
St Blaise Ltd	sm ss	112
Stoneguard	cr sm tc	113
Stoneworks of Bath Ltd	sm	113
W J Haysom & Son	sm	114
Wells Cathedral Stonemasons	sm ss	113
Western Stone	cr sm	113

KEY
- br brick services
- cr concrete repairs
- dw dry stone walling
- ea earth and cob
- mm mould makers
- sm stone masons
- ss statuary and stone carving
- tc architectural terracotta

Regional designation is according to office location. Many firms operate nationally.

MASONRY PRODUCTS

A D Calvert Architectural Stone Supplies Ltd	sq	104
Abbey Heritage Ltd	cs pv	104
Ace Demolition – The Reclamation Centre	bk pv sq	159
Ajeer Limited	bk pv sq	159
Allison Stone (Wigan) Limited	sq	106
Alpine Reclamation	bk pv sq	159
Antique Buildings Limited	bk pv sq	162
Artisan Oak Buildings Ltd	pv	142
Ashby Stone Masonry Limited	pv	105
The Bath Stone Group	sq	105
Boden & Ward	sq	106
Bovingdon Brickworks Ltd	bk	99
Broadmead Cast Stone	cs	116
The Bulmer Brick and Tile Co Ltd	bk	99
Burleigh Stone Cleaning & Restoration Co Ltd	cs	97
Cameron (UK) Plc	sq	107
Centreline Architectural Sculpture	ge sq ss	107
Chelwood Brick Ltd	bk	99
Chilstone	cs ss	116
Classical Flagstones	pv	114
Cobar Services	pv	160
Conservation Building Products Ltd	bk pv sq	160
Cranborne Stone	cs	116
David Ball Restoration (London) Limited	cs	109
The Downs Stone Company Ltd	pv sq	108
Drummonds Architectural Antiques Limited	bk pv sq	160
Nicholas Durnan	ss	114
Graciela Ainsworth	ss	114
H G Clarke & Son	sq	108
Haddonstone Limited	cs ss	117
Hammill Brick Ltd	bk	100
Hanson Brick	bk pv	100
The Hopton Wood Stone Firms Ltd	pv sq	109
Ibstock Brick Limited	at bk pv	100
Ibstock Hathernware	at	97
Keim Mineral Paints Ltd	cr	184
Ketley Brick Company Limited	bk pv	100
Keyline	cs mx ss	115
Lambs Bricks and Arches	at bk	101
LASSCO	pv	160
Leander Architectural	ss	115
McMarmilloyd Limited	sq ss	96
Mosaic Marble and Granite Ltd	sq	96
Mott Graves Projects Ltd	ss	110
Original Architectural Antiques Company Ltd	pv	160
Original Oak	at	208
Retrouvius Architectural Salvage	pv	160
Robus Architectural Ceramics	at	97
S & J Whitehead Ltd	sq	66
Shaws of Darwen	at	99
Smithbrook Building Products Ltd	bk	91
Solopark Plc	bk pv	161
South West Reclamation	bk pv sq	162
The Stone Doctors	sq	112
Stonecraft	sq	113
Stoneguard	cs	113
Sussex Brick Ltd	bk	100
Symonds Salvage	bk pv sq	162
Thorverton Stone Company	cs	116
W J Haysom & Son	pv sq	114
Walcot Reclamation	bk pv sq	162
Weldon Stone Enterprises Ltd	sq	114
Wells Cathedral Stonemasons	sq	113
Wells Masonry Services Ltd	sq	114
West Meon Pottery	at bk	98
Westland	ss	190
William Taylor Stonemasons	sq	112

KEY
- at architectural terracotta suppliers
- bk brick suppliers
- cs cast stone
- ge stone identification and sourcing
- mx mould making materials
- pv flagstones and paving
- sq stone suppliers
- ss statuary and sculpture suppliers

SHINING STONES

Britain's native 'marbles'

GRAHAM LOTT AND DAVID SMITH

Britain's historic buildings can often prove to be a treasure trove for the marble enthusiast. High status buildings like royal palaces or stately homes may have colourful marble-lined halls, grand marble staircases, or extravagantly carved fireplaces. Our cathedrals and churches can show an even more diverse use of marbles from simple memorial tablets, intricately carved tombs, pulpits, fonts, decorative columns, colourful marbled floors to loud and extravagant Victorian graveyard statements. In the 19th century marbles became decorative status symbols in the newly established banks and commercial headquarters of our major cities and towns.

In the main, the marbles on view are imported varieties, a trade which has grown since Roman times, but mingled amongst them particularly in our parish churches and lesser buildings are many equally attractive native 'marbles'. Strictly defined, marbles are metamorphosed limestones that in their raw state tend, like most rocks, to look rather drab and uninteresting. However, if they are cut and carefully polished they are transformed into the extravagantly colourful and patterned 'shining stones' (the term marble derives from the Greek word for shining or sparkling) that have been coveted for decoration and ornament since first exploited and displayed by the Greeks and Romans.

Anyone who has travelled in Europe, particularly to Greece or Italy, and visited the many great churches and cathedrals cannot fail to be attracted, or at the very least distracted, by the sheer range and magnificence of the decorative marble-work on display. One imagines that the same impact was felt by earlier travellers from Britain for whom it was considered an essential part of their education to spend a period on the continent of Europe. Among these visitors were many of the great architects, artists and intellectuals who subsequently influenced styles and tastes throughout British life. How disappointed some of them must have been to find that on returning to Britain they could find no marbles available to rival those of the Mediterranean area; no white Carrara (Tuscany), no Verde Antico (Thessaly), no Rouge Languedoc (Carcassone) and no Port d'Oro (Gulf of Spezia).

A shaft made of Frosterley Marble, in the North Porch of Bristol Cathedral

WHY DOESN'T BRITAIN HAVE ITS OWN RANGE OF METAMORPHIC MARBLES?

The simple answer is that Britain's geological history has been very different from that of the Mediterranean area. The marbles of the Mediterranean region formed by the alteration of beds of sedimentary limestones over long periods of geological time. They began life as sediments, formed from the skeletal remains of calcareous fossils, shells and corals fragments, in ancient tropical seas. Subsequently these limestones were deeply buried and subjected to pressures and temperatures high enough to cause all of this skeletal material to recrystallize, destroying any signs of the original sedimentary fabric. The extravagantly coloured marbles we see today are now fine crystal mosaics of calcium or magnesium carbonate, sometimes veined and fractured, but showing little sign of their sedimentary origins.

Iona

Purbeck

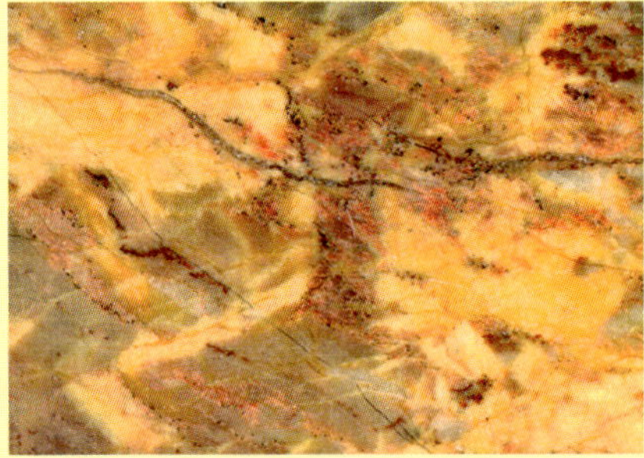

Petitor

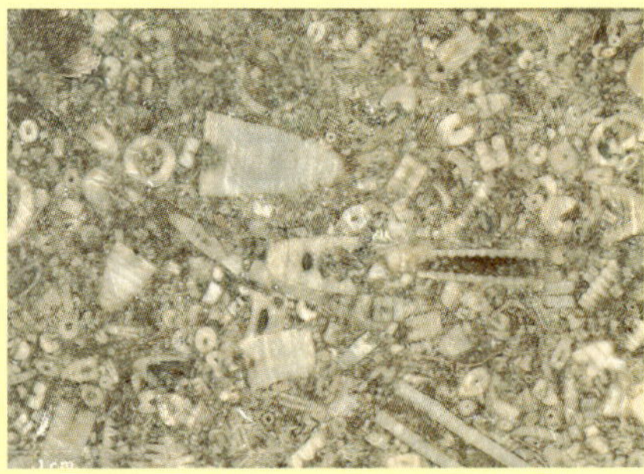

Derby Fossil

Britain's geological succession also has many thick beds of limestone. They principally occur in the Pre-Cambrian, Cambrian, Devonian, Carboniferous, Jurassic and Cretaceous rocks *(see table)*. However, with the exception of the Pre-Cambrian and Cambrian limestones of Scotland and North Wales, they have not been subjected to the same pressures and temperatures over geological time and, consequently, are less altered. Though now hard and cemented, internally they remain today much as when they were deposited – tropical marine sediments packed with unaltered calcitic fossil fragments.

The colour variations so characteristic of true marbles are part of this same alteration process (metamorphism) which redistributes the various chemical compounds present in the original limestone throughout the new crystalline fabric. Reds and yellows are a result of the presence of various iron compounds, blacks contain organic carbon, greens include copper compounds and whites are almost pure calcium or magnesium carbonate.

BRITISH 'MARBLES'

Historically, the only true marbles produced in Britain were quarried in north-west Scotland on the isles of Iona, Tiree and Skye. Iona Marble is predominantly white with greenish veining and mottling and though known to have been worked since the 13th century, for Iona Cathedral, it was never a large industry. Iona Marble can still be seen in a number of churches in southern Scotland and was one of many marbles used to decorate Westminster Cathedral in London. Today the metamorphosed Durness Limestone is quarried near Ullapool to produce Britain's sole remaining true marble, the variegated greenish white *Ledmore Marble.*

From earliest times this scarcity of indigenous marble was overcome by importing European marbles for high quality decorative work. Such European marbles are well known from many Roman sites in Britain. Presumably the high cost of such a trade meant that alternative sources of decorative stone were soon sought out and the local hard sedimentary limestones were quickly exploited for decorative and ornamental work. Although rarely as extravagantly coloured as their Mediterranean counterparts, Britain's native 'marbles' were available in a variety of colours and textures, and by medieval times were extensively used in cathedrals, churches and great houses.

MARBLES OF GREAT BRITAIN

GEOLOGICAL AGE	'MARBLE'
QUATERNARY	
TERTIARY	
CRETACEOUS, UPPER	
CRETACEOUS, LOWER	Sussex, Petworth, Bethersden, Charlwood, Small Paludina, Large Paludina
JURASSIC, UPPER	Purbeck, Melbury
JURASSIC, MIDDLE	Forest, Alwalton, Yeovil, Bowden, Crackemont, Stamford, Weldon Rag, Raunds, Stanwick
JURASSIC, LOWER	Ammonite, Marston, Banbury
TRIASSIC	Cotham, Draycott, Alabaster
PERMIAN	
CARBONIFEROUS	Frosterley, Dent, Nidderdale, Poolvash*, Swaledale, Orton Scar, Halkyn, Penmon, Pembroke, Mumbles, Ashford, Furness, Duke's Red, Rosewood, Birdseye, Muscle Shoal, Hopton Wood*, Monyash, Derby Fossil, Tournai
DEVONIAN	Ashburton*, Plymouth, Petitor, Ipplepen, Radford, Ogwell, Bradley Wood
SILURIAN	Ledbury
ORDOVICIAN	
CAMBRIAN	Ledmore*, Skye, Tiree, Swithland Slate
PRE-CAMBRIAN	Iona, Mona

KEY
Native 'marble' (mostly limestones) True marble * active quarries

The best known is undoubtedly *Purbeck Marble* a dark greenish grey, reddish or dark grey fossiliferous limestone that is found only in thin beds on the Isle of Purbeck in Dorset. First exploited by the Romans, there are few medieval cathedrals and churches of southern England that do not have some Purbeck Marble decoration in the form of columns, coffin lids, tombstones or fonts (such as Salisbury, Ely, Llandaff, Winchester). Cathedrals as far afield as Lincoln, York, Beverley and Durham and a number of churches in Leeds also have Purbeck Marble decorative stonework.

Purbeck Marble is, however, very much a southern stone. Elsewhere, other limestones were often exploited for such decorative work. In northern England, at Weardale, a beautiful black limestone studded with large white corals, known as *Frosterley Marble*, appears in Durham, York and Norwich cathedrals and in a number of churches in Yorkshire, such as St Mary's in Beverley, St John's in Leeds. In more recent times, Frosterley Marble was used in Bristol and St Albans cathedrals and to provide a pulpit for Bombay Cathedral.

Black limestones were always very much sought after and other later regional examples include the *Ashford Black* from Derbyshire (worked from the 17th century), *Pembroke* (coralline), *Nidderdale* (crinoidal) and *Poolvash* (black) from the Isle of Man. These native 'marbles' had to compete not only with Mediterranean varieties but also with a vibrant trade in the famous *Tournai Marble* (Belgian Black or Touchstone). Belgian 'marbles' were extensively imported around the 12th century for use as fonts and grave slabs in many of our cathedrals and larger churches (Winchester, Lincoln and Ely cathedrals). Black 'marbles' from Ireland were also imported, such as *Kilkenny Black* (crinoidal) and *Galway Black* (pure black). The expense of importing such limestone meant that occasionally other materials were substituted, for example polished dark grey Swithland Slate memorial slabs are common in some churches in the East Midlands.

More colourful native 'marbles' were also widely worked for decorative purposes. Many of the great houses, stately homes and palaces of Britain contain some marble decoration perhaps as flooring or commonly for elaborately decorated fire surrounds. The Dukes of Devonshire, over many centuries, exploited a variety of colourful limestones from their estates in Derbyshire. Houses like

Ashburton

Penmon

Alabaster

Images courtesy of Natural History Museum

Chatsworth, Haddon Hall, Hardwick Hall and Bolsover Castle all contain decoration carved from local limestones such as the *Duke's Red* (as at Great Longstone Church, Derbyshire and St John's Chapel, Cambridge), *Birdseye* (crinoidal), *Rosewood* (finely layered) and *Muscle Shoal* (Bolsover Castle; a shelly limestone from the Coal Measures). *Furness Marble*, a grey-brown crinoidal variety, was also produced from the Devonshire estates in Lancashire.

The Carboniferous limestones of the Derbyshire area are notable for another famous limestone, the *Hopton Wood Stone.* This pale, buff-grey crinoidal limestone is still produced and has a long history of use for memorial and ornamental work. It was particularly widely used for the construction of war memorials after both World Wars, and was one of a limited number of stones selected for the manufacture of stone crosses to commemorate the war dead in tens of thousands of graves across Europe and further afield in North Africa.

Other local 'marbles' include the buff coloured, coarsely fossiliferous crinoidal limestone beds such as *Monyash, Derby Fossil* (Derbyshire), *Swale Dale Fossil*, (Yorkshire) and *Orton Scar* (Cumbria); the grey to buff, white veined *Mumbles* variety from Swansea; the grey-brown, veined *Penmon* and *Halkyn* (crinoidal) marbles from North Wales.

During the 19th century some of the most important sources of native 'marbles' were the limestones outcropping in the Plymouth, Ipplepen and Torquay areas of Devon. Though still not true marbles they commonly show fabrics and textures which suggest they have locally been subjected to high pressures and temperatures associated with earth movements in this area. These compact limestones show a wide range of colours from light grey to black with shades of red, cross-cut by veins of white, yellow and red. They are often characterized by a variety of large fossils (corals, crinoids, bivalves, stromatoporoids and ammonoids) and consequently sometimes termed *Madrepore* marbles (coral-rich), but may also be veined, fractured or brecciated and when polished produce a wide range of distinctive marble-like textures and patterns. They are known by a variety of local names (26 or more in all) including *Plymouth* (black, grey and red), *Petitor* (yellow pink and grey), *Ipplepen* (reddish grey and white), *Radford, Ogwell* and *Bradley Woods.* The best known are probably those of *Ashburton* (dark grey to black, coral-rich with red and white veining). Fine examples of their use can be seen in Keble College Chapel at Oxford, Chichester Cathedral (Ashburton) and Brompton Oratory (Radford).

Locally important in some parts of Britain were thin beds of fossiliferous limestones that were hard enough to work for decorative or ornamental purposes. The obvious added value of a good polished stone meant that many such limestones formed the basis of small local industries and commonly appear in local churches and houses. Some examples include *Ledbury Marble* (mottled red, purple white and blue) a coral-crinoid rich limestone outcropping in the Malverns and *Cotham Marble* from Somerset. Cotham is a buff coloured limestone with dark, dendritic, tree-like mineralized growths, hence its alternative name *Landscape Marble.* From the Lower Jurassic rocks came the *Ammonite* or *Marston Marble* at Yeovil and *Banbury Marble* in Oxfordshire; from the Middle Jurassic rocks *Stamford Marble* and *Weldon Rag* (Lincolnshire Limestone) and *Raunds* or *Stanwick Marble* (Blisworth Limestone) in Northamptonshire. *Alwalton Marble* was produced from thin shelly limestone beds in the Middle Jurassic succession of the Peterborough area, the best examples of which can be seen in the 12th century tomb of Abbot Benedict at Peterborough Cathedral. The so-called *Forest Marble* (*Yeovil Marble* in Somerset; *Bowden* or *Crackement* marbles in Dorset) is also a hard, thin, shelly limestone, once extensively used for paving and roofing in Wiltshire and Gloucestershire, but also polished to provide decorative fire surrounds. Fractured limestone nodules (septaria) from the Oxford Clay in Dorset, known as *Melbury Marble*, were once cut and polished for decorative slabs.

A number of thin bands of blue-grey limestones outcrop in the Weald of south-east England. They were known by a variety of local names including, *Sussex, Petworth, Charlwood, Bethersden Small and Large Paludina* marbles. These limestones, like the Purbeck Marble, formed in freshwater lakes, and because they contain numerous coiled gastropod shells are commonly confused with it, despite the larger size of the fossils. Unlike Purbeck, however, these Wealden 'marbles' were only used locally (as at Canterbury and Chichester cathedrals, and at churches in Arundel, Burton, Horsham, Lavant, Pulborough and Stopham) and are rarely found further afield.

OTHER BRITISH 'MARBLES'

A number of stones which carry the epithet 'marble' cannot even be classified as limestones. *Draycott Marble* (brecciated) was quarried in Somerset. This reddish coloured rock originally formed as an accumulation of coarse, angular limestone and sandstone fragments subsequently cemented together by carbonate and known by geologists as a breccia. The original quarries were recently re-opened to provide new stone for the conservation of Bristol Temple Meads railway station. The metamorphic alteration of some igneous rocks, particularly those of basic composition i.e. those rich in the green mineral olivine and lacking quartz, produces the rock type commonly termed *serpentinite.* The *Mona Marble* from the Pre-Cambrian Mona Complex at Roscolyn in Anglesey is variegated dark green, white veined serpentinite (a metamorphised gabbro) that was apparently the basis of a small London-based industry in the early 19th century. *Polyphant* from Launceston in Cornwall is also a dark green serpentinite (a metamorphosed picrite), which when polished, has a rich, marble-like finish. It was used decoratively in many local churches and can be seen in Truro, Canterbury and Exeter cathedrals.

Alabaster is often confused with marble. Alabaster is a hard, compact, finely crystalline form of gypsum (calcium sulphate) which is usually white, translucent or mottled red in colour. Although beds of gypsum are common in the Triassic successions of Britain, the alabaster variety, used for carved ornamental work is much rarer. The most important sources of alabaster for monumental work were along the Nottinghamshire-Derbyshire border at Chellaston and Red Hill, and at Fauld in Staffordshire. Alabaster from the Derbyshire pits was carved into a large number of fine medieval sepulchral monuments, for export around England and to France. Local churches in the East Midlands include many fine alabaster figures and carvings as tomb monuments (Swarkestone, Radcliffe-on-Soar and Bottesford for example) as do several of our cathedrals (Lichfield, Canterbury, Worcester and Southwell). Alabaster from these pits was also used for the massive white and red mottled columns of Holkham Hall (the so-called Marble Hall) in Norfolk and Kedleston in Derbyshire. In the 19th century architects like Robert Adam produced fire surrounds carved from white alabaster.

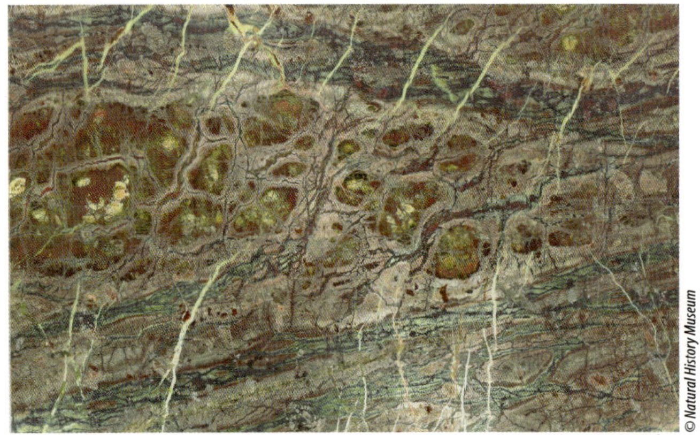
© Natural History Museum

Serpentine

MARBLE COLLECTIONS

The wide range of colours and rock fabrics that characterise British marbles make their identification something of a problem to the untrained eye. There are fortunately, however, a number of outstanding collections of European marbles in Britain which should be the first port of call if identification is a problem. The largest collections are held by the Natural History Museum in London and the Oxford Museum of Natural History. In addition to these collections, both buildings have used marbles for decorative effect in the original Victorian display areas. The Natural History Museum's Collection of Building and Decorative Stones includes over 4,000 polished samples from international sources, representing the use of marble since the mid 19th century. It is hoped that a searchable database of digital images of these specimens will be available on the Internet by 2002. The Oxford Museum houses the famous Corsi Collection of European marbles and the Watson Collection that includes many British varieties. As an added bonus, the museum also has a fine display of clearly labelled marble columns lining the corridors surrounding the main gallery. Both these collections can be viewed by appointment with the curators. Though not strictly a collection, the wide range of marbles used to decorate Westminster Cathedral provide one of the few British examples of decorative marble-work on a scale to rival the great churches in Rome.

RECOMMENDED READING

Austin, RL; *Mumbles Marble and its Association with Swansea and District.* Minerva, Journal of Swansea History, 1999

Ashurst, JA and Dimes, FG; *Conservation of Building and Decorative Stone. Vol 1* Butterworth-Heinemann, 1990

Blacker, JG and Mitchell, M; *The Use of Nidderdale Marble and other Crinoidal Limestones in Fountain's Abbey, North Yorkshire.* The Leeds Philosophical and Literary Society, 1998

Tomlinson, JM; *Derbyshire Black Marble.* Peak District Mines Historical Society Special Publication No 4, 1996

Young, J; *Alabaster.* Derbyshire Museum Service, 1990

This article is published with the permission of the Director of the British Geological Survey, NERC.

Dr GRAHAM LOTT is a sedimentary petrologist with the British Geological Survey, Nottingham E-mail g.lott@bgs.ac.uk

DAVID A SMITH is the Loans Manager and Petrology Curator, Mineralogy Department, at The Natural History Museum, London E-mail d.a.smith@nhm.ac.uk

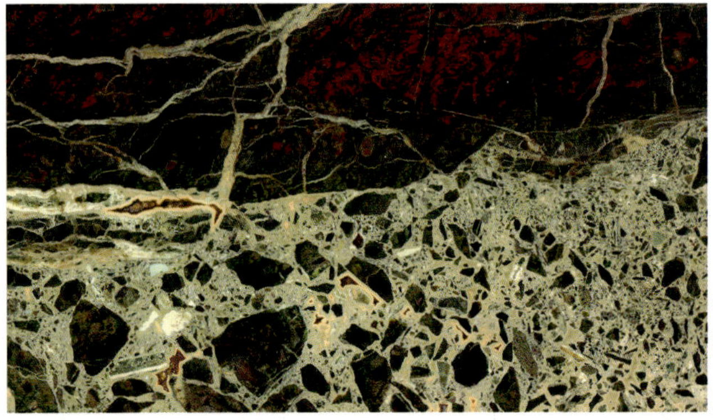
© Natural History Museum

Serpentine E3873

MARBLE & GRANITE

▶ H W POULTER & SON

279 Fulham Road, London SW10 9PZ
Tel 020 7352 7268 Fax 020 7351 0984

FIREPLACES: *See also: profile entry in Fireplaces section, page 189.*

▶ McMARMILLOYD LIMITED

Brail Farm, Wilton Road, Great Bedwyn, Marlborough, Wiltshire SN8 3LY
Tel 01672 870227 Fax 01672 870053
E-mail info@mcmarmilloyd.co.uk
Website www.mcmarmilloyd.co.uk

SPECIALISTS IN RARE AND ANTIQUE MARBLES: Based in beautiful rural Wiltshire, McMarmilloyd, for ten years has specialised in the procurement of rare, antique and archaeological marbles for restoration and refurbishment. Regular granites, marbles and slates are supplied as are ex-stock, the finest statuary marble for sculpture work. For further details or to arrange a visit please contact Ian McDonald. *See also: display entry above.*

▶ MOSAIC MARBLE AND GRANITE LTD

Dartford House, Two Rivers, Station Lane, Witney, Oxon OX8 6BH
Tel 01993 708899 Fax 01993 709959

SPECIALISTS IN MARBLE, GRANITE, MOSAIC RESTORATION AND CLEANING TECHNIQUES: As specialists, consultants and contractors, MMG offer a comprehensive service and are able to advise on, and implement professional cleaning and restoration to external facades and cladding, as well as all internal and sensitive areas. They supply and install for new-build and maintenance projects, and work closely with their clients on their commissions, which include historic, national and public buildings.

▶ NOSTALGIA

Holland's Mill, Shaw Heath, Stockport, Cheshire SK3 8BH
Tel 0161 477 7706 Fax 0161 477 2267
Website www.nostalgiafireplaces.co.uk

RECLAIMED ANTIQUE FIREPLACES: Nostalgia supplies and restores fine marble chimneypieces. *See also: profile entry in Fireplaces section, page 189.*

▶ BURLEIGH STONE CLEANING & RESTORATION CO LTD

The Old Stables, 56 Balliol Road, Bootle, Merseyside L20 7EJ
Tel 0151 922 3366 Fax 0151 922 3377
E-mail info@burleighstone.co.uk
Website www.burleighstone.co.uk

ARCHITECTURAL TERRACOTTA: A high quality comprehensive service on terracotta: cleaning, repair, restoration and replacement. *See also: display entry in Masonry Cleaning section, page 172.*

▶ CAMERON (UK) PLC

Cockshades, Wybunbury, Nantwich, Cheshire CW5 7HA
Tel 01270 841122 Fax 01270 841520 E-mail cameron@dial.pipex.com

RESTORATION AND CONSERVATION OF NEW/ REPLACEMENT TERRACOTTA AND FAIENCE: Cameron are highly experienced in the fixing of terracotta and faience. Examples of their work can be seen on the restored Fenwicks, Newcastle-Upon-Tyne where over 3,500 blocks were fixed to the extended facades. *See also: display entry in Stone section, page 107.*

▶ IBSTOCK HATHERNWARE

Station Works, Rempstone Road, Normanton on Soar, Loughborough, Leicestershire
Tel 0870 903 4016
Website www.hathernware.co.uk

ARCHITECTURAL TERRACOTTA AND FAIENCE: *See also, display entry above.*

▶ LAMB'S TERRACOTTA & FAIENCE

Nyewood Court, Brookers Road, Billingshurst, W Sussex RH14 9RZ
Tel 01403 785141 ext 223 Fax 01403 784663
E-mail sales@lambsbricks.co.uk

TERRACOTTA: Lamb's Bricks & Arches have been involved with the manufacture and supply of decorative traditional terracotta for over 100 years. If you are looking for replacement materials to match those found on any listed building, Lamb's Bricks & Arches can satisfy all your current terracotta requirements. *See also: display entry in this section, page 98, and entries in Brick section, page 100 and 101.*

▶ MAYSAND LIMITED

109-111 Windsor Road, Oldham, Lancs OL8 1RH
Tel 0161 628 8888 Fax 0161 627 0996
E-mail sales@maysand.co.uk

MASONRY REPAIR, CLEANING, RESTORATION AND REPLACEMENT: *See also: display entry in Stone section, page 110.*

▶ ROBUS ARCHITECTURAL CERAMICS

Evington Park, Hastingleigh, Ashford, Kent TN25 5JH
Tel/Fax 01233 750330
E-mail robusceramics@excite.co.uk
Website www.robusceramics.co.uk

CLAY BUILDING PRODUCTS: Specialist in clay building products. Robus manufactures a wide range of architectural ceramics for exterior and interior use, which includes mathematical tiles, roof tiles and fittings, chimney pots, floor and wall tiles. Robus offers a specialist service of replacement terracotta for renewal projects and the range also includes garden statuary, urns, terracotta gazebos and follies.

▶ SHAWS OF DARWEN

Waterside, Darwen, Lancashire BB3 3NX
Tel 01254 775111 Fax 01254 873462

ARCHITECTURAL TERRACOTTA AND FAIENCE: Manufacturers for over a century, Shaws offers traditional craftsmanship combined with the latest production technology for architectural restoration. For advice, surveys or quotations, please contact Jon Wilson. A division of Shires Limited. *See also: display entry in this section, page 99.*

▶ STONEGUARD

St Martins House, High Street, Ruislip, Middlesex HA4 7AU
Tel 01895 675577 Fax 01895 679125
Website www.stoneguard.co.uk
▶ with offices also in Bath, Birmingham, Bradford, Manchester and Stirling

BUILDING RESTORATION AND CONSERVATION: *See also: profile entry in Building Contractors section, page 66.*

ARCHITECTURAL TERRACOTTA

▶ STONEWEST LTD

Lamberts Place, St James's Road, Croydon CR9 2HX
Tel 020 8684 6646 Fax 020 8684 9323
E-mail stonewest@cwcom.net
Website www.stonewest.co.uk

BUILDING CONTRACTORS AND STONE MASONS: *See also: display entry in Building Contractors section, page 64.*

▶ WEST MEON POTTERY

Church Lane, West Meon, Petersfield, Hampshire GU32 1JW
Tel 01730 829434

ARCHITECTURAL CERAMICS: This country workshop has the resources to reproduce practically any piece of architectural ceramic for conservation projects. Hand thrown chimney pots up to 5 feet tall are a speciality. Also, special bricks, terracotta blocks and mouldings, finials and ridge tiles, floor tiles and large garden pots. They can mix different clays and aggregates to match the finish of an original example. Recent work includes pieces for The British Museum, Uppark House, Kensington Palace, Watts Memorial Chapel, Ishiya project, Japan; Kirby Hall; Kirklees Terrace, Glasgow; Shetland Amenity Trust, Ham House and other National Trust properties.

BRICK SERVICES

▶ BECKWITH TUCKPOINTING

24 The Meadowway, Billericay, Essex CM11 2HL
Mobile 0958 727208 Home 01277 659949

TUCKPOINTING, BRICK AND STONE RESTORATION: *See also: profile entry in Pointing section, page 99.*

▶ GERARD C J LYNCH, LCG, Cert Ed, MA (Dist)

10 Blackthorn Grove, Woburn Sands, Milton Keynes MK17 8PZ
Tel/Fax 01908 584163
E-mail glynch@cwcom.net
Website www.brickmaster.co.uk

HISTORIC BRICKWORK CONSULTANT: Gerard Lynch is an internationally acknowledged historic brickwork consultant and master bricklayer, offering a comprehensive consultancy and on occasions, a specialist bricklaying service. He lectures on historic brickwork and its conservation, and runs master classes embracing his extensive practical skills, covering historic, modern, gauged and decorative brickwork and traditional pointing methods. Mr Lynch is frequently relied on by heritage bodies, architectural practices and lay people to provide reports on brickwork status, methods of sympathetic repair and conservation, and training for craftsmen. Author of *Gauged Brickwork a Technical Handbook*, and *Brickwork: History, Technology and Practice volumes 1 & 2* (Donhead).

▶ MATHIAS BUILDERS

5 Elmside, Kensworth, Dunstable, Beds LU6 3RR
Tel 01582 873418 Fax 01582 517837

HISTORIC BRICKWORK SPECIALISTS: Mathias is a well-established father and son partnership specialising in the repair and restoration of historic brickwork. Work includes rubbed and gauged arches and flintwork using traditional methods and materials, and the use of lime mortars. Work includes brickwork to many National Trust properties, on the Ashridge Estate, Tithe Barn in Edlesborough, Goodwood House and to many other listed buildings. Recent major contracts include repair and restoration of the Tudor chimneystacks within Chequers Estate. Most work is completed in Bedfordshire, Hertfordshire and Buckinghamshire but Mathias will travel within an 80-mile radius of Dunstable.

ARCHITECTURAL TERRACOTTA

Terracotta
&
Faience
MANUFACTURERS FOR OVER A CENTURY,
SHAWS OFFER TRADITIONAL CRAFTSMANSHIP
COMBINED WITH THE LATEST TECHNOLOGY
FOR ARCHITECTURAL RESTORATION
AND CONSERVATION
SHAWS OF DARWEN
For advice, surveys & quotations, please contact Jon Wilson
Shaws of Darwen, Waterside, Darwen, Lancashire.BB3 3NX
Telephone: 01254 775111 Fax: 01254 873462

BRICK SERVICES

▶ STONEWEST LTD

Lamberts Place, St James's Road, Croydon CR9 2HX
Tel 020 8684 6646 Fax 020 8684 9323

BUILDING CONTRACTORS AND STONE MASONS*: See also: display entry in Building Contractors section, page 64.*

▶ TIMOTHY J SHEPHERD HISTORIC BRICKWORK SPECIALIST

101 Moffat Road, Thornton Heath, Surrey CR7 8PZ
Tel/Fax 020 8653 2438
E-mail tim@histbric.demon.co.uk

POINTING

▶ BECKWITH TUCKPOINTING

24 The Meadowway, Billericay, Essex CM11 2HL
Mobile 0958 727208 Home 01277 659949

TUCKPOINTING, BRICK AND STONE RESTORATION: Keith Beckwith runs a small team of skilled and dedicated craftsmen which has over 40 years experience in specialist pointing, the main type being tuckpointing. All types of brick and stone construction can also be restored, or completed as part of a larger restoration or sympathetic new-build project. Keith Beckwith has worked on many historic buildings including, Marlborough House, The Royal Mews and St James's Palace. Beckwith Tuckpointing's craftsmen take great pride in their work, and offer samples free of charge for projects of 25 sq m or larger. Please ring Keith Beckwith to discuss your requirements.

BRICK SUPPLIERS

▶ BOVINGDON BRICKWORKS LTD

Pudds Cross, Ley Hill Road, Bovingdon, Nr Hemel Hempstead, Herts HP3 0NW
Tel 01442 833176 Fax 01442 834539

BESPOKE BRICKMAKERS: Bovingdon Brickworks is the last working brickworks in Hertfordshire. Brickmakers skills are still to be found manufacturing in the best bespoke manner to duplicate colours, shapes and sizes of bricks made years ago. Their traditional skills, born of long experience, are also reflected in the wide and varied range of hand and machine made facing bricks they produce.

▶ THE BULMER BRICK & TILE CO LTD

Bulmer, Nr Sudbury, Suffolk CO10 7EF
Tel 01787 269232 Fax 01787 269040

CLAY BRICKS AND TILES: Situated on one of the finest seams of clay in Britain, with bricks, tiles and pots having been made at Bulmer since the Bronze Age. With this long history, the traditions and knowledge passed down to today express themselves in the specialist work of restoration. Many of the company's 19th century moulds are being used to provide replacement bricks for the original. New patterns are being added to the 3,000 plus patterns they already hold. The company produces a range of quality rubbing bricks from fine washed clays and has a cutting service for special shapes and gauged arch work.

▶ CHELWOOD BRICK LTD

Adswood Road, Cheadle Hulme, Cheadle, Cheshire SK8 5QY
Tel 0161 485 8211 Fax 0161 488 4827

CLAY BRICKS AND TILES FOR REFURBISHMENT: Chelwood offers a wide range of clay bricks and tiles for refurbishment projects and conservation work. The range includes traditional stock bricks, genuine handmade and modern wirecut facings. The company also offers authentic handmade roof tiles which are also suitable for vertical tile hanging. Chelwood's handmade facings can be made to any size to perfectly tie in with existing brickwork, whilst a range of 73mm wirecut facings offers a simple solution to match brickwork on older properties. A full range of BS and bespoke specials is also available for the truly professional finish.

BRICK SUPPLIERS

▶ HAMMILL BRICK LTD

Woodnesborough, Nr Sandwich, Kent CT13 0EJ
Tel 01304 617613 Fax 01304 611036

SPECIALIST STOCK BRICK MANUFACTURERS: The brickworks was established at the Hamlet of Hammill, near Sandwich in 1926. Production has concentrated on quality stock facing bricks, which closely match those used for centuries in Southern and Eastern England. The range consists of the Light Red Stock, Medium Multi Stock and the Selected Light Multi Stock, all produced with a fine hand-made texture. The brickworks will also blend different ranges to suit a customer's preference. The company stocks an excellent range of British Standard specials and has a CAD service to assist clients in the design of special shapes and detail work. The bricks have been used on many environmentally sensitive schemes and are ideally suited to restoration, matching of old work and new projects in conservation areas.

▶ HANSON BRICK

Marketing Department, Stewartby, Bedford MK43 9LZ
Tel 08705 258258 Fax 01234 762040
Website www.hanson-brickseurope.com

FACING BRICKS: Hanson Brick are the UK's leading manufacturers of quality facing bricks. The product portfolios of London, Kempston, Butterley and Desimpel include over 200 different colours and textures from handmades and stocks through to wirecut rustics. Hanson Brick also manufactures a range of clay paviors in complementary colours and surface textures. A range of 73mm bricks in tones selected to aid the renovation and extensions of old buildings is also available. A special service is the manufacture of decorative brickwork plaques, using individually carved components. Special shapes are a available along with a technical advisory service, with CAD facilities, to assist in the creation of brickwork features.

▶ IBSTOCK BRICK LIMITED

Ibstock, Leicestershire LE67 6HS Tel 01530 261999 Fax 01530 261888

FACING BRICKS: Ibstock Brick has probably the widest range of quality facing bricks in the country. The wide range of clays ensures that there is always an Ibstock brick to blend in with existing brickwork and a sensitive environment. An Ibstock speciality is its ability to supply any special shapes or sizes whether to British Standard or to the designer's own specification. Anything that has been made in clay can be matched by Ibstock. Ibstock has a team of design advisers ready to advise on any brickwork detail, which is supported by a highly advanced CAD system. *See also: Ibstock Hathernware display entry in Architectural Terracotta section, page 97.*

▶ KETLEY BRICK COMPANY LIMITED

Dreadnought Works, Pensnett, Brierley Hill, Staffs DY5 4TH
Tel 01384 78361/77405 Fax 01384 74553

BRICKS: Ketley Brick is a specialist producer of Staffordshire Blue, Brown Brindle and Red Engineering bricks, in both metric and Imperial sizes. British Standard and purpose made specials are offered, as well as matching creasing tiles in all three colours.

▶ LAMB'S BRICKS & ARCHES

Nyewood Court, Brookers Road, Billingshurst, W Sussex RH14 9RZ
Tel 01403 785141 ext 223 Fax 01403 784663 E-mail sales@lambsbricks.co

MANUFACTURERS OF TRADITIONAL BRICKS AND RED RUBBERS: Lamb's are able to manufacture yellow, gault or red rubbers using authentic brick earth and traditional methods that produce exact replicas of original details. Their CAD department produces drawings which, after site approval, their masons use as templates. Experience has shown that only traditional brick earth can produce the necessary quality red rubbers for matching purposes including specials with the TLB mark that is still in use by Lamb's Bricks & Arches today. They will design and draw arches, for new openings or as replacements, using their specialist computer programmes. They produce a wide range of handmade and machine made facing bricks in Imperial sizes which are suitable for colour selecting to match existingwork. *See also: display entry in this section, page 101, and entries in Architectural Terracotta section, page 97 and 98.*

Figure 1 Surface degradation of sandstone caused by salt crystallisation

STONE CONSERVATION AND REPLACEMENT

IAN SIMS

The modern architect has access to an unprecedented variety of building stone from world-wide sources, whereas in the past best use had to be made of materials available in the local vicinity. Although some regions produced granites and other types of highly durable stone, much of the UK had only limestones or sandstones, which were usually less durable but easier to quarry and process with primitive tools. These sedimentary stone types exhibit a wide range of properties and vary greatly in durability from source to source, but all are prone to some degree of deterioration on buildings in the face of natural weathering and pollution damage. Modern conservation seeks to save and stabilise decaying stone fabric whenever feasible, but it is sometimes necessary to identify matching or sympathetic replacement material.

'Stone' is the term applied to rock used in building construction. Accounts of the geology of stone materials may be found in several books (see Recommended Reading below). Basically there are three groups of rocks, classified according to their mode of formation: igneous rocks that solidify from the molten state (such as granite, gabbro, dolerite and basalt); sedimentary rocks that form by the consolidation of sediments, variously created by the erosion of earlier rocks, chemical precipitation or the accumulation of plant and animal remains (such as sandstone and limestone); and metamorphic rocks that derive from the recrystallisation of existing igneous and sedimentary rocks, for example during mountain building (such as gneiss, schist, slate and marble). The performance and durability of a rock used as building stone depends largely upon its basic composition and texture, but is often crucially influenced by its geological history since formation, especially secondary alteration and weathering.

All limestones and sandstones are relatively porous stones. By comparison igneous and metamorphic materials typically have very low porosity in their unaltered and unweathered conditions. This relatively high porosity causes the limestones and sandstones used widely in the UK to be generally vulnerable to decay, although the nature of the pore structure also plays a critical role in their durability. Porosity allows rainwater and resultant solutions to penetrate into the stone, causing the integrity of the microstructure to be degraded by such actions as acid dissolution, wetting and drying cycles, freezing and thawing cycles and organic colonisation. Another major cause of degradation is the complex mechanism associated with the crystallisation of salts (sulphate salts in particular) within porous stones, which has been especially associated with industrial and urban pollution *(Figure 1)*. There are many other potential mechanisms of decay that can compromise durability.

Conservation requires an assessment of damaged or deteriorating stone to establish the cause(s) and extent of its degradation, as well as its residual capability to continue to perform its structural, protective or aesthetic role in the building concerned. Petrographic[1] examination is an essential means of making this assessment. Owing to the recent reductions in some types of pollution, much of the existing damage to historic stone fabric might have taken place in the past, so that present deterioration is either relatively slow or different in character. It is also necessary to recognise that, in some circumstances, the cleaning or removal of soiled or damaged stone surfaces can accelerate deterioration of the remaining stone, which was effectively being protected by the removed material. Chemical treatments can be successfully applied in the right circumstances, but caution is recommended *(see page 185)* as they may do more harm than good. Eventually, stone units that are disfigured or weakened to an unacceptable extent will need to be sensitively replaced by new or reclaimed material.

In stone replacement work, it may be necessary to specify a particular stone material. In such cases it is essential to specify the characteristics required of the stone as well as its petrographic identity, whilst still ensuring that the materials scientist has the scope to assess and control the quality of new stone or the condition of reclaimed material. Even the supply of specifically named stone materials can be made conditional upon acceptable quality and condition, providing that a pre-selection process has established that the preferred stone is realistically available with the properties required. Geological names for stone, especially the traditional and/or exotic trade names that are commonly used throughout the stone industry, are very unreliable indications of quality and engineering properties.

Detailed microscopical examination of thin-section specimens *(Figure 2)*, augmented by some appropriate chemical analysis, is an invaluable technique for assessing the nature, provenance and condition of an existing stone material, also for diagnosing the cause of any deterioration. Petrography should similarly be part of the fundamental assessment of replacement stone and is particularly effective for assessing weathering durability. Submission of detailed petrographic analyses should be specified for candidate stones. These should have been conducted within the previous six months by an independent and accredited petrographer experienced in building stone. Samples examined must be wholly representative of the range of materials that would be supplied.

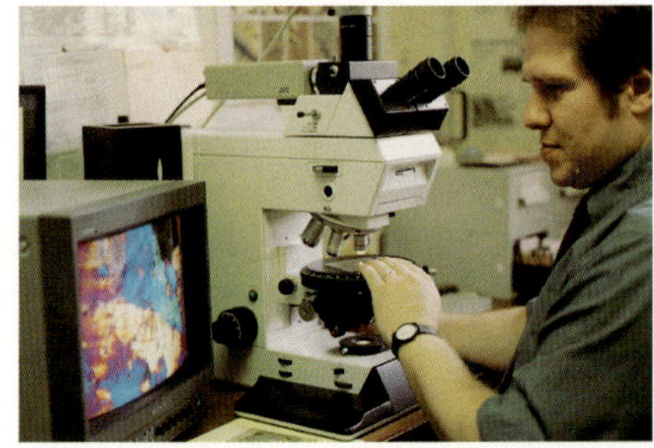

Figure 2 High-power petrological microscopy

[1] *Petrography: the description of rocks in terms of mineralogy and texture.*

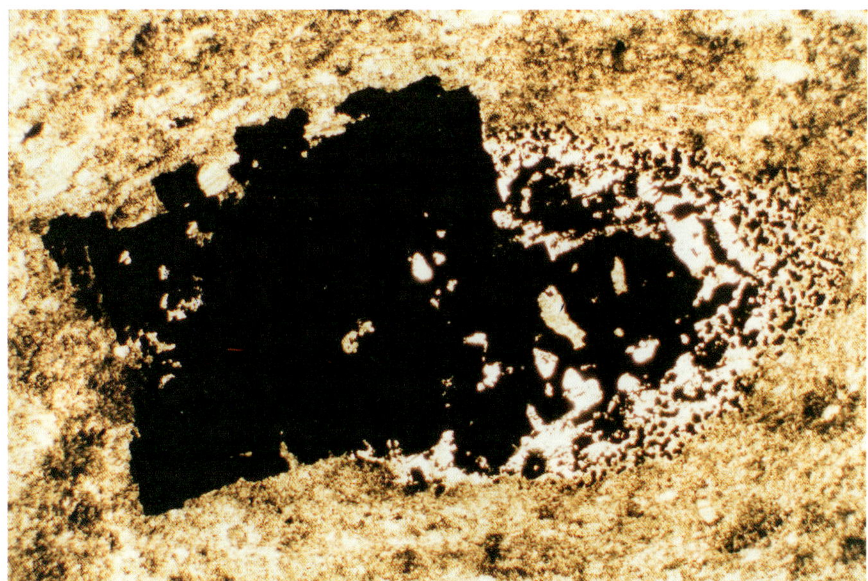

Figure 3 *Thin-section microscopical view: potentially reactive pyrite in slate*

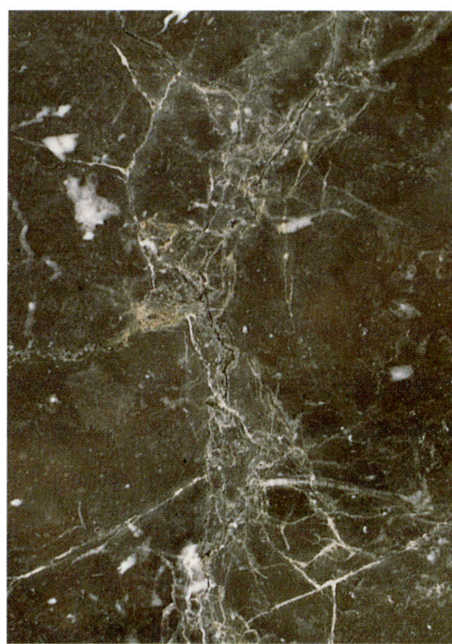

Figure 4 *Deleterious stylolitic features in a black limestone*

Petrography can identify the presence of any constituents and/or textural features that could be susceptible to weathering. Examples include acid soluble components, iron compounds that could oxidise causing expansion and staining *(Figure 3)*, clay-bearing constituents that could cause moisture movements and/or weaken the stone in its wetted condition, pore structures that would be susceptible to freeze-thaw or salt crystallisation and pre-existing micro-fracturing or incipient delamination.

Porosity and pore structure are critical properties that influence the performance of stone, especially changes in appearance on ageing and weathering. Pore size and the degree of pore interconnection are usually more important than total porosity. In freeze-thaw resistance, two candidate stones might have similar total porosity, but one with a large proportion of micro-pores could be less durable than one with a smaller proportion of larger pores. Many methods have been devised for the objective assessment of porous stones.

One of the more popular is 'saturation coefficient', which is the ratio of water absorption under normal conditions to that achieved under vacuum. A porous limestone with a higher proportion of fine pores accessible by capillary action would yield a higher saturation coefficient, up to the maximum possible value of 1.0. Saturation coefficients greater than 0.85 indicate limestone of relatively low durability, whereas values of less than 0.65 indicate durable stone. This leaves a band of uncertainty (0.65–0.85) in which the measure is unreliable and in practice many materials yield values within this range. Moreover, saturation coefficient is only meaningful for limestones with a minimum porosity value of five per cent.

Compressive strength seems an obvious requirement for stone masonry, yet historically it was rarely quantified, mainly because material that could survive the rigours of extraction, dressing and transportation to site could reasonably be expected to withstand the relatively low loading requirements of traditional construction on large stone blocks.

Strength in tension will often dominate the mechanical performance of stone. This is typically evaluated by a bending test, wherein a relatively thin specimen is loaded to failure whilst supported at its extremities. Two test options are generally considered: one loaded at a single central point (modulus of rupture), the other loaded at two points (flexural strength). Different determinations must be established, variously perpendicular or parallel to any layered structure or grain orientation (rift) and in dry or saturated condition.

Building exteriors are subjected to wetting in service, so that the saturated basis will be the appropriate criterion. The relationship between increasing moisture content and reducing strength is by no means linear and, particularly with stronger stones, only very modest moisture contents can induce significant strength reductions. Recent research involving 30 sandstones found losses in compressive strength on wetting to range from as little as eight per cent to as much as 82 per cent.

Stylolites are unique to limestones and some marbles and are a source of concern to materials scientists owing to their implications for strength performance and durability. These irregular suture-like boundaries form in limestones as the result of pressure-controlled dissolution during diagenesis and undoubtedly form a physical discontinuity within the resultant stone. Most limestones contain stylolites to some degree and clearly they are infrequent causes of bending failures or premature deterioration. However, the inappropriate concurrence of stylolites with positions of stress concentration must always be avoided, also some forms of stylolite can be the cause of premature deterioration *(Figure 4)*.

Stone exhibits dimensional changes in response to variations in temperature and moisture condition. Usually these changes are reversible, contracting or shrinking on cooling or drying and expanding or swelling on heating or wetting, but failure can occur, particularly if the stone is fixed too rigidly or is inappropriately jointed, restricting movement.

Figure 5 *Thin-section microscopical view: unsound clay minerals in a basalt*

Figure 6 *Thin-section microscopical view: carbonate slate*

It is much less common for stone, other than roofing slates, to be tested for dimensional stability in the face of wetting and drying cycles, although these are likely to be a significant factor in exterior service in many parts of the world. Some clay-bearing stones, including slate, impure limestones or marbles and some geologically altered igneous rocks *(Figure 5)*, can exhibit prodigious dimensional changes on wetting and drying in service.

Externally exposed stone must be durable to the weathering effects that prevail in the climate and circumstances pertinent to the building in question. However, durability is a concept that must be conditioned by the nature of the threat to endurance and the length of time for which that endurance is required. Certain types of slate, for example, contain high proportions of carbonate (carbonate slate) that are vulnerable to rapid decay in polluted acidic conditions, but might well be entirely durable in comparatively benign unpolluted environments *(Figure 6)*. Also, the poorer grades of roofing slate might nevertheless be classified as 'durable' if a life of less than 40 years is acceptable, whereas superior grades will otherwise be required.

An additional factor of particular importance in restoration and conservation is the extent to which gradual deterioration is tolerable, or even desirable. Slates and other roofing stones will often display a visible degradation with time, so that the materials take on a weathered appearance, whilst still performing their function. Architects may favour this visible aspect of natural weathering for replacement stone in conservation work.

Weathering threats to stone durability vary with climatic conditions, modified by local meteorological and atmospheric pollution factors, and must be established for each building. Many tests have been devised individually to address the resistance of stone to various threats, but actual exposure conditions are impossible to reproduce in the laboratory and in reality stone materials are exposed to complex combinations of factors. Durability can be assessed by inspecting the actual performance of the same type of stone on existing buildings. There is undoubted value in carrying out such reconnaissance, providing suitable examples exist and can be accessed, but the approach is not entirely reliable.

There are four test categories for weathering durability. Firstly, those that directly measure properties of the stone that relate to its performance in resisting weathering (eg water absorption and porosity). Secondly, tests that create artificial physical demands as surrogates for natural weathering conditions and thus provide an indirect measure of durability, calibrated by experience and/or testing reference materials (eg salt crystallisation). Thirdly, tests that simulate an aspect of real exposure, but accelerated by increased severity and/or greater frequency of cycles, and thus obtain comparatively direct information on resistance (eg acid immersion and cycles of freezing and thawing, wetting and drying or heating and cooling). Finally, applying mechanical tests before and after a regime of simulated weathering.

RECOMMENDED READING

Ashurst, J, Dimes, FG, 1990, *Conservation of building and decorative stone*, Volumes 1 and 2, Butterworth-Heinemann, London

Grossi, C, Hunt, B, Smart, S, 1999, Urban pollution and stone decay, *Natural Stone Specialist*, Volume 34 (7) 22-32

Sims, I, Miglio, BF, Richardson, B, 2002 (in preparation), *Stone for building – selection and technology*, E and F N Spon, London

Smith, MR, 1999, *Stone: building stone, rock fill and armourstone in construction*, Geological Society, London

Dr IAN SIMS graduated as a geologist in 1972, following eight seasons in archaeological excavation. His doctoral research concerned concrete, and he also worked on Roman stone and mortar. In 1975 he joined a London consultancy, establishing its geomaterials department. Joining STATS Limited, St Albans in 1996, he is director responsible for geomaterials and expert services.

STONE

▶ A D CALVERT ARCHITECTURAL STONE SUPPLIES LTD

Smithy Lane, Grove Square, Leyburn, North Yorkshire DL8 5DZ
Tel 01969 622515 Fax 01969 624345
E-mail stone@calverts.co.uk
Website www.calverts.co.uk

STONE MASONRY AND ARCHITECTURAL STONE SUPPLIER. From the heart of the Yorkshire Dales A D Calvert supplies stone sawn six sides, profiled, lathed, carved and hand finished walling, using eight types of sandstone and two limestone. This includes the company's own local Witton Fell medium grained buff coloured sandstone which is highly recommended for new and renovation work. For further information contact Andrew Calvert who will be pleased to discuss your requirements.

▶ A F JONES (STONEMASONS)

33 Bedford Road, Reading, Berkshire RG1 7EX
Tel 0118 957 3537 Fax 0118 957 4334

MASTER STONEMASONS: Established in 1865 by Arthur F Jones and continued today by G A and A G Jones, this firm has accumulated experience and expertise gained by five generations of stone masons. There are currently 20 highly skilled craftsmen many who have spent their working lives with A F Jones, restoring, carving, cleaning and conserving stone facades. A F Jones offers a complete service from consultancy and specification to production and site fixing. Their style of management is non-confrontational and above all, fair and honest. A F Jones is an approved contractor for The Churches Conservation Trust, English Heritage the National Trust and Historic Royal Palaces. *See also: display entry in Stone section, page 105.*

▶ ABBEY HERITAGE LTD

Dartford House, Two Rivers, Station Lane, Witney, Oxon OX8 6BH
Tel 01993 709699 Fax 01993 709959
E-mail stone@abbeyh.co.uk
▶ York Office Tel 01904 567332
▶ Edinburgh Office Tel 0131 228 2281

MASONRY, RESTORATION, FACADE CLEANING AND LASER CLEANING: Specialists in facade cleaning, anti-graffiti treatments, natural stone replacement and restoration, new stone projects, terracotta, faience, granite, marble and all aspects of brickwork, Abbey Heritage works closely with leading architects, surveyors, local authorities and consultancies throughout the UK on historic and prominent national buildings. Dedication to detail and a high level of managerial input has earned Abbey Heritage a reputation for integrity and reliability, generating consistent repeat business. As principal contractor Abbey provides associated leadwork, roofing, decoration and joinery services. Leaders in the field of laser cleaning for buildings, artefacts and artworks. They offer full national and international consultancy services.

▶ ABBEY MASONRY AND RESTORATION

Heol Parc Mawr, Cross Hands Business Park, Cross Hands, Llanelli, Carmarthenshire SA14 6RE
Tel 01269 845 084 Fax 01269 831 774

CRAFTSMEN IN STONE MASONRY AND SPECIALISTS IN CARVING CONSERVATION: As recognised specialists in the conservation and restoration of ecclesiastical and historical buildings, Abbey Masonry have restored over 80 churches and cathedrals in England and Wales. Purpose built masonry works accommodate state of the art technology enhanced by skilled, dedicated craftsmen which combine to offer an exceptional service. They are also approved installers of Cintec anchors and registered operatives of the Jos and Doff stone cleaning systems. Contact Abbey Masonry today to discuss your requirements.

▶ ANELAYS

William Anelay Limited, Murton Way, Osbaldwick, York YO1 5UW
Tel 01904 412624 Fax 01904 413535
E-mail office@williamanelay.co.uk
Website www.williamanelay.co.uk

BUILDING AND RESTORATION CONTRACTORS: *See also: profile entry in Building Contractors section, page 58.*

▶ ASHBY STONE MASONRY LIMITED

Ashby Yard, Gibbs Wharf, Hedley Avenue, Grays, Essex RM20 4EZ
Tel 01375 397001 Fax 01375 397002

STONEMASONRY AND RESTORATION CONTRACTOR: Ashby Stone Masonry Limited provides a full restoration service as principal contractor or sub-contractor from consultancy, survey and design, through procurement and production to site installation and repairs. All aspects of the conservation of buildings are encompassed to meet the project requirements to the highest standards. Recent restoration projects include: St Margaret's Church, Barking, Port East, Docklands, and Cavendish Laboratories, Cambridge. ASM also carry out new build projects using in-house CAD design and production facilities with site management and fixing completing the range of services offered.

▶ THE BATH STONE GROUP

Stoke Hill Mine, Midford Lane, Limpley Stoke, Nr Bath BA3 6JR
Tel 01225 723792 Fax 01225 722129
E-mail bsg@bath-stone.co.uk
Website www.bath-stone.co.uk
Contact Elaine Dickerson

STOKE GROUND BATH STONE: The Bath Stone Group, producing award winning Stoke Ground Bath Stone, offers an integrated service from design through to installation on site. The company's highly skilled masons and carvers, can fabricate and finish stone to exacting specifications. Stoke Ground is regularly specified by English Heritage and the National Trust, and used in restoration and new-build projects including: Buckingham Palace and Windsor Castle; Oxbridge colleges: Merton, Manchester, Brazenose and Pembroke; Stations: Liverpool Street, London and Temple Meads, Bristol; Development: Wessex Water, Bath and Britannic Assurance HQ, Birmingham; Conservation: Temple of Concorde, Stowe and Waddesdon Manor, Berks. Bath Stone Group offers cost effective natural stone building solutions.

▶ BODEN & WARD STONEMASONS LTD

Ox-House Farm, Brington Road, Flore, Northants NN7 4NQ
Tel 01327 349081 Fax 01327 349290

CRAFTSMEN IN STONE: Masons and stone carvers, highly experienced in all aspects of stone, repair, cleaning and restoration. Projects undertaken extend from repairs to parish churches to major restoration projects. One such, recently completed, involved the restoration of the central pavilion at Woburn Abbey. This won the company a millennium Natural Stone Award from the Stone Federation of Great Britain. Boden & Ward also supply worked and sawn stone to other masonry companies. They now produce hand made fireplaces individually designed as well as bespoke carvings for enhancing garden design in all situations from the smallest local project. *See also: display entry in this section, page 106*.

▶ BURLEIGH STONE CLEANING & RESTORATION CO LTD

The Old Stables, 56 Balliol Road, Bootle, Merseyside L20 7EJ
Tel 0151 922 3366 Fax 0151 922 3377
E-mail info@burleighstone.co.uk
Website www.burleighstone.co.uk

STONE SPECIALISTS: A comprehensive high quality service in stone masonry, cleaning, restoration, carving and replacement. *See also: display entry in Masonry Cleaning section, page 172.*

▶ BURROWS DAVIES LIMITED

The Stoneyard, West End, Strensall, York YO32 5WH
Tel 01904 491849 Fax 01904 491910

MASONRY AND RESTORATION SPECIALISTS: Burrows Davies Limited carries out high quality masonry, conservation and restoration works on historic properties and listed buildings including churches, historic houses and monuments throughout the Midlands, North of England and Scotland. The company works primarily as principal contractor and occasionally as specialist sub-contractor and over the past few years has successfully completed various prestigious projects ranging in value from a few thousand pounds to £1 million.

STONE

HERALDIC BEAST, THE GREAT HALL – STIRLING CASTLE

▶ JOHN BYSOUTH – STONE CONSULTANT

90 Selborne Road, Southgate, London N14 7DG
Tel/Fax 020 8882 1554

NATURAL STONE CONSULTANT: John Bysouth uses his knowledge of stone and the experience of a lifetime in the stone industry to give independent advice to architects, surveyors, engineers and property owners on the best methods of using and dealing with natural stone, on both old and new buildings. Services include, surveys and reports, recommendations for repairs and restoration, sourcing of suitable material, budget estimates and project management and supervision. Contracts large or small undertaken nationwide. Technical advisor to Stone Federation, Great Britain.

▶ C GINN BUILDING RESTORATION

89 Lunedale Road, Fleet Estate, Dartford, Kent DA2 6LW
Tel/Fax 01322 290505
E-mail c.ginnbuildingrestoration@btinternet.com
Website www.stonecleaning-restoration.com

RESTORATION AND CLEANING SPECIALISTS: C Ginn Building Restoration has a proven track record for the repair and cleaning of exterior fabrics such as stone, stucco, faience, terracotta and brickwork. They are also experienced at supplying and fixing natural and precast stone details, and are always willing to undertake surveys together with costs for all aspects of work within their field. Whilst undertaking large projects the company is always available for the smaller projects and offers the same level of service throughout. If your project requires assistance in 'taking the weight' please contact C Ginn Building Restoration for further discussions and information required.
See also: display entry in Masonry Cleaning section, page 172.

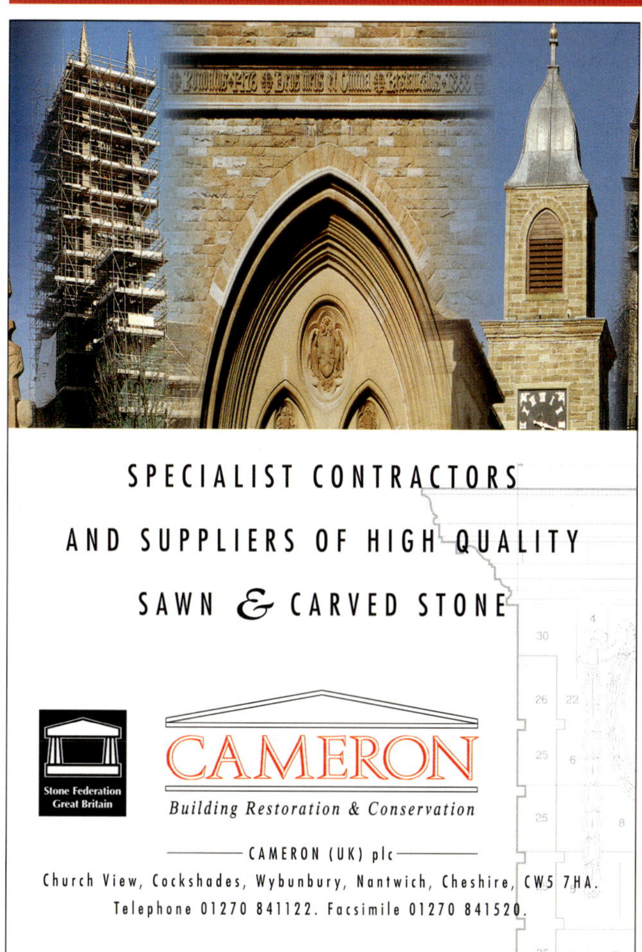

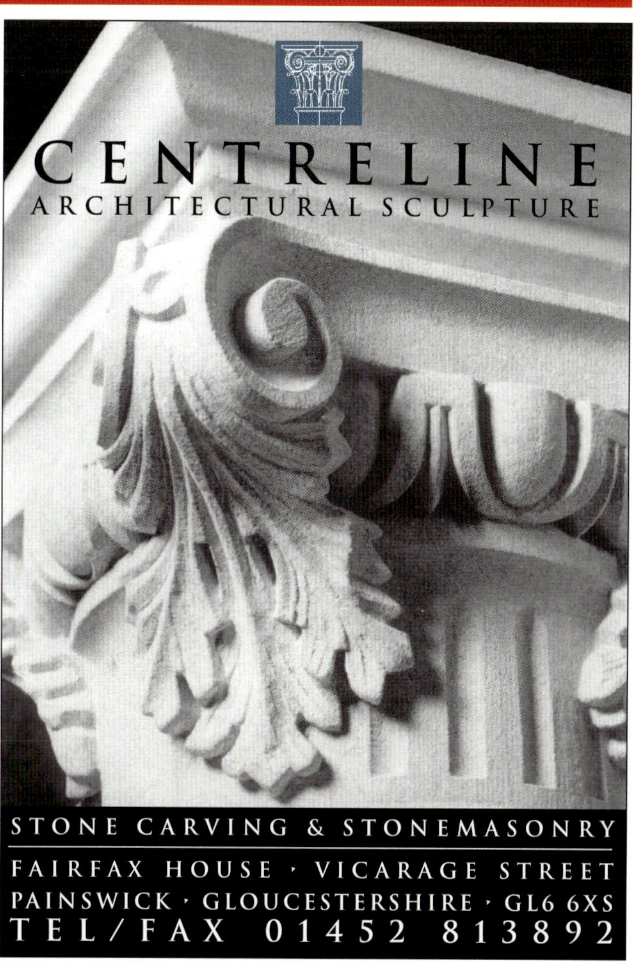

▶ CAMERON (UK) PLC
Cockshades, Wybunbury, Nantwich, Cheshire CW5 7HA
Tel 01270 841122 Fax 01270 841520
E-mail cameron@dial.pipex.com

SUPPLY AND FIX OF NATURAL AND CAST STONE: Production of sawn and carved stone in both modern and traditional techniques. Cameron's skilled fixer masons are experienced in handling and installing stonework of the most intricate detail. *See also: display entry above.*

▶ CARREK LTD
Mason's Yard, Wells Cathedral, Wells, Somerset BA5 2PA
Tel 01749 689000 Fax 01749 689089
Website www.carrek.co.uk

HISTORIC BUILDING REPAIR COMPANY: *See also: profile entry in Building Contractors section, page 60.*

▶ CARTHY CONSERVATION LTD
18 Alexandria Road, London W13 0NR
Tel/Fax 020 8840 3294
E-mail deborahcarthy@btclick.com

CONSERVATION OF ARCHITECTURAL DETAIL AND SCULPTURE: Carthy Conservation offers an established team which carries out high quality conservation and consultancy projects on stone, terracotta, plaster, mosaic and wood. Gilding, polychrome and monochrome surfaces are also specialities. The company has established strong links with analytical laboratories and scientists both in this country and abroad who have established reputations in conservation. Work is undertaken across the United Kingdom for private clients, architects and building consultants, construction main contractors, cathedrals, churches and government agencies. This conservation practice is included on the Conservation Register maintained by the United Kingdom Institute for Conservation.

▶ CATHEDRAL WORKS ORGANISATION (CHICHESTER) LIMITED
Terminus Road, Chichester, West Sussex PO19 2TX
Tel 01243 784225 Fax 01243 813700
E-mail C.W.O@osborne.co.uk
Website www.stone-federationgb.org.uk

SPECIALIST STONEMASONRY, RESTORATION, CLEANING AND SUPPLY: Royal warrant holder CWO has the skills, experience and resources to handle all types of stonework, restoration and conservation projects throughout South and South East England including Greater and Central London. CWO offers a complete service from specification and drawings, to production at its Chichester workshops and completion by masons on site. CWO is the first stonemasonry company to be accredited as an Assessment Centre for National Vocational Qualification levels 2 and 3. Recent contracts include: The Royal Albert Hall, South Steps reinstatement; The Queens Gallery, Buckingham Palace; All Saints Church, Ryde; Christchurch Priory, Christchurch; Paternoster Square, Portland stone column, London. CWO offers a supply-only service in a range of limestone and sandstones for hand finished stonework. Specialist Small Works department carries out work up to £100,000.

▶ CENTRELINE ARCHITECTURAL SCULPTURE
Fairfax House, Vicarage Street, Painswick, Gloucestershire GL6 6XS
Tel 01452 813892 Fax 01452 813892

ARCHITECTURAL SCULPTURE: A dynamic organisation of specialist stone carvers and stonemasons based in Gloucestershire, able to tackle large carving contracts on a national level. The team benefits from graduate, postgraduate and international training with a combined 70 years professional experience working with all types of stone on buildings ranging from vernacular to stately. *See also: display entry above.*

▶ CLIVEDEN CONSERVATION WORKSHOP LTD

The Tennis Courts, Cliveden Estate, Taplow, Maidenhead, Berkshire SL6 0JA
Tel 01628 604721 Fax 01628 660379

SCULPTURE, STONE, PLASTER AND WALL PAINTINGS CONSERVATION: Established 1983 as the statuary workshop for the National Trust, independent since 1991 serving English Heritage and the Royal Palaces. Experienced in carving, modelling and conservation to museum standards, particularly in the country house. Recently re-instated Jacobean plaster ceilings at Edinburgh Castle, and Rococo and Neo-classical plasterwork at Uppark. Replicated chimneypieces at Abbey Leix, and repaired stonework and monuments, Ely and Winchester Cathedrals. Repair of statuary at Stowe Landscape Gardens, Osborne House and Aphrodisias, Turkey and reinstatement of sculpture carved in marble and stone at Belton House and Castletown Cox, Eire. Conservation and consultancy services, survey and recording, advice on security, drawings and specifications. By appointment to the National Trust, advisors for the conservation of stone and plaster.

▶ DAVID BALL RESTORATION (LONDON) LIMITED

104A Consort Road, London SE15 2PR
Tel 020 7277 7775 Fax 020 7635 0556
E-mail mail@dbr.uk.com

CLEANING AND REPAIR SPECIALISTS: David Ball Restoration Limited specialises in the cleaning and repair of stone, brick, stucco and terracotta facades. The company has the management and craft skills to undertake major contracts, but also acts as a specialist sub-contractor in their core business activity of cleaning and repairing commercial and public buildings. The company's comprehensive knowledge of the repair of historic buildings enables them to offer advice on traditional forms of construction as well as the latest conservation techniques. Ancillary craft operations, such as leadwork, ironwork, woodwork, roofing, painting and decoration, may also be undertaken as part of a managed contract. *See also: display entry in this section, page 109.*

▶ NICHOLAS DURNAN

The Flaxmill, Lopen, South Petherton, Somerset TA13 5JS
Tel/Fax 01460 241770 Mobile 07790 027012
E-mail durnan@globalnet.co.uk
Website www.stone-studio.co.uk

STONECARVER AND SCULPTOR, STONE AND SCULPTURE CONSERVATION CONSULTANT: *See also: profile entry in Sculpture section, page 114.*

▶ H & W SELLORS LTD

Milford Works, Bakewell, Derby DE45 1DX
Tel 01629 812058 01629 815138

ROOFING, RESTORATION AND BUILDING CONTRACTORS: A family firm which was established in 1850 as traditional stone slating contractors has expanded to undertake work in building conservation and conversion. The company has been involved in a programme of maintenance work at Chatsworth House, Derbyshire including restoration of the Cascade and Cascade House. Work is undertaken on both residential and commercial properties, and a recent conversion of a timber mill to offices within the Peak District National Park has been nominated for an award.

▶ H G CLARKE & SON

Weston Underwood, Olney, Bucks MK46 5JS
Tel 01234 711358/712047

TOTTERNHOE LIMESTONE SUPPLIERS: H G Clarke & Son supplies original Totternhoe Clunch limestone, used throughout the southern half of Great Britain on historic buildings which have chalk stone connections. Supplied in quarry block and slabbing especially to builders in the Valley of The White Horse including Oxford, Gloucester and Swindon areas, it is also recognised as similar to stone used in Jerusalem.

STONE

▶ THE HOPTON WOOD STONE FIRMS LTD

New Road, Middleton, Wirksworth, Derbyshire DE4 4NA
Tel 01629 822216 Fax 01629 824348
E-mail info@lowesmarble.com
Website www.lowesmarble.com

SUPPLIERS OF HOPTON WOOD LIMESTONE: This famous hard limestone appreciated by masons, sculptors and lettercutters is available as sawn six sides or fully worked. Since the reintroduction of Hopton Wood the material has been used for restoration and new projects throughout the country and overseas. Some examples are Pershore Abbey, Barlaston Hall, The Albert Memorial, Axa Equity & Law head office IoM, The Gilbert Collection at Somerset House and many of the country's most noted public buildings. The combination of modern plant and a skilled workforce enables Francis N Lowe Ltd and Longcliffe Quarries Ltd as joint owners of The Hopton Wood Stone Firms to offer an efficient and reliable service. Contact their office for further information, samples or a company brochure.

▶ J JOSLIN (CONTRACTORS) LTD

Southrah Quarry, Lower Road, Long Hanborough, Witney, Oxfordshire OX8 8LN
Tel 01993 882153 Fax 01993 882960
Contact Managing Director, Roy Kelly

CONSERVATION, RESTORATION, AND SPECIALIST MASONRY CLEANING: Joslins of Oxfordshire is one of Britain's leading stonemasons, with over 100 years experience. Recently awarded the Royal Warrant Joslins has an enviable reputation for quality of workmanship and reliability. Current projects include the restoration of Chichester Cathedral and Waddesdon Manor. Whether a project involves restoration, conservation or new build Joslins' skilled craftsmen can provide a complete service of the highest standards including hand carving stone to specification. Joslins successfully combines the traditional skills of hand-crafted stonemasonry with the latest technology, and are also approved users of the Jos, Doff and Torc methods of specialist stove cleaning. *See also: display entry on page 108.*

▶ MAYSAND LIMITED

109-111 Windsor Road, Oldham, Lancs OL8 1RH
Tel 0161 628 8888 Fax 0161 627 0996
E-mail sales@maysand.co.uk

MASONRY REPAIR, CLEANING, RESTORATION AND REPLACEMENT: *See also: display entry in this section, page 110.*

▶ KIM MEREDEW

5 Neville Road, London N16 8SH
Tel 020 7254 9155 Mobile 0976 903678
E-mail kim.meredew@virgin.net
Website kimmeredew.co.uk

BUILDING AND DETAIL RESTORATION: Kim Meredew is an English Heritage trained stone carver and mason with 20 years building and design experience. He specialises in stone, stucco and gauged brick restoration, working to meticulously high standards in both private and public projects including listed buildings, churches, banks etc. Please visit the above website to see examples of his work.

▶ NEGUS

(Ken Negus Ltd), 90 Garfield Road, London SW19 8SB
Tel 020 8543 9266 Fax 020 8543 9100
E-mail enquiries@kennegus.co.uk
Website www.kennegus.co.uk

STONE CLEANING AND CONSERVATION SPECIALISTS: *See also: profile entry in Masonry Cleaning section, page 173.*

STRUCTURE & FABRIC
Masonry

▶ NIMBUS CONSERVATION LTD

Eastgate, Christchurch Street East, Frome, Somerset BA11 1QD
Tel 01373 474646 Fax 01373 474648
E-mail enquiries@nimbusconservation.com
Website www.nimbusconservation.com

STONE CONSERVATORS: Nimbus has an established reputation for conservation on historic buildings, monuments, churches and memorials as well as archaeological sites, sculptures and museum artefacts. With particular expertise in limestone, marble and plaster, Nimbus also conserves brick, terracotta, sandstone, gilding and polychromy. In addition, Nimbus provides sensitive carving, letter cutting, specialist cleaning, technical advice and consultancy. Clients include English Heritage, the National Trust, the CCT, HRP, dioceses, councils and individuals. Recent contracts include Ardfert Cathedral in Ireland, Bath Abbey, Cobham Hall, Westminster Cathedral, Brodsworth Hall and Ightham Mote. Nimbus also supplies TRASS, a pozzolanic additive for mortars, and ceramic dowels.

▶ RICHARD NOVISS

63 North Street, Middle Barton, Chipping Norton, Oxon OX7 7BH
Tel 01869 347062

SCULPTOR AND STONEMASON: Specialist in carving and replacement of figurative and decorative stonework. Richard Noviss trained as a stonemason at Weymouth College and worked in the Cathedral workshops at Wells and Winchester before becoming self-employed in 1988. He is based at his workshop near Oxford, but undertakes commissions widely in the southern half of England. Works undertaken: re-carving and repairs to historic statuary; portraits in stone; repairs to stonework of historic buildings using lime based techniques; individually designed and carved fireplaces, memorials and headstones.

▶ PAYE STONEWORK & RESTORATION LTD

44-46 Borough Road, London SE1 0AJ
Tel 020 7928 4000 Fax 020 7928 4004 Website www.payestone.co.uk

MASONRY CONSTRUCTION AND REPAIR SPECIALISTS: With its own labour resources skilled in traditional techniques, PAYE is able to undertake new construction and conservation repairs of stone, brickwork, stucco, terracotta and faience to the highest standards. They frequently work on historic monuments, national public buildings, Royal Palaces, churches and commercial projects both as a principal contractor and specialist subcontractor. Recent contracts include Tower of London, Waterloo Block, Marlborough House, H M Treasury, Whitehall, Royal Marsden Hospital, Osborne House Isle of Wight. PAYE Stonework is pleased to undertake surveys, provide cost budgets and consultancy advice. Contact David Manktelow for further information and a company brochure. *See also: display entry on page 111.*

▶ PRIEST RESTORATION

96 Moyser Road, London SW16 6SH
Tel 020 8677 5660 Fax 020 8677 2550
E-mail enquiries@priestrestoration.co.uk Website www.priestrestoration.co.uk

STONE SPECIALISTS: Priest Restoration is able to offer a comprehensive service providing a fully detailed survey identifying the cause of defects and most appropriate methods of cleaning, restoration and repair. They have recently restored prestigious buildings such as: The Mansion House, The Lyceum Theatre, Aldwych Theatre, Hatchlands Park Surrey, Aquascutum Regent Street, Langham Hotel Regent Street, Eton College, and The Royal Brompton Hospital. They have also been responsible for the following new-build projects: Louis Vuitton and Tommy Hilfiger stores in New Bond Street, stone paving and flooring at Columbus Courtyard and HSBC Tower Canary Wharf, and the Sackler Library stonework at Oxford University. Priest's management team has considerable experience within the stone and restoration industry offering expertise in cleaning and restoring; stonework, brickwork, terracotta, faience, stucco, marble and granite. *See also: display entry on page 111.*

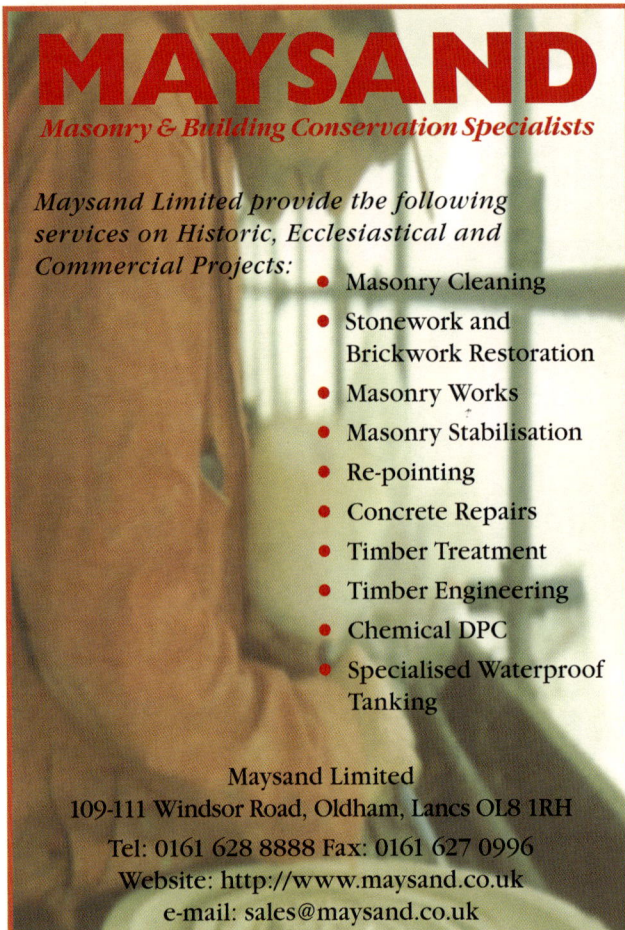

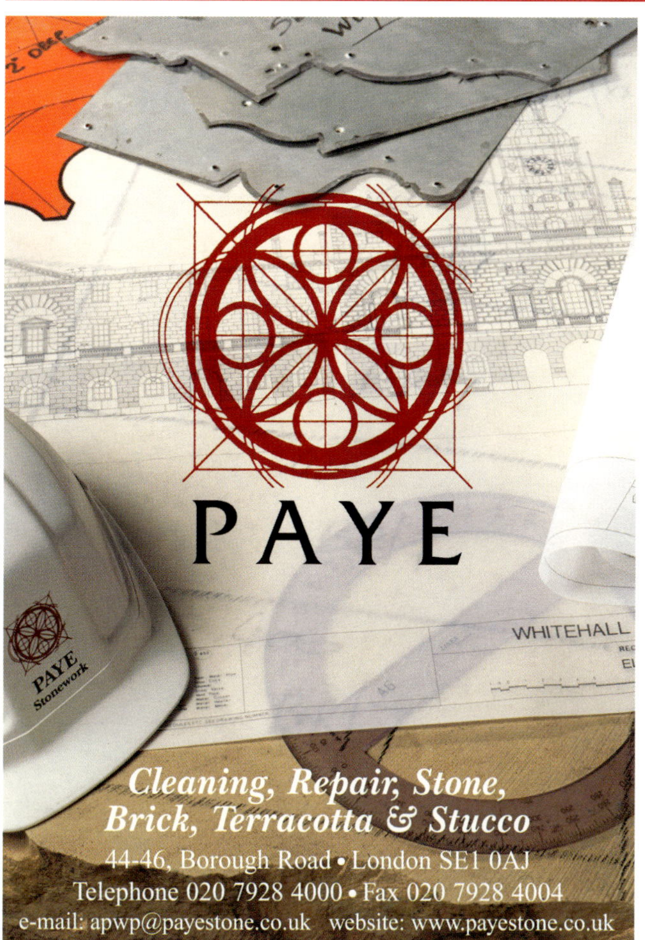

PRIEST
Stonework & Restoration

PRIEST RESTORATION LTD PRIDE THEMSELVES ON PROVIDING A COMPLETE PROFESSIONAL SERVICE AND MEETING AN EXACTING STANDARD OF EXCELLENCE IN ALL ASPECTS OF THE STONEWORK AND RESTORATION INDUSTRY.

- New Stonework
- Stone & Brickwork Restoration
- Stone & Brick Cleaning
- Marble & Granite
- Design & Survey Service

96 Moyser Road, London SW16 6SH
Tel: 020 8677 5660
Fax: 020 8677 2550
Email: enquiries@priestrestoration.co.uk
Web: www.priestrestoration.co.uk

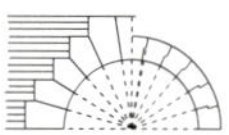

R.M.H. EATON
STONEMASONRY

VAT NO. 509 3290 50

- specialist in natural stone restoration and conservation
- consultancy service to fixing
- work undertaken as main contractor or sub-contractor

R.M.H. Eaton Stonemasonry has built up an excellent reputation with large contractors and local authorities. Past work includes Heeley Bridge, Sheffield; Breadsall Priory and St.Pierre Marriot Hotels; stonework supplied to riverside/wharf regeneration scheme, Newark. Also ongoing restoration work at Haddon Hall, Derbyshire.

R.M.H. Eaton Stonemasonry
Harebell House, Birchover, Matlock, Derbyshire DE4 2BL
Tel 01629 636541 Fax/Tel 01629 650085

Specialists in the repair and conservation of historical buildings and monuments

- Conservation
- Restoration
- Stonemasonry
- Specialist Cleaning
- Lime Technology
- Stucco Repairs
- Analysis
- Reports & Surveys

The Barn, Wood Lane, Eastcote, Warwickshire B92 0JL
Tel. 01675 446123 Fax. 0121 706 3886
www.russcott.co.uk

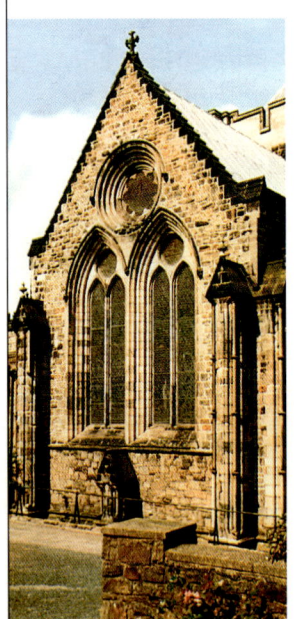

▶ RUSSCOTT CONSERVATION AND MASONRY

The Barn, Wood Lane, Eastcote, Warwickshire B92 0JL
Tel 01675 446123 Fax 0121 706 3886

SPECIALISTS IN THE REPAIR AND CONSERVATION OF HISTORIC BUILDINGS AND MONUMENTS: *See also: display entry in this section, page 111.*

▶ SOMERSET STONE MASONS LTD

37B Broadway East, West Wilts Trading Estate, Westbury, Wiltshire BA13 4JA
Tel 01373 858159 Fax 01373 827205 E-mail charlie@somersetstonemasons.co.uk
Website www.somersetstonemasons.co.uk

STONEWORK RESTORATION: Somerset Stone Masons Limited is a well established traditional masonry company with a reputation for high quality craftsmanship. An experienced and sympathetic approach to the restoration, repair and cleaning of historic and ecclesiastical stone buildings and monuments is provided. Fully equipped workshops enable Somerset Stone to provide a full service to the trade and private sector in supplying, carving and fixing stone components. The fixing service covers the entire South West, South Wales, Gloucestershire and the M4 corridor. A full advisory and report service is available on all aspects of masonry. Please contact the company for further information and visit the website address shown above.

▶ ST BLAISE LTD

Westhill Barn, Evershot, Dorchester, Dorset DT2 0LD
Tel 01935 83662 Fax 01935 83017 E-mail stblaise@compuserve.com

STONE, MASONRY, CARVING, FIXING, CLEANING AND CONSERVATION, CINTEC ANCHORS, DESIGN: *See also: entry in Building Contractors section, page 65.*

▶ STATS CONSULTANCY (STATS Limited, founded 1974)

Porterswood House, Porters Wood, St Albans, Herts AL3 6PQ
Tel 01727 833261 Fax 01727 835682 E-mail ian.sims@stats.co.uk
Contact Dr Ian Sims

SPECIALIST ENGINEERING, MATERIALS AND ENVIRONMENTAL CONSULTANTS: *See also: profile entry in Materials Analysis section, page 45.*

▶ STONE CENTRAL LIMITED
Unit 4, Sedgley Park Trading Estate, Prestwich, Manchester M25 9WD
Tel 0161 773 1349 Fax 0161 773 4405
E-mail stonecentral@btclick
Website www.stonecentral.co.uk
BUILDING RESTORATION AND STONE MASONRY: *See also: display entry on page 112.*

▶ STONECRAFT
Beeches, The Hill, Merrywalks, Stroud, Glos GL5 4EP Tel 01453 765776
RECLAIMED AND NEW BUILDING AND WALLING STONE, NATURAL STONE FIREPLACES AND DRY STONE WALLING

▶ STONEGUARD
St Martins House, High Street, Ruislip, Middlesex HA4 7AU
Tel 01895 675577 Fax 01895 679125
Website www.stoneguard.co.uk
▸ with offices also in Bath, Birmingham, Bradford, Manchester and Stirling
BUILDING RESTORATION AND CONSERVATION: *See also: profile entry in Building Contractors section, page 66.*

▶ STONEWEST LTD
Lamberts Place, St James's Road, Croydon CR9 2HX
Tel 020 8684 6646 Fax 020 8684 9323
E-mail stonewest@cwcom.net
Website www.stonewest.co.uk
BUILDING CONTRACTORS AND STONE MASONS: *See also: display entry in Building Contractors section, page 64.*

▶ STONEWORKS OF BATH LTD
Old Orchard, Walcot Street, Bath BA1 5AX
Tel 01225 311136 Fax 01225 481119
Website www.stoneworks.co.uk
MASTER STONE AND MARBLE MASONS, SPECIALISTS IN CHIMNEYPIECES AND FIRE SURROUNDS, DESIGN AND CONSULTANCY: *See also: entry in Fireplaces section, page 189.*

STONE

▶ W J HAYSOM & SON

St Aldhelm's Quarry, Worth Matravers, Dorset BH19 3HL
Tel 01929 439217 Fax 01929 439215
▶ Landers Quarries Ltd, Kingston Road, Langton Matravers, Dorset BH19 3JP
E-mail haysom@purbeckstone.co.uk or landers@purbeckstone.co.uk
Website www.purbeckstone.co.uk

PURBECK LIMESTONE AND MARBLE: has, for centuries, been used for general masonry and dimension work. W J Haysom & Son have two masonry workshops supplying Purbeck stone for both restoration and new build. The Purbeck and Purbeck-Portland beds each exhibit different colour and fossil content and are available in a range of finishes, including polished and non-slip. Projects range from royal houses, cathedrals and parish churches to small individual commissions. Recent work includes Windsor Castle, Hampton Court, Portsmouth, Salisbury and Chichester Cathedrals, Amersham and Sevenoaks parish churches, Radnor and Westbury Homes, Merrill Lynch UK Headquarters and Blandford Town Centre.

▶ WELDON STONE ENTERPRISES LTD

106 Kettering Road, Weldon, Corby, Northants NN17 1UE
Tel 01536 261545 Fax 01536 262140

CRAFTSMEN IN STONE: Weldon Stone is a well established company with experience in the restoration and conservation of churches, stately homes, and other listed buildings. The company takes pride in its ability to produce accurately sawn and worked masonry, which is fixed by teams of masons who understand the importance of careful handling and traditional techniques. Recent contracts include Stoneleigh Abbey, Warks. (East Wing and Laundry House); St Mary Magdalene, Enfield, London; Moseley Grammar School, Birmingham; and St Botolph's, Ratcliffe-on-the-Wreak, Leicestershire.

▶ WELLS MASONRY SERVICES LTD

Ilsom Farm, Cirencester Road, Tetbury, Glos GL8 8RX
Tel 01666 504251 Fax 01666 502285

STONE RESTORATION SERVICES: A privately owned specialist masonry company, Wells Masonry Services has been involved in many prestigious contracts with its team of expert operatives capable of handling all the diverse aspects of structural and decorative natural stonework, particularly in the field of repair and conservation of historic buildings. Projects like the new entrances, doorways and windows for the Palace of Westminster and the extensive restoration at Balliol College, Oxford have ensured that the company's services are always in demand where quality workmanship is of paramount importance. Wessex Stone Fireplaces, a subsidiary company, is a market leading producer of natural stone fire surrounds.

PAVING

▶ CLASSICAL FLAGSTONES

Lyncombe Vale Farm, Lyncombe Vale, Bath Avon BA2 4LT
Tel 01225 316759 Fax 01225 482076
Website www.classical-flagstones.com

REPRODUCTION FLAGSTONES: Classical Flagstones produce a range of good-looking flagstone and cobblestone reproductions that are ideal for gardens, courtyards, driveways, swimming pools, roof terraces and conservatories. Deliveries throughout the UK and abroad.

▶ KETLEY BRICK COMPANY LIMITED

Dreadnought Works, Pensnett, Brierley Hill, Staffs DY5 4TH
Tel 01384 78361/77405 Fax 01384 74553

PAVERS: Ketley has over a century of experience in producing traditional Victorian patterned pavers in Diamond Chequer 8 panel and 2 panel. They are ideal for breaking up larger more mundane areas of paving.

For Masonry Cleaning, see page 171

SCULPTURE

▶ CENTRELINE ARCHITECTURAL SCULPTURE

Fairfax House, Vicarage Street, Painswick, Gloucestershire GL6 6XS
Tel 01452 813892 Fax 01452 813892

ARCHITECTURAL SCULPTURE: A dynamic organisation of specialist stone carvers and stonemasons based in Gloucestershire, able to tackle large carving contracts on a national level. The team benefits from graduate, postgraduate and international training with a combined 70 years professional experience working with all types of stone on buildings ranging from vernacular to stately. *See also: display entry in Stone section, page 107.*

▶ CLIVEDEN CONSERVATION WORKSHOP LTD

The Tennis Courts, Cliveden Estate, Taplow, Maidenhead, Berkshire SL6 0JA
Tel 01628 604721 Fax 01628 660379

SCULPTURE, STONE AND WALL PAINTINGS CONSERVATION: *See also: profile entry in Stone section, page 108.*

▶ NICHOLAS DURNAN

The Flaxmill, Lopen, South Petherton, Somerset TA13 5JS
Tel/Fax 01460 241770 Mobile 07790 027012
E-mail durnan@globalnet.co.uk
Website www.stone-studio.co.uk

STONECARVER AND SCULPTOR, STONE AND SCULPTURE CONSERVATION CONSULTANT: Nicholas Durnan, a leading architectural and stone sculpture conservation consultant, has advised on major conservation projects at Salisbury, Wells, Exeter, Canterbury and Rochester cathedrals and Westminster Abbey. Specialising in limestone and lime-based conservation techniques, he has helped develop a range of repair mortars and conservation techniques for the many limestones used in southern England. Nicholas Durnan also works as a carver and sculptor, producing a range of work including a life size reclining Buddha (carved, painted and gilded), relief statuary, foliage and ornamental work. As a lecturer he regularly presents his conservation projects and has taught short courses on stone conservation and carving.

▶ GRACIELA AINSWORTH

Units 4 & 10 Bonnington Mill Business Centre, 72 Newhaven Road, Leith, Edinburgh EH6 5QG
Tel 0131 555 1294 Fax 0131 467 7080
E-mail graciela@graciela-ainsworth.com

SCULPTURE CONSERVATION AND RESTORATION, SCULPTURE CARVING COMMISSIONS: Dedicated to the conservation and restoration of sculpture, ornaments, monuments, memorials, museum artefacts, architectural ornament and historic building fabric. Work is conducted in specially designed workshops or on-site in museums, galleries, churches, and historic buildings and gardens. Scale of projects undertaken range from fine pieces from the Oriental Gallery in Durham, marble busts from the Wallace Monument and Playfair Library, through to the statuary of Drummond Castle Gardens and Glamis Castle; and up to the 32 foot William Wallace statue in Drybourgh, St Giles Cathedral and the decorative façade of Jenners in Edinburgh. Regular clients include The National Galleries of Scotland, The University of Edinburgh, Glasgow Museums, Perth and Kinross Council, and notable private clients.

▶ HANNA CONSERVATION

1 The Poplars, Crooked Lane, Kirk Hammerton, York YO26 8DG
Tel/Fax 01423 330228
E-mail hanna.conservation@talk21.com
Partners SB Hanna, H Hanna

SCULPTURE AND HISTORIC BUILDINGS CONSERVATION AND CONSULTANCY: Established in 1988 and led by an accredited conservator, this practice is committed to excellence and provides a wide range of consultancy and practical conservation services. These include: condition surveys; environmental monitoring; conservation plans; investigation of decay mechanisms; scientific analysis; conservation and repair of sculpture, monuments, decorative elements of buildings in all stone types; painted surfaces and mosaics. Clients include: English Heritage; the National Trust; Carlisle, Durham, Ely, Lichfield, Norwich and St Albans Cathedrals, Southwell and York Minsters; churches; museums; art galleries and private collections.

▶ NICOLAS BOYES STONE CONSERVATION

9 Palmerston Place Lane, Edinburgh EH12 5AE
Tel 0131 255 4438 Fax 0131 226 3673 Website www.nb-sc.co.uk

STONE CONSERVATION SERVICES: *See also: profile entry in Masonry Cleaning section, page 173.*

STATUARY

▶ CHILSTONE

Victoria Park, Fordcombe Road, Langton Green, Tunbridge Wells, Kent TN3 0RE
Tel 01892 740866 Fax 01892 740249
STONEMASONRY: Handmade reconstructed stone statuary. *See also: profile entry in Cast Stone section, page 116.*

▶ EURA CONSERVATION LTD

Unit H3, Halesfield 19, Telford, Shropshire TF7 4QT
Tel 01952 680218 Fax 01952 585044 E-mail mail@eura.co.uk
CONSULTANTS AND CONSERVATORS FOR SCULPTURE, ARCHITECTURAL DECORATION AND GARDEN ORNAMENT: Eura Conservation's projects have included the ornamental cladding of the Albert Memorial, the Hubert Fountain, the Canada, Australia and South Africa Gates and many municipal monuments. The company has also dealt with computerised conservation documentation systems, security fixings, packing and transportation of museum objects and EU funded research. Clients include English Heritage, the National Trust, the Department of Culture, Media and Sport, many local authorities, museums and private estates. The company acts as main, principal and sub-contractor.

▶ HADDONSTONE LIMITED

The Forge House, East Haddon, Northampton NN6 8DB
Tel 01604 770711 Fax 01604 770027 Website www.haddonstone.co.uk
STANDARD RANGE INCLUDES BUSTS AND FULL FIGURES IN CLASSICAL AND TRADITIONAL STYLES: *See also: display entry in Cast Stone section, page 117.*

▶ HERITAGE ENGINEERING

22-24 Carmyle Avenue, Foxley, Glasgow G32 8HJ
Tel 0141 763 0007 Fax 0141 763 0583 E-mail lndherco@aol.com
STATUARY RESTORATION: *See also: profile entry in Millwrights section, page 80.*

▶ HIRST CONSERVATION

Laughton, Sleaford, Lincolnshire NG34 0HE
Tel 01529 497449 Fax 01529 497518
MONUMENTS AND SCULPTURE CONSERVATION: *See also: display entry on the inside front cover and profile entry in Building Contractors section, page 62.*

▶ LEANDER ARCHITECTURAL

Fletcher Foundry, Halstead Close, Dove Holes, Buxton, Derbyshire SK17 8BP
Tel 01298 814941 Fax 01298 814970
SCULPTURAL AND BAS RELIEF CASTINGS: *See also: display entry in Signage section, page 157.*

▶ McMARMILLOYD LIMITED

Brail Farm, Wilton Road, Great Bedwyn, Marlborough, Wiltshire SN8 3LY
Tel 01672 870227 Fax 01672 870053
E-mail info@mcmarmilloyd.co.uk Website www.mcmarmilloyd.co.uk
FINE CARVED MARBLE WORK: Suppliers of the very finest statuary marble for refurbishment and the production of garden statuary. *See also: display entry in Marble & Granite section, page 96.*

▶ PLOWDEN & SMITH

190 St Ann's Hill, London SW18 2RT Tel 020 8874 4005 Fax 020 8874 7248
E-mail info@plowden-smith.com Website www.plowden-smith.com
CONSERVATION AND RESTORATION: *See also: display entry in Fine Art Conservation section, page 202.*

▶ ST BLAISE LTD

Westhill Barn, Evershot, Dorchester, Dorset DT2 0LD
Tel 01935 83662 Fax 01935 83017 E-mail stblaise@compuserve.com
CARVING, MODELLING, CONSERVATION, IN STONE, WOOD, PLASTER AND CLAY: *See also: entry in Building Contractors section, page 65.*

▶ TAYLOR PEARCE RESTORATION SERVICES LIMITED

Fishers Court, Besson Street, London SE14 5AF
Tel 020 7252 9800 Fax 020 7277 8169
STATUARY AND ARCHITECTURAL ORNAMENTS CONSERVATORS: Specialists in the restoration, conservation and installation of stone statuary and ornaments, architectural statuary, architectural ceramics, mosaic work and church monuments. Taylor Pearce has a reputation for professional conservation practice executed to the highest standards. Projects have been carried out for the V & A, The Royal Academy, English Heritage, National Trust, Imperial War Museum, Lincoln Cathedral (great frieze and judgement porch), Westminster Abbey (Chapter House annunciation group) and Westminster Cathedral. Because the company's projects are executed to museum standards within commercial parameters it also numbers top architectural and surveying practices amongst its regular clients.

MOULDING MATERIALS

STRUCTURE & FABRIC
Masonry

CAST STONE

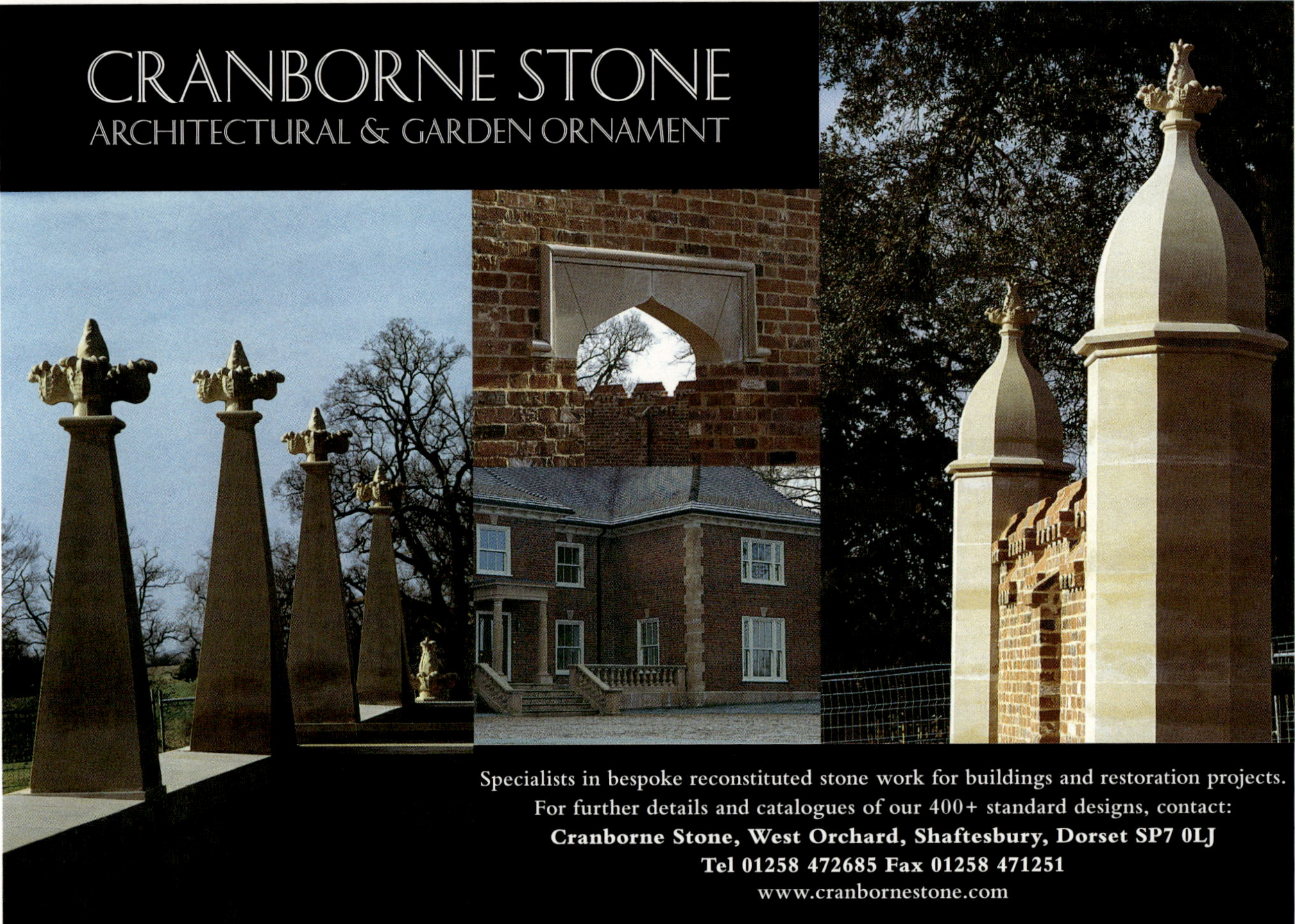

CRANBORNE STONE
ARCHITECTURAL & GARDEN ORNAMENT
Specialists in bespoke reconstituted stone work for buildings and restoration projects.
For further details and catalogues of our 400+ standard designs, contact:
Cranborne Stone, West Orchard, Shaftesbury, Dorset SP7 0LJ
Tel 01258 472685 Fax 01258 471251
www.cranbornestone.com

▶ THE BATH STONE GROUP

Stoke Hill Mine, Midford Lane, Limpley Stoke, Nr Bath BA3 6JR
Tel 01225 723792 Fax 01225 722129
E-mail bsg@bath-stone.co.uk Website www.bath-stone.co.uk
Contact Elaine Dickerson

STOKE GROUND BATH STONE: The Bath Stone Group, producing award winning Stoke Ground Bath Stone, offers an integrated service from design through to installation on site. The company's highly skilled masons and carvers, can fabricate and finish stone to exacting specifications. Stoke Ground is regularly specified by English Heritage and the National Trust, and used in restoration and new-build projects including: Buckingham Palace and Windsor Castle; Oxbridge colleges: Merton, Manchester, Brazenose and Pembroke; Stations: Liverpool Street, London and Temple Meads, Bristol; Development: Wessex Water, Bath and Britannic Assurance HQ, Birmingham; Conservation: Temple of Concorde, Stowe and Waddesdon Manor, Berks. Bath Stone Group offers cost effective natural stone building solutions.

▶ BROADMEAD CAST STONE

Broadmead Works, Hart Street, Maidstone, Kent ME16 8RE
Tel 01622 690960 Fax 01622 765484

RECONSTRUCTED STONEWORK: Broadmead Cast Stone has since 1920 manufactured reconstructed stonework and is an acknowledged leader in the market, undertaking both refurbishment and new build projects. The ability of Broadmead to recreate high quality stone units by the use of natural aggregates with extremely accurate surfaces and details is the basis of its market strength.
By integrating their specialist knowledge with that of the designer's objectives, they are able to provide a satisfactory solution to all aesthetic and construction requirements. They have an unparalleled reputation for the quality of their work, and have attained accreditation as a Quality Assured Company to BS 5750 Part 2. *See also: Wallis display entry in Building Contractors section, page 55.*

▶ CHILSTONE

Victoria Park, Fordcombe Road, Langton Green, Tunbridge Wells, Kent TN3 0RE
Tel 01892 740866 Fax 01892 740249

RECONSTRUCTED STONE PRODUCTS: Handmade architectural stonework for classical buildings and gardens. Range includes columns, coping, balustrade, cornice, temples, statues, urns, porticos, birdbaths, sundials, fountains, gate piers and finials. Used in numerous historic gardens – Warwick and Hever Castles, Kew, Regents Park and Buckingham Palace. Also 'specials' to your design. Fully illustrated catalogue available.

▶ HADDONSTONE LIMITED

The Forge House, East Haddon, Northampton NN6 8DB
Tel 01604 770711 Fax 01604 770027 Website www.haddonstone.co.uk

RECONSTRUCTED STONE: An extensive standard range of architectural and ornamental stonework in Haddonstone and Haddon-Tecstone materials. Custom-made stonework a speciality: *See also: display entry in this section, page 117.*

▶ KEYLINE EAST LONDON

Atlas Wharf, Hackney Wick, Berkshire Road, London E9 5NB
Tel 020 8533 3499 Fax 020 8985 9861

PLASTERWORK MATERIALS DISTRIBUTOR: *See also: display entry in Moulding Materials section, page 115.*

▶ THORVERTON STONE COMPANY

Seychelles Farm, Upton Pyne, Exeter, Devon EX5 5HY
Tel 01392 851822 Fax 01392 851833
E-mail caststone@thorvertonstone.co.uk Web www.thorvertonstone.co.uk

RECONSTRUCTED STONE: Manufacturers of standard and bespoke architectural masonry in reconstructed cast stone, for use in all types of building and landscapes: typical products include mullion window surrounds, cills, heads, columns, arches, porticos, steps, balustrading, coping, pier caps, ball finials, fireplaces, barbecue kits and cobble slabs. Thorverton's reconstructed stone is manufactured from natural aggregates and crushed stone from many regions of the country. Each piece is carefully finished by skilled craftsmen. Specialists in customer designed specials.

METALWORK

Abbey Heritage Ltd	wr	104
Ace Demolition – The Reclamation Centre	ci	159
Ajeer Limited	ci	159
Albion Manufacturing Ltd	bs ci wr	145
Alpine Reclamation	ci wr	159
Antique Buildings Limited	ci	162
Architectural Metalwork Conservation	wr	145
Barr & Grosvenor Ltd	ci	145
Boshers (Cholsey) Ltd	wr	58
Brassart Limited	bs	123
Broxap Dorothea	ci	145
Britannia Architectural Metalwork Ltd	ci	145
C E L Architectural Metal Roofing	pb	86
C J Blacksmiths	wr	146
Calder Group	ci wr	59
Cambrian Castings (Wales) Ltd	ci	146
Carrek Ltd	wr	146
The Cast Iron Company	ci wr	152
Castaway Cast Products and Woodware	ci	146
Casting Repairs	ci wr	146
Chris Topp & Co Blacksmiths	ci wr	146
Conplex	pb	88
Conservation Building Products Limited	ci	160
Crowncast Ltd	ci	147
Don Barker Ltd	wr	147
Dorothea Restorations Ltd	ci wr	147
Drummonds Architectural Antiques Limited	ci	160
Eura Conservation Ltd	wr	148
Fabco	ci wr	148
Fine Iron	bs ci wr	147
Fred Brodnax Architectural Blacksmiths	wr	148
George James & Sons	wr	148
Glasgow Steel Nail Co Ltd	wr	179
Hartbro Weld Engineers	ci	148
Hodgsons Forge Decorative Metalwork & Restoration	wr	149
J & J W Longbottom Limited	ci	87
J H Porter & Son Ltd	ci wr	149
James Hoyle & Son	ci	150
Joseph Giles	bs ci wr	124
L C Jay Limited	ci	149
LASSCO	ci wr	160
Leander Architectural	ci	151
Malborough Forge	ci pb wr	149
Marsh Brothers Engineering Services Ltd	ci wr	151
Mather & Smith Ltd / M J Allen Group	ci	149
Norgrove Studios	wr	130
Northwest Lead	pb	86
Nostalgia	ci	189
Original Architectural Antiques Co Ltd	ci wr	160
Peter S Neale Blacksmiths	ci wr	150
The Real Wrought Iron Company	wr	150
Retrouvius Architectural Salvage	ci	160
The Round Wood Timber Co	wr	141
Rupert Harris Metalwork Conservation	bs ci gs pb	150
Saint-Gobain Pipelines	wr	87
Sarum Metal Craft	wr	151
Shepley Engineers Limited	ci wr	151
Smith of Derby	bs ci wr	152
South West Reclamation	ci	162
Structural Perspectives	ci wr	151
Symonds Salvage	ci	162
Victorian Classic Style	ci	152
Walcot Reclamation	ci wr	162
Walter MacFarlane & Co Ltd	ci wr	152

KEY
- bs brass
- ci cast iron
- gs gold and silversmiths
- pb decorative leadwork
- wr wrought ironwork

STREET AND GARDEN

A D Calvert Architectural Stone Supplies Ltd	ss	104
A F Jones (Stonemasons)	ss	105
Abbey Heritage Ltd	pv ss	104
Abbey Masonry and Restoration	ss	104
Acanthus Associated Architectural Practices	la ur	24
Ace Demolition – The Reclamation Centre	pv	159
Ajeer Limited	pv	159
Albion Manufacturing Ltd	gt	145
Allison Stone (Wigan) Limited	ss	106
Alpine Reclamation	pv	159
Anderson and Glenn	la	39
Andrew Martin Associates	ur	38
Anthony Blacklay & Associates	la	25
Antique Buildings Limited	pv	159
Architectural Metalwork Conservation	gt	145
Architectural Reclaim	gt	159
Arrol & Snell	la	25
Artisan Oak Buildings Ltd	pv	142
Ashby Stone Masonry Limited	pv ss	105
Benjamin Tindall Architects	ur	26
Boden & Ward Stonemasons Ltd	ss	106
Britannia Architectural Metalwork Ltd	gt sf	145
Broadway Malyan Cultural Heritage	la	26
Broxap Dorothea	sf	152
Building Design Partnership	ur	27
Carthy Conservation Ltd	ss	107
The Cast Iron Company	gt sf	152
Castaway Cast Products and Woodware	gt si	146
Casting Repairs	gt	146
Cathedral Works Organisation (Chichester) Limited	ss	107
Centreline Architectural Sculpture	ss	107
Chilstone	cn sf ss	156
Chris Topp & Co Blacksmiths	gt	146
Clague	la	27
Classical Flagstones	pv	114
Cliveden Conservation Workshop Ltd	ss	108
Cobar Services	pv	160
Conservation Building Products Limited	pv	160
Cranborne Stone Limited	gt sf	152
Crowncast Ltd	gt sf	147
The Cumbria Clock Company	cl	157
David Ashton Hill Architects	la	28
Don Barker Ltd	gt	147
Donald Insall Associates Ltd	ur	29
Dorothea Restorations Ltd	gt	147
The Downs Stone Company Ltd	pv	108
Drummonds Architectural Antiques Limited	pv	160
Eura Conservation Ltd	sf ss	115
Exitex	cn	156
Fabco	gt	125
Fine Iron	gt sf	147
George James & Sons	gt	148
Gibberd Conservation	la	29
Haddonstone Limited	cn sf ss	152
Hanna Conservation	ss	114
Hanson Brick	pv	100
Hearns (Specialised) Joinery Ltd	cn sf	140
Heritage Engineering	cn gt ss	115
Hill Leigh Ltd	cn	156
Hodgsons Forge Decorative Metalwork & Restoration	gt	149
Holden Conservation Ltd	ss	115
The Hopton Wood Stone Firms Ltd	pv	109
I J P Building Conservation	gt ss	61
Ibstock Brick Limited	pv	100
J H Porter & Son Ltd	gt	149
James Hoyle & Son	gt si	150
Julian Harrap Architects	la	32
Ketley Brick Company Limited	pv	100
L C Jay Limited	gt sf	149
LASSCO	pv	160
Latham Architects	la ur	32
Leander Architectural	si sf ss	157
Malborough Forge	gt	149
Malbrook Conservatories	cn	156
Mansfield Thomas & Partners	la	33
Marsh Brothers Engineering Services Ltd	gt	151
Mather & Smith Ltd / M J Allen Group	gt sf si	149
McMarmilloyd Limited	ss	96
Morgan & Co (Strood) Ltd	fe	143
Mowlem Rattee & Kett	ss	62
MRDA	la	32
Nimbus Conservation Ltd	ss	110
Original Architectural Antiques Co Ltd	pv	160
Peter S Neale Blacksmiths	gt sf	150
Retrouvius Architectural Salvage	gt pv	160
Richard Noviss Sculptor & Stonemason	ss	110
Robus Architectural Ceramics	ss	97
Rupert Harris Metalwork Conservation	gt	150
Shepley Engineers	gt	151
Smith of Derby	gt cl	157
Solopark Plc	gt pv sf ss	161
South West Reclamation	pv	162
Spencer & Richman	ss	190
St Blaise Ltd	ss	65
The Stone Doctors	ss	112
Stuart Page Architects	la	35
Symonds Salvage	pv	162
T J Crump Oakwrights Limited	cn	78
Taylor Pearce Restoration Services Limited	ss	115
Thwaites & Reed	cl	157
Vale Garden Houses	cn	156
Victorian Classic Style	gt sf	152
W J Haysom & Son	pv	114
Walcot Reclamation	pv	162
Wells Cathedral Stonemasons	ss	113
West Meon Pottery	sf	98
Westland	gt ss	190
Westoncraft	cn	137
Wildwood Joinery	cn	142

KEY
- cl clocks
- cn garden buildings and conservatories
- fe timber fencing
- gt gates and railings
- la landscape architects
- ln landscaping
- pv flagstones and paving
- si signage
- sf street and garden furniture
- ss statuary and stone carving
- ur urban designers

GLASS

Albion Manufacturing Ltd	gp	145
Bursledon Brickworks Conservation Centre	wg	123
C J L Designs	gp ll sg	129
The Cathedral Studios	sg	129
Chester Masonry Group	sg	59
David Bowman Stained Glass	sg	129
Fabco	gp	125
Fine Iron	gp	147
Goddard & Gibbs Studios Ltd	ll sg	130
Illumin Glass Studio	ll sg wg	129
Linley Stained Glass Studios	sg	130
The London Crown Glass Company	wg	129
Nick Bayliss Architectural Glass	sg ll	130
Norgrove Studios	sg wg	130
Norman & Underwood	ll	86
Riverside Studio	gp ll sg	129
The Stained Glass Specialist	sg wg	129
Tatra Glass	sg wg	129

KEY
- gp window grilles
- ll leaded lights
- sg stained glass
- wg window glass

WINDOWS AND DOORS

KEY

- cs cast stone
- gp window grilles
- mw metal windows
- rl roof lights
- sk secondary glazing
- td timber doors
- tw timber windows
- wf door and window furniture
- wg window glass

JOINERY

KEY

- cb cabinet makers
- jo joinery
- wc wood carvers and turners

TIMBER SUPPLIERS

KEY

- ti timber suppliers
- tp timber and parquet flooring
- ve veneers

Figure 1. A beautiful scrolled turnbuckle of circa 1700 from Blanchworth Gallery, Alkington, Glos

Fixed Light
Fixed Light
Mullion
Saddle bar
Stanchion bar
Transom
Mullion (section)
Turnbuckle (from above)
Handle (from side)
Casement (opening light)
Fixed Light

Figure 2. A typical cross-window. This window typifies many of the features of the 17th and early 18th century transomed window, divided by a stone mullion and a transom into four lights, only one of which opens. Based on an attic window of 1707 from a farm house at Pilning, Gloucestershire, the window catches and their 'turnbuckle' handles are embellished with a typical Gloucestershire open heart design and the handle at the bottom of the casement is simply ornamented with a knob.

EARLY CASEMENT WINDOW FURNITURE

LINDA HALL

Windows are one of the most important items for giving character to an old house, and the more original window furniture that survives, the more character the house will have. Put in the wrong type of windows and the house, be it ever so old, will look modern. Retain the correct window details, and preferably the original windows, and the house will look its correct age.

THE DESIGN OF WINDOWS

Medieval houses generally had unglazed windows; it was not until glass became less expensive during the reign of Elizabeth I that the ordinary householder could begin to afford glazing. Elizabethan windows were usually divided into several small 'lights' by mullions (the vertical dividers of wood or stone) and sometimes also by transoms (horizontal dividers), but they rarely had more than one opening casement. This could be quite small in the late 16th and early 17th century, often taking up only about two thirds of the height of the window. 18th and 19th century casements tend to be much larger, and always take up the full height of the window. In the late 17th and 18th centuries, cross-windows *(Figure 2)* were common. These have a central mullion and a horizontal transom set somewhere above the mid point, and the casement takes up the whole of one of the lower lights. The fixed lights were held in place with wire which attached the leadwork either to vertical stanchions or to horizontal bars known as 'saddle bars'. Stanchions were slender bars of either iron or wood, square in section, which were set diamond-wise in the centre of each light. Commonly they were set into mortises in the window frame, but some iron ones had broad flattened ends which were nailed to the wooden window frame *(Figure 6)*. This design appears to be a later development, often appearing in late 17th century and 18th century windows. Occasionally windows have two stanchions per light: these are for security, rather than for fixing the leaded lights, and generally seem to date from the later 17th century. Saddle bars were either square in section or, in the 18th and 19th centuries, round, and there were two, three or more in each light. Windows with saddle bars will commonly also have a stanchion to provide security in the opening light.

The glass panes of leaded lights were diamond-shaped in the early period and this shape continued in common use throughout the 17th century. This is a far more economical shape than the later square or rectangular panes, as oddments of glass – which was still a relatively expensive material - could be used up in the small triangles at the edges of the window. The Elizabethans sometimes used an assortment of geometric shapes for their leaded lights, making wonderful patterns.

In the later 17th century square or rectangular panes came in. Letters written between 1692 and 1695 from the steward to the owner of Levens Hall in Westmorland refer to the use of 'quarry' glass (meaning diamond panes) in various buildings around the estate, including the stables and the clock house. The accounts for Levens Hall itself contain the

Figure 3. A wrought iron casement with horizontal saddle bars and diamond leaded lights in Cheddington, Bucks. Note the spiral handle at the base of the casement and the plain hook stay. Date uncertain.

item *"put into severall windos 14 squers of new glass att 2d the squer".* Presumably the now old-fashioned diamond panes were acceptable in the socially inferior buildings but not in the main house. The accounts also refer to *"naylls to sett the glas in the wood windos"*; these must be the 17th century equivalent of the modern glazing sprig.

In probate inventories of the early 17th century the opening casements were regarded as moveable items, to be valued with the other

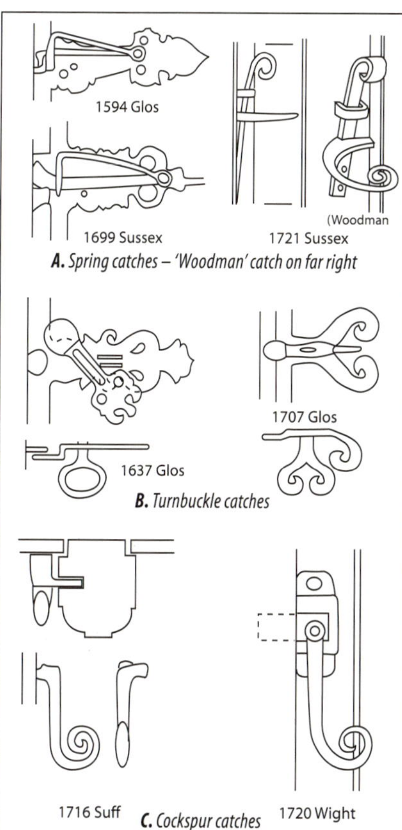

Figure 4. *Window catches from dated houses (from 'Fixtures and Fittings in Dated Houses')*

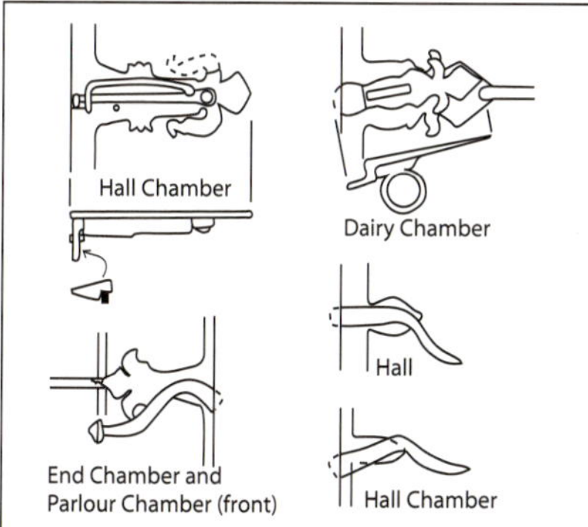

Figure 5. *Window catches from 38, Latimer, Bucks illustrating the great variety of window catches found in this house which dates from the second half of the 16th century. Only the spring catch in the hall chamber is likely to be original. The dairy chamber catch is late 17th century, the parlour chamber catch probably early 18th century, and the others any date up to the early 20th.*

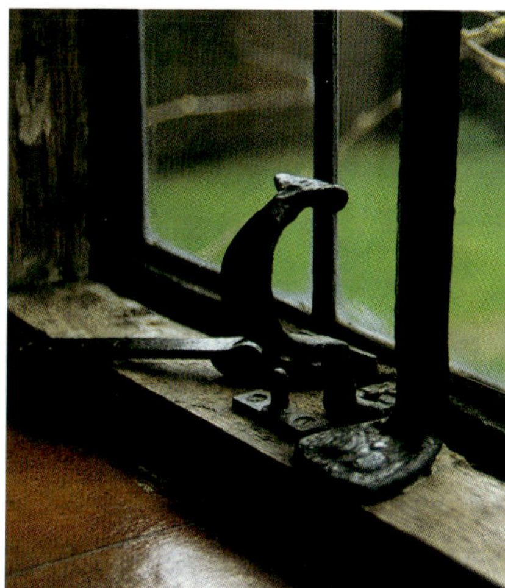

Figure 6. *Church Farm, Haddenham, Bucks. An iron stanchion with an expanded foot. The casement has a very unusual handle and a modern bar stay.*

Figure 7. *Crossways Farm, Abinger Hammer, Surrey. Vertical 'Woodman' catch. The window has lost its leaded lights.*

contents. They are hung in the same way as a door, on iron pintles protruding from the frame, and can easily be lifted off. The casements are made of wrought iron and have a variety of ironwork fittings; catches to hold them shut, stays to hold them open, and handles for moving them.

Iron casements continued in use throughout the 18th century and even into the 19th century, often being used for attics and service rooms while the rest of the house was given fashionable sashes.

WINDOW CATCHES

Catches took two distinct forms, the spring catch and the turnbuckle.

The spring catch seems to have been popular throughout the 17th century. This design has a baseplate, usually decorative, attached to the iron frame of the casement *(Figure 4A)*. The catch itself is a horizontal bar which engages a small iron plate set into the window frame; the spring is a second bar which forces the latch into place and holds it shut. There were several variations on this design including vertical spring catches such as the Woodman catch which is found in the South East. In one version of the Woodman catch the spring bar is attached to the window frame; in another the spring is attached to the casement *(Figure 7)*.

Far more common is the turnbuckle catch, in which the baseplate attached to the casement carries a swivelling catch or turnbuckle *(Figure 4B)*. If the surround is of stone, the catch can simply fasten against the metal frame of the window, but if it is timber the frame is usually omitted and there is either a projecting metal plate or a slot to receive the catch.

The baseplate is simply there to provide an anchorage for the turnbuckle, and may be small and plain, but in many instances it was used as a vehicle for the most elaborate decoration *(Figure 1 for example)*. The designs vary enormously and probably each blacksmith had his own favourites; a double scroll or open heart design is common in south Gloucestershire *(see Figure 2)*, but as yet not enough recording or research work has been done in other areas to be able to note regional variations.

Sometimes the actual turnbuckle has a design which matches or complements that of the baseplate *(as in Figure 2)*, but quite commonly it is fairly plain, often comprising little more than a loop and the expanded end. Unfortunately turnbuckles are easily broken and there are many examples of baseplates with either no turnbuckle at all or with what is obviously a much later replacement in a contrasting style.

The most simple of all the window catch designs is the cockspur catch *(Figure 4C)*, a small latch attached to a handle with a spiral end and commonly fixed directly to the outer rim of the casement. These are found with both iron casements and with the later wooden ones, which in Gloucestershire seem to have started being used at the end of the 17th century. Cockspur catches have been found in houses dated 1716 in Suffolk, 1720 on the Isle of Wight and 1721 in Surrey, and in 18th century houses in East Sussex.

Dating can be difficult, but many iron casements with turnbuckles belong to the later 17th century and the early 18th century. The elaborate use of decoration occurs on many items of furnishings after the Restoration of Charles II in 1660; people seem to have reacted against the enforced plainness of the Puritan regime by looking back to the great age of Elizabeth I, when decoration was used in great abundance. Even if the date of the house is known, the casements may have been replaced at a later date. This is especially true in a house which has a number of different designs; if all the window details match, they are more likely to be original to the house.

One house with a huge variety of window fittings is 38, Latimer in Buckinghamshire *(Figure 5)*. It has one spring catch, matching turnbuckles in two of the bedrooms and a different one in the third. Other windows have the simplest variety of catch, with a plain turnbuckle comprising a flat-sectioned 'latch' and a round-sectioned handle, all slightly different and carried on varying small undecorated baseplates. Such catches are difficult to date, but are unlikely to be earlier than the 18th century and could be considerably later, even early 20th century vernacular revival. It is tempting to imagine that a previous owner bought a job lot of casements and inserted them in the house, but as they all fit the windows it may simply be that casements were replaced at different dates as wear and tear made necessary.

RIM HANDLES

To avoid putting undue strain on the catch when opening and shutting the window, casements were usually provided with a handle attached to the bottom rim. Some very elaborate designs are known, but most take one of three forms; the tulip leaf, the spiral or the knob *(Figure 8)*. As with the catches, the date range of each type is large and they cannot as yet be used for more precise dating. Tulip leaf handles are known from 1620 to 1687, spirals have been found in 1645 and 1721 and knobs from 1624 to 1727, but they may all have a much wider period of use.

STAYS

Many handles incorporate a hole in the base for a hook stay, a long hooked rod

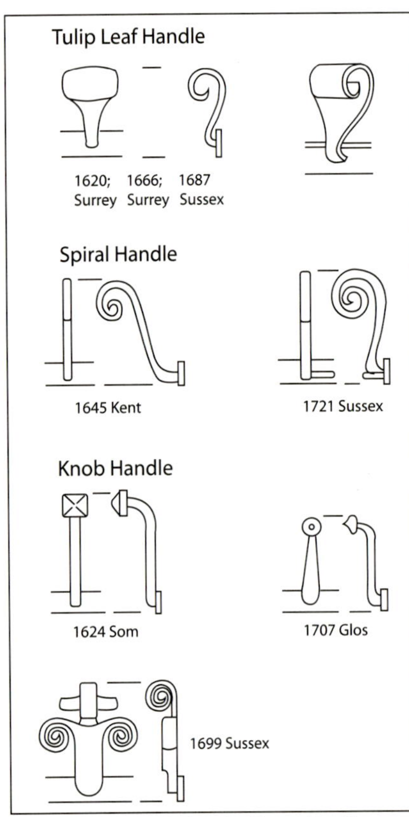

Figure 8. Window handles from dated houses (From Fixtures and Fittings in Dated Houses).

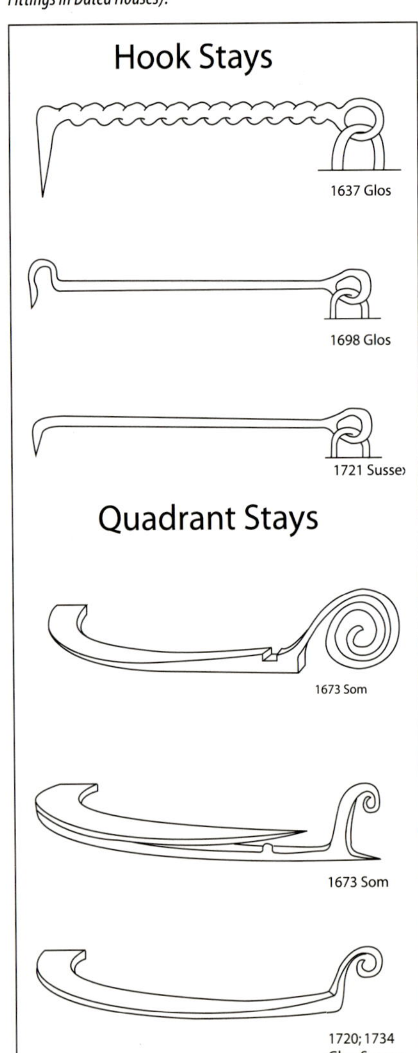

Figure 9. Window stays from dated houses (From Fixtures and Fittings in Dated Houses).

Figure 10. Manor Farm, Compton Greenfield, Glos. Wooden-mullioned window of 1637. It has an iron stanchion, a decorative turnbuckle which engages an iron plate set into the mullion, and an external quadrant stay.

Figure 11. Church Farm, Haddenham, Bucks. A tulip leaf handle with a hole in the base for the hook stay. Alongside is the modern bar stay, which clearly would not be able to hold the window this far open.

located either outside or inside the window which holds the casement open *(Figure 9)*. Alternatively there may be a separate iron loop next to the handle. The hook stay is usually very plain, but some have a twisted shaft.

This type of stay can only hold the window open at a fixed distance, unlike the more modern version with a series of holes in the bar and a peg on the sill. In order to provide more variation, the house at Latimer has three stays for each window, all of different lengths, so that the casement can be fully open, half open or almost closed. In each case the longest stay is fixed to the outer sill, the other two to the inside of the window frame.

The alternative way to hold a casement open was the exterior quadrant stay *(Figure 9)*. This is a flat iron bar in a quarter circle which is fixed to the outer window sill. Usually it has some form of decorative scroll at the end. Many have a notch near the end to hold the casement in place, or else the bar gets thicker towards the end so that the casement wedges against it. The disadvantage of this type of quadrant stay is that it will only hold the window fully open; there is no halfway position. The problem was overcome by using a split bar, so that the upper portion acts as a spring and will wedge against the casement at any point along its length.

Again, dating is difficult. Many of the quadrant stays are in 17th century or early 18th century buildings, whereas hook stays are found at much later dates as well.

The fact that whole casements can be replaced so easily means that original window furniture is not as common as it might be. The owner of a house in Endon, Staffordshire for example, remembers that every window in the house had beautiful and highly elaborate window catches before her father removed all but one. Elsewhere decorative catches show the evidence of daily wear and tear, and many are damaged in some way. It is therefore all the more important to retain any that still remain; even partial or broken catches and handles tell us something of the house's history and should be kept. The catches in particular, with their wonderful variety of forms, are a marvellous reminder of man's basic creativity and ingenuity, and are worthy of more attention than they usually receive.

RECOMMENDED READING

Ayres, James; *The Shell Book of the Home in Britain*. Faber and Faber, 1981

Calloway, Stephen (ed); *The Elements of Style; an encyclopedia of domestic details*. Mitchell Beazley, 1991

Hall, Linda and Alcock, NW; *Fixtures and Fittings in Dated Houses 1567-1763*. Council for British Archaeology 1994

Lloyd, Nathaniel; *History of the English House. 1931*, reissued by The Architectural Press, 1975

McCann, John; *Antique Ironwork. Period Home*, June/July 1982

Martin, David and Barbara; *Domestic Building in the Eastern High Weald, Part 2 – Windows and Doorways*. Hastings Area Archaeological Papers, 1991

LINDA HALL read archaeology and history at Southampton University and has studied vernacular architecture, particularly in south Gloucestershire, for many years. She is a member of the Vernacular Architecture Group, The Institute of Field Archaeologists, and the Regional Furniture Society. Published works include *Rural Houses of North Avon and South Gloucestershire 1400-1720* (Bristol Museum, 1983) and, with Dr Nat Alcock, *Fixtures and Fittings in Dated Houses 1567-1763* (Council for British Archaeology, 1994). She lectures widely and is available to record buildings for either planning or research purposes. Her earlier article on door furniture for *The Building Conservation Directory 1999* is available on-line on the BCD's website, www.buildingconservation.com. Contact Linda Hall, 38 Hawthorn Way, Shepperton, Middx TW17 8QH, Tel 01932 787836.

DOOR & WINDOW FITTINGS

▶ BRAMAH

31 Oldbury Place, London W1U 5PT
Tel 020 7486 1739 Fax 020 7935 2779 E-mail jlbramah@globalnet.co.uk

SASH WINDOW LOCKS AND HIGH SECURITY DOOR LOCKS: Bramah first made locks for historic buildings when they were being built in the early 1800s. Today they still make high security locks, and where they can, tailor them to suit the situation. This applies to their high security key registered deadlocks for secure areas, cabinet locks and cylinder locks. All Bramah locks may be mastered or keyed alike and are available in a variety of finishes, with black and bronze often being used. In addition Bramah manufacture the Rola range of window locks and bolts, where they have a niche expertise in the manufacture of sash window locks.

▶ BRASSART LIMITED

Attwood Street, Lye, Stourbridge DY9 8RY
Tel 01384 894814 Fax 01384 423824 E-mail sales@brassart.co.uk

PERIOD BRASS IRONMONGERY AND HERITAGE FITTINGS: Manufacturing at Lye in the Midlands, each piece of door furniture, bathroom accessory or electrical fitting made in the factory is crafted from local materials and hand finished and polished by craftsmen using timeless techniques. Brassart's accurate period ranges have evolved over many decades and their archive of patterns now dates back over 200 years. Renowned for flexibility, their Bespoke Products Division has been involved in a number of notable projects including The Houses of Parliament and St Paul's Cathedral.

▶ BURSLEDON BRICKWORKS CONSERVATION CENTRE

Bursledon Brickworks, Coal Park Lane, Swanwick, Southampton SO31 7GW
Tel/Fax 01489 576248 E-mail bursledon@ndirect.co.uk

TRADITIONAL CAST IRON DOOR AND WINDOW FURNITURE: *See also: profile entry in Mortars & Renders section, page 167.*

▶ CLAYTON MUNROE LIMITED

Kingston, Staverton, Totnes, Devon TQ9 6AR
Tel 01803 762626 Fax 01803 762584
E-mail mail@claytonmunroe.co.uk
Website www.claytonmunroe.co.uk

DOOR AND WINDOW FITTINGS AND ACCESSORIES: Formed in the early 80s in response to demand for authentic restoration hardware, Clayton Munroe Limited now offers an extensive selection of innovative door furniture, cabinet fittings and accessories, including a new range of stair rods. The 'Rough at the Edges' range, based on 18th century designs is hand forged and supplied with traditional fixings and addresses the search for authentic hardware to retain the true character of a period building. 'Rough at the Edges' products are used in historic buildings throughout the world as well as film sets for period dramas such as Jane Eyre. The company is also well known for its trade marked 'Patine' finish (similar to pewter in colour) and offers over 3,500 items featured in four mail order brochures. *See also: display entry on this page.*

▶ DON FORBES SASH FITTINGS

Cotterton, Logiealmond, Perthshire PH1 3TJ
Tel/Fax 01738 880329

MANUFACTURING AND WHOLESALE IRONMONGERS: Specialist suppliers of all types of sash window fittings including: axle pulleys in a range of sizes and finishes, sash cords and chains, sash openers and easyclean fittings, a large selection of sash fasteners, lifts, stops, etc in a variety of styles and finishes. Don Forbes can refurbish existing fittings and supply replacement items to match existing. A comprehensive range of cast iron and lead sash weights and makeweights are available ex-stock, or made to specification. Kame lead and other lead products available. Brochure and price lists on request.

▶ JOSEPH GILES

The Old Dairy, 51a St Peters Street, South Croydon CR2 7DG
Tel 020 8680 2602

SUPPLY AND REPLICATION OF IRONMONGERY: *See also: display entry on page 124.*

STRUCTURE & FABRIC
Metal, Glass & Wood

DOOR & WINDOW FITTINGS

▶ MACKINNON & BAILEY

Floodgate Street, Birmingham B5 5SL
Tel 0121 773 5827 Fax 9121 766 6072
E-mail sales@mackinnons.telme.com Website www.mackinnons.co.uk

ARCHITECTURAL HARDWARE AND VENTILATOR MANUFACTURERS: *See also: display entry on page 124.*

▶ MBL

55 High Street, Biggleswade, Bedfordshire SG18 0JH
Tel/Fax 01767 318695 E-mail mid-beds.locksmiths@talk21.com

SPECIALIST ARCHITECTURAL IRONMONGERS AND LOCKSMITHS: MBL is a specialist company involved in bespoke, unique and specialist architectural ironmongery and locking systems. Founded in 1990, MBL has provided a quality service involving refurbishment, replication and restoration of historic and period fittings for all types of properties. From single commissions to full design, scheduling and supply of complete packages, MBL integrates modern security systems to complement historic buildings. Recent and on-going projects are Chicksands Priory and Woburn Abbey, Bedfordshire; Waddesdon Manor Stables, Buckinghamshire; Somerset House, London; and various ecclesiastical projects and private houses. *See also: display entry in this section, page 124.*

▶ QUALITY LOCK CO

Leve Lane, Willenhall, West Midlands WV13 1PS
Tel/Fax 01902 602942

HAND CRAFTED TRADITIONAL AND MODERN LOCKS: Quality Lock Company manufactures locks in the traditional way from original patterns and drawings and refurbishes and re-key locks dating from the early 1800s to the present time. Using your lock as a pattern, Quality Lock will produce new locks to exactly match the original. Their stock range of brass cased rimlocks harmonises well with historic buildings. They can produce master keyed locks and extensions to existing master key suites and are also able to produce short runs of specially designed locks to meet customers' individual locking needs.

▶ REDDISEALS

The Furlong, Droitwich, Worcestershire WR9 9BG
Tel 01905 779961 Fax 01905 770131
E-mail reddiseals@reddiplex.co.uk
Website www.reddiseals.com

See also: display entry on page 124.

WINDOW PROTECTION

PLAIN GLAZING

BEN SINCLAIR

New cylinder glass set into restored cast iron window frames at Tardebigge, Worcestershire.

Sash windows restored with new cylinder glass at Maryvale, Birmingham.

Perhaps the most noticeable and yet misunderstood feature of an historic building is that of its glass and the way that glass influences both the atmosphere and ambience of the building.

The influence of glass can be seen both internally and externally, working in a reflective and refractive capacity. Externally, the reflective quality of glass will be most noticeable, varying from the crude slab-like appearance of float glass to the subtle shimmer of period cylinder or crown glass. Internally, handmade glass will influence the softness and warmth of incoming light, as well as the appearance of external views through that glass. The difference between a cold clinical light seen through modern float glass and a warm refracted light through early Victorian cylinder glass is profound.

EARLY GLASS PRODUCTION

Glass used for glazing was made by heating together sand (silica) and lime (calcium oxide) at a high temperature with a flux of wood ash or soda lime to reduce the temperature at which the particles fuse. Modern clear float glass is manufactured from silica (70-74%), lime (5-12%), and soda (12-16%), varying between manufacturers.

Before the industrial revolution, furnaces used to make glass were fuelled with charcoal. The industry was nomadic because production depleted forests for fuel. Silica was taken straight from rivers or quarried sand, with little regard for contaminants. Sands often contained metal oxides which discoloured the glass, and early glasses would rarely have been pure 'white' or uncoloured. Beechwood ash, soda lime, and other materials used in the smelting process also added colour to the glass, which was most commonly green or straw coloured. Regular production of white glasses was uncommon until the early 19th century, when consistent materials became available, and more efficient coal- then later gas-fired furnaces were introduced.

It is a commonly held misconception that all historic windows were glazed with crown glass. Both cylinder (or broad) glass and crown glass methods of handmade flat glass production survived alongside each other until the 20th century, when crown glass declined into obscurity. Often harsh taxation regimes tended to promote one method over another, causing variations in production techniques in differing periods. Much mid and late 18th century glass was crown, where thin glass attracted less duty being taxed by weight. However, by the mid 19th century, glass production was again dominated by the cylinder method. Paxton's 1851 Crystal Palace was built using cylinder glass, made by Chance Bros. of Birmingham. This building used just over 80,000m^2 glass, approximately one third of the English annual production at that time. Packington Hall, at Meriden in the Midlands, glazed *c*1840 has a good mix of both white crown and white cylinder glass. St James's Church, on the same estate, *c* 1792 is glazed with predominantly crown glass, very active with obvious curvilinear striations, bubbles, and a straw tint.

WINDOW GLASS TYPES & TERMINOLOGY

Slab Glass

The earliest and most primitive form of window glass, made by casting molten glass as a thick slab onto a bench or flat surface, often within a mold. Also later used as the base material for mirror and plate glass production, prior to hand/machine polishing and finishing.

Cylinder Glass

A handmade mouth-blown glass made by blowing a bottle-shaped cylinder and removing the two ends. The cylinder is then cut,

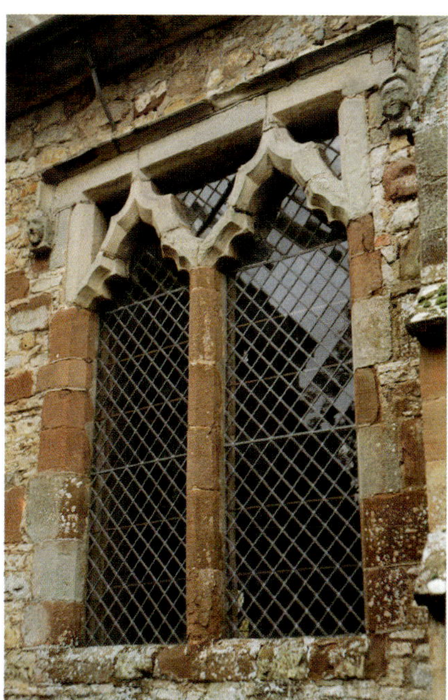

Leaded glazing at Leamington Hastings Parish Church with historic glass replaced by leaded float glass. Period glass survives in the heads and traceries only.

Leaded lights with 17th century glass at Compton Wynyates, after restoration. Almost all of the original glass was re-used (approximately 98%), with new glass made to match.

reheated and unrolled to give a flat piece of glass which is then annealed slowly. Recognisable by naturally distorted movement without pattern. Archaic names: broad glass, forest glass and occasionally, muff glass.

Crown Glass

A handmade mouth-blown and spun glass, made by first blowing a sphere which is then opened out at one end and spun into a flat disc. Several panes of crown glass may be cut out of each disc. Recognisable by curvilinear, naturally distorted movement. Often panes are slightly bowed as a result of a secondary annealing process. Archaic name: Normandy glass.

Polished Plate Glass

Slab glass, ground and polished on both sides. Used initially for mirrors then grand windows, it was a prohibitively expensive 19th century glass, but was truly flat. Easily replaced today by float glass.

Drawn Sheet Glass

First produced in the early 20th century, this machine-made glass was drawn vertically out of a glass furnace, or Lehr, in sheet form between rollers. Recognisable by some movement in the glass in the vertical plane, but of consistent thickness and flatness. Modern drawn sheet is usually imported from Eastern Europe, for the greenhouse trade. Often called 'horticultural' or 'dutch lights'. Used as an inferior alternative to replicate handmade glass.

Float Glass

The ubiquitous glass of today, made by floating molten glass over a bed of liquid tin. Available since the late 1950s, it is a truly flat, functional glass without movement or interest, visually ugly in historic settings.

RE-USING HISTORIC GLASS

The removal of historic glass from timber sash windows and cast iron frames is fraught with difficulty, usually as a result of hard putties and rigid frames. Where new windows are required, the original frames can be sacrificed to save the glass for re-use, but if the frames are to be repaired, extracting the panes requires care and a great deal of time. Even once the face putty is removed, the thin line of putty between the glass and the frame is usually more than enough to keep the glass immoveable. It is sensible to leave glass *in situ,* reface any lost or damaged putty with new, and refrain from disturbing surrounding framework if at all possible.

The removal of historic glass from leaded lights is not so difficult because the lead cames which support it are flexible, and all surviving glass can and should be re-used when releading historic windows. Nevertheless, there is a considerable risk of accidental damage. Loss of historic glass in the releading process is indicative of poor workmanship. Where historic glass is partially missing, then very similar glasses should be used to complement the originals. The use of float glass alongside handmade glass looks ugly and is again indicative of poor workmanship, resulting in a 'spotted dick' appearance.

MODERN GLASS ALTERNATIVES

The beauty of original handmade glass and the way in which it disperses and reflects light is due to the method of manufacture. Minor imperfections, inclusions, and tints of colour impart life and vitality. Selecting new handmade glass which exhibits the same qualities can be difficult, particularly as crown glass glazing is no longer made. Modern, machine-made glass on the other hand, is dull, flat and lifeless, and as inappropriate in a historic façade as concrete blocks in a handmade brick wall.

MACHINE-MADE GLASS

Float glass, due to its flat perfection is not an appropriate alternative. Float glass that has been distorted in a kiln has an individual appearance markedly different to handmade glass. It is bland to view, has a 'bendy' appearance, lacks surface brilliance, and is quite unlike handmade glass. Its use should

The West window of the fine parish church of Kilpeck, Herefordshire church was spoiled by the use of machine-made float glass leaded lights. All the earlier glass was lost.

Lodge Park seen from its former deer course, and (below) a detail of the new cylinder glass glazing.

CASE STUDY: LODGE PARK

Lodge Park, near Aldworth in Gloucestershire, is an ornate Jacobean pavilion built *c*1634 to view the sport of deer coursing. Over the next 300 years it was converted into a small, two-storey house, and substantially altered several times. By the time the National Trust acquired the building in 1982 little of its original interior was to be seen and many of its original fittings had been lost, including all early glazing.

The building has been recently restored under the direction of the architect Michael Reardon, in collaboration with Jeffrey Haworth of the Trust's Severn Region office, and the archaeologist Warwick Rodwell.

The Glass used to glaze the 13 principal windows was purpose-made as a traditional cylinder, or broad glass. This was a deliberate attempt to replicate English cylinder glass typical of the mid 17th century, with new glass manufactured in France. Primitive surviving glasses were copied from this period, based largely on plain broad glass dated 1667 in the private church at Compton Wynyates, Tysoe, Warwickshire. Although deliberately active, straw green tinted, and with numerous bubbles, inclusions and imperfections, this glass was made as it was 350 years ago. It is not excessively distressed or over active, as a modern art glass might appear; neither is it as perfect as a 19th century handmade cylinder, or crown glass. The colouring is typical of a glass manufactured from contaminated sands and potash within a charcoal furnace; the oxides and impurities creating a sparkling, lightly coloured and imperfect glass with tremendous life and movement.

The resultant light in Lodge Park is much warmer than the clinically clear white light one might expect of a more modern building, emphasising the natural colour of fine stone, timber and lime plasterwork now found within the building.

be discouraged, as the visual impact of this 'economy' glass on a building is false and obvious.

Drawn sheet glass, commonly called 'horticultural', has some linear movement, though still dull and relatively lifeless. Its only advantage, along with float glass, is price. It is worth remembering that cheap glass looks exactly that: cheap.

Cathedral glass is an obscured textured rolled glass with smooth and rough faces, commonly used in lavatories. Originally hand rolled, it is now machine-made in Germany, America and China.

Pressed patterned glasses are produced in a similar fashion to cathedral glass. Some Victorian patterns are available, mainly as imports from America.

Machine-made 'antique' glass is made by wire-stroking drawn sheet glass whilst still plastic. Produced mainly for the art glass market, its busy striations make the glass artificially active.

HANDMADE GLASS ALTERNATIVES

Cylinder glass is a handmade mouth-blown glass, readily available commercially. There are many different types, some produced for restoration purposes, but most are art glasses. Most handmade glass for the stained glass market is produced by this method.

Crown glass is a handmade mouth-blown spun glass which is only available today in the form of small, heavy bullions. True crown glass replicating the fine thin crown glasses typical of the mid 19th century is unavailable, and any modern glass sold today as 'crown' glass is likely to be cylinder glass of one type or another. Fine clear handmade cylinder glass is the only compatible glass alternative today.

Care must be taken in the selection of handmade glass: there are many types of cylinder glass, and specifying that term alone is not enough. The exaggerated appearance of many contemporary types of 'antique', 'reamy' or 'seedy' cylinder glasses produced for the art glass market make them poor substitutes for historic window glass.

There are good imported glasses from Germany, France and Poland that are extremely close to old glasses that still survive, and it is these that must be sought out. There is now no handmade window glass made commercially in the UK.

BEN SINCLAIR is a traditional glazier working in building restoration. He runs Norgrove Studios, whose work involves new and restored stained glass and historic plain glass restoration and facsimile.

SOURCES OF WINDOW GLASS

Norgrove Studios
Tel 01527 541545
See page 130.

The London Crown Glass Company
Tel 01491 413227
See page 129.

Tatra Glass (UK)
Tel 01509 235387
See page 129.

Machine-drawn sheet glass can be obtained from Bursledon Brickworks Conservation Centre
Tel 01489 576248
See page 167.

WINDOW GLASS

DECORATIVE & STAINED GLASS

▶ C J L DESIGNS

Pant y Ffynnon, Felindre, Llandysul SA44 5XS
Tel 01559 371670 Mobile 07968 175069

STAINED GLASS AND LEADED LIGHTS: Christopher and Bronwen Worrall offer a design service for new windows in all settings, and will manufacture new windows or rebuild and repair the existing ones of your period building. They advise on the maintenance needs of your leaded glass and frames and offer refurbishment of metal frames and window furniture including Crittal-type. New hoppers and other frame types can be made to order and window protection is available using polycarbonate and metal grills. Their customers include many churches of the Oxford and Winchester Dioceses dating from the 12th Century, private houses and listed buildings such as Littlecote House, Spring Cottage, Cliveden and the Falkland Islands Chapel, Pangbourne. CJL Designs is a member of the Guild of Master Craftsmen.

▶ THE CATHEDRAL STUDIOS

Dean and Chapter of Canterbury, 8a The Precincts, Canterbury, Kent CT1 2EG
Tel 01227 865265 Fax 01227 865222 E-mail stainedglass@canterbury-cathedral.org

SPECIALISTS IN STAINED GLASS CONSERVATION AND RESTORATION, BSMGP Accredited Studio – Categories 1-5: The Cathedral Studios look back on 29 years of successful conservation of the medieval, late medieval and Victorian heritage in Canterbury Cathedral and in many other churches and buildings throughout the country. The team of seven highly trained and specialised conservators is headed by Dr Sebastian Strobl, UKIC accredited conservator and an internationally renowned stained glass expert, who will be pleased to advise on the conservation and protection of your glass.

▶ DAVID BOWMAN STAINED GLASS

10 Bernard Close, Cuddington, Bucks HP18 0AJ
Tel/Fax 01844 292813 Mobile 07880 702450

DECORATIVE AND STAINED GLASS: David Bowman Stained Glass is a family team of craftsmen with a combined experience of over 100 years in the restoration of historic glass. A modern and traditional design and repair service is offered by Stewart Bowman. The company is also experienced in the field of window protection.

▶ ILLUMIN GLASS STUDIO

82 Bond Street, Macclesfield, Cheshire SK11 6QS Tel 01625 613600

STAINED GLASS WINDOWS AND LIGHTING: Illumin manufacture and repair stained glass windows and lighting; and supply antique and other sheet glass. They also provide to order sandblasting, brilliant cutting, etching, etc. Established in 1979, Illumin cover all aspects from design to installation. They cover a 30 mile radius of Macclesfield and will travel further for an interesting and viable project. Past work has included: The Millennium Window at Great Brington Church, Northampton; Grade I listed Chethams Library, Manchester; renovating windows of a 16th century farmhouse; repairs to local churches, and; repairs and new windows for private houses of various ages.

▶ RIVERSIDE STUDIO

9b Curzon Street, Hull, East Riding of Yorkshire HU3 5PH
Tel 01482 563742 Fax 01482 575900 E-mail riverside.studio@ic24.net

STAINED AND LEADED GLASS WINDOWS: Riverside Studio are specialists in all aspects of the conservation, restoration and repair of stained and leaded glass windows. Wire guards and polycarbonate window protection are also manufactured and fitted. Working nationally, the studio undertakes work of the highest calibre for its clients who include ecclesiastical authorities, historic building owners and the top national conservation organisations. On its team Riverside has two UKIC accredited conservators. For free estimates and advice please contact Richard Green.

▶ THE STAINED GLASS SPECIALIST

35 Leigh Lane, Wimborne, Dorset BH21 2PW
Tel/Fax 01202 882208 E-mail ssherriff@yahoo.com Website www.lead-windows.co.uk

STAINED AND DECORATIVE GLASS: A full comprehensive stained glass design studio offering repair, restoration and new design including *in situ* repairs of lead, tiffany and kiln work. This family run company offers sound advice on the repair and maintenance of all windows and their frames including the supply and fitting of all methods of protection. Operating south of a line between Bristol and London, a reliable dependable service and top quality workmanship is always provided. Recent projects include the extensive repair and restoration of Tarrant Rushton Church windows, comprehensive renovation of a listed school buildings leaded light windows and fabrication and fitting of 'real' millennium commissions.

***On the left**, a well-painted window, showing reasonable overpainting of the glass. **Above**, a similar example, but here the painter did not understand the difference between the frame and the sash: there is an arbitrary line across the top of the muntins (which are in the same plane as the other outer linings). He has also not returned the frame colour round the lining returns, which has resulted in a slightly insubstantial appearance.*

SASH WINDOWS: PAINTING AND DRAUGHT-PROOFING

DAVID WRIGHTSON

The sliding sash window has been with us for over three centuries and the operating principle has remained almost unchanged throughout that time. Some of the components, such as the staff bead and the parting bead can still be bought off the shelf.

Sash windows are made in such a way that they can be easily dismantled for repair or for replacing broken sash cords. Many people fail to realise this when they encounter problems and think that the only sensible option is replacement – in some cases, with plastic ones – which is simply not the case. Some timber windows have lasted for centuries because they have been properly maintained and painted regularly. Plastic windows (PVCu) by comparison, cannot be dismantled and repaired so easily, and the components cannot be made by any competent joiner. They have not been tested by time and there are already signs of failure. Plastic windows, which usually come with double glazed sealed units are generally only guaranteed for ten years and are expensive to replace if they fail. Furthermore, they are constructed in such a way that they cannot reproduce the mouldings and details characteristic of traditional timber windows. They are, almost without a single exception, completely unsuitable for use in any historic context.

The problems most likely to be encountered with traditional timber sashes are sticking, failure of joints, failure of putty, wet rot, rattling, and draughts. The first four of these are the result of poor maintenance and the lack of a good protective coat of paint. Rattling and draughts can be dealt with in a number of ways, which we will examine later.

MAINTENANCE

You should aim to inspect your windows every year (and, ideally, get a qualified professional to inspect the whole house every four or five years). Typical sash window problems likely to be encountered include:

- Cracked and flaking paintwork: the outside of the windows should be repainted at intervals of five to eight years, normally.
- Sticking windows: usually the result of either careless replacement of staff bead, following repair or re-cording, which is easily remedied, or a build up of paint which needs to be removed.
- Failed putty and broken glass panes: these are relatively easy to replace.
- Broken cords: in former times people re-corded their own windows – the cords and sash weights were available at any ironmongers (and still are at some).
- Timber decay, particularly to the bottom rail: fillers are invaluable for minor decay and surface imperfections where the strength of the timber is unaffected; loose corner joints can be strengthened by means of corner brackets which can then be painted over; and more significant repairs can be carried out by any competent joiner.

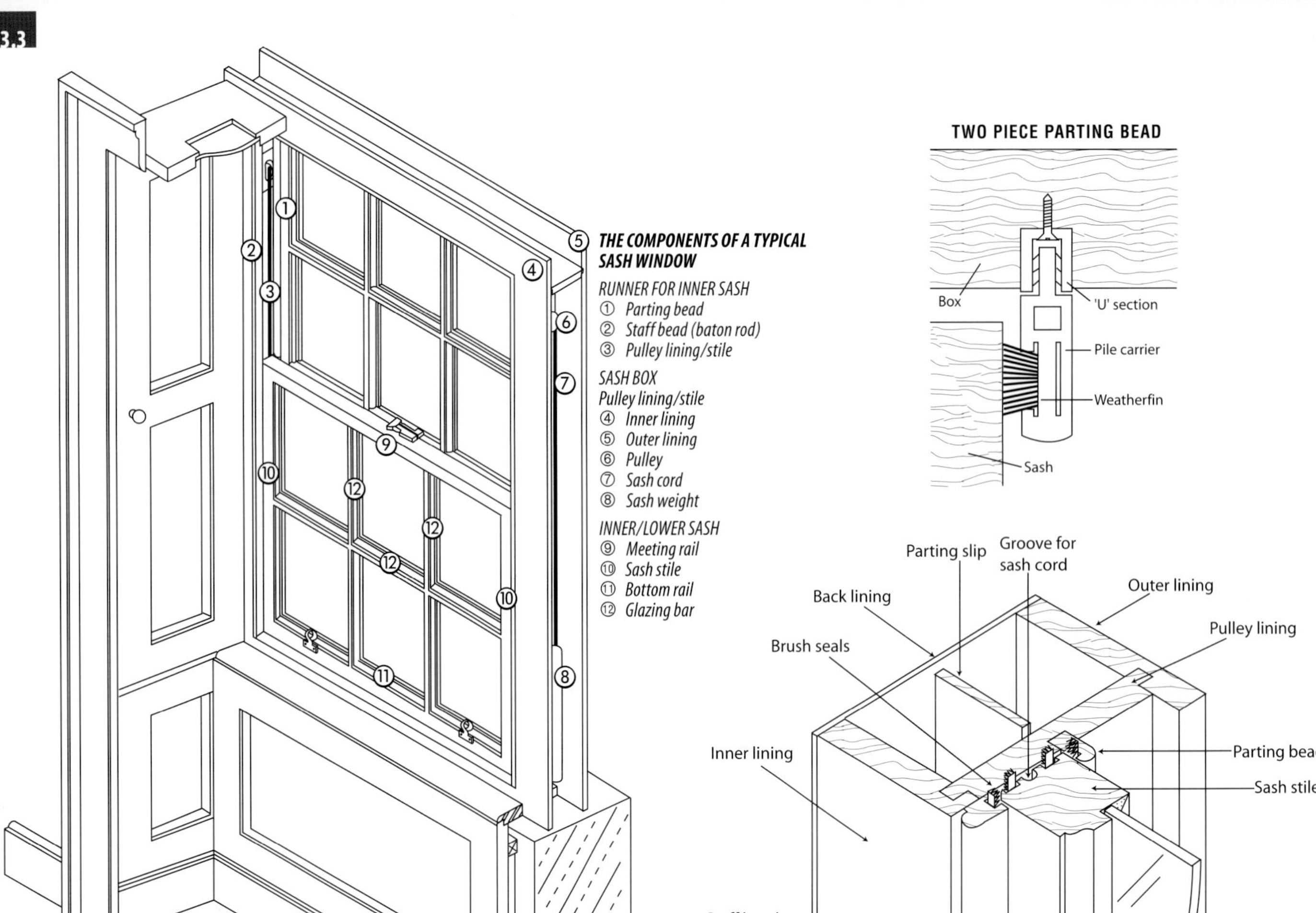

Illustrations reproduced with permission of Historic Scotland and Simpson and Brown Architects from Historic Scotland's Technical Advice Note 3: Performance Standards for Timber Sash and Case Windows 1994.

PAINTING

Preparation

Carry out any repairs, rot treatment, re-glazing or re-cording. Remove all sash furniture (window fasteners and sash lifts or handles). Wash down the previous paintwork with sugar soap, soda or detergent, and rinse off. Rub down with pumice stone, pumice block or sandpaper and brush away all dirt and grit. If your windows were painted with a lead based paint, you must use a wet or damp process to prevent the release of toxic dust into the air. Use a hot air paint stripper only if necessary (if a build up of paint is causing sticking or completely obscuring mouldings). Protect glass by using a suitable shield. Use a shaped shavehook to strip mouldings. A respirator should be worn when burning off old lead paint, because of the toxic fumes. Never use a blowlamp or propane torch; they have been responsible for too many fires in the past. Next, spot prime any areas of exposed bare wood. (If you have taken out the sashes for repair or re-cording you will notice the edges of the stiles are unpainted, leave them this way; that's how they're meant to be).

Paints and equipment

Provide yourself with a safe ladder, protective dustsheets, appropriately sized brushes, a paint kettle, and the correct type of paint. In most cases you will find it easiest to use a one-inch brush for the glazing bars, and a larger brush for the other areas (professionals tend to use the largest brush they can in any given circumstances, but this requires experience). A paint kettle holds a manageable amount of paint at a time, is easy to hold, and helps prevent grit from dirty brushes getting back into the paint tin. The correct type of paint depends on what you are painting onto. Some paints are not compatible with others, especially earlier paints, and can fail quite rapidly. The make up of modern paints has changed so much that you will need the advice of a reliable supplier. Unless the chosen paint is thixotropic (a thickish gel usually marketed as a single-coat paint) make sure you stir it well, since the various constituents quickly separate out during storage, and the solid pigment goes to the bottom of the tin. Don't paint in very hot or in very cold weather; painting outside in wet weather is not advisable.

Glazing bars

Some people find these difficult to paint and rely on proprietary masking devices or on masking tape. It is much better to learn how to do it properly, using a brush of the right size. Try to develop a technique that allows you to just cover the junction between the glass and the putty. This prevents water from getting in and stops the putty from deteriorating. If you accidentally paint onto the glass this can be cleaned up later using a special scraper. Some painters will overdo this principle and paint (very neatly) much too far onto the glass. Try to avoid this, as it increases the visual weight if the glazing bars often in an inappropriate way. (Broadly speaking the width of the glazing bars gives an instant clue as to the period of a building).

Procedure for internal painting

1 Slide the outer sash down a little and raise the inner sash slightly, leaving a gap top and bottom.
2 Paint the glazing bars and the surfaces of the inner sash including the top surface of the meeting rail and the underside of the bottom rail.
3 Then paint what you can see of the outer sash, including the face (but not the underside) of the meeting rail.
4 Now slide the outer sash up, but not quite closed, and lower the inner sash by gripping it from the outside so that you don't touch the wet paint.
5 You can now complete the painting of the outer sash, omitting the top surface of the top rail. Ideally you should leave the sashes in this position until dry.
6 According to how your colour scheme relates to the outside paint colour, you can either paint the pulley stile at the same time as the sashes, or later. If painting at the same time, then the stile, parting bead and staff bead can be painted in sections as the sashes are moved up and down. Pull the sash cords out when you are working behind them (painting the cords makes them more likely to break); paint the pulley housing but not the pulley itself, otherwise it will jam and the sashes will not run easily.
7 Finish off by painting any inside linings, frames, and shutters. Tackle one component at a time if possible, and

On this window the inside colour stops at the staff bead and the runners are both white, which is the outside colour of the windows.

As the internal frame and the sashes are different colours the upper half of the inner runner has been painted the same light colour as the frame. The lower half of the outside runner is the same as the outside colour of the window (white).

complete it before moving on to the next. Any framed elements such a shutters or lining should be covered in the following order; mouldings and panels (at the same time), muntins, top rail, middle rail, bottom rail, and stiles. Remember to 'lay off' (the final brush strokes) in the same direction as the grain of the wood.

Procedure for external painting

1. Reverse the sashes as described above and paint all visible surfaces except the top surface of the top rail of the inside sash. The pulley stiles can be painted at this time also.
2. Return the sashes to an almost closed position and complete the painting of the inner sash and the pulley stiles, together with the cill.

When finished, all exposed wood should be covered; any unprotected wood can be affected by moisture, providing entry points for rot and allowing unwanted expansion with consequent sticking of the sashes. At the same time, the coating of paint should not be so thick as to cause its own problems, jamming the sashes in their runners, or preventing them from moving at all.

Paint colours

The question of where colours should finish often arises, particularly when dealing with the runners (the inner faces of the staff bead and parting bead that the sash stile runs between). There are no hard and fast rules, but the problem becomes acute when sharply contrasting colours are in use. A certain amount of judgement is called for, but generally it is simplest if the runners are the same colour as the sashes, although it is sometimes the case that inner and outer runners are different colours. To complicate matters further, the upper and lower halves of the inner runners can be different *(see photographs)*. Don't forget that outside painting in a new colour may need listed building approval if it significantly alters the character and appearance of the building.

DRAUGHT-PROOFING AND WEATHER-STRIPPING

Take any combination of the words draught/weather and proofing/stripping/sealing – they all mean exactly the same thing: anything done to reduce air leakage. More heat is lost in this way than can be saved by trying to double-glaze your existing windows (which in many cases will be impossible anyway). The gaps around an average sash window can be the equivalent of an aperture measuring ten or more square inches. Therefore it is important to reduce leakage as far as possible. Doing so will increase your comfort and reduce your heating bills. The cost of draught-proofing will generally be recouped within 5–25 years. It will take 60 to 100 years for new plastic double glazed windows to start saving you money, and added to this will be the cost of replacing the sealed units if they fail (remember that most guarantees are only for ten years). You might consider secondary glazing (an internal light metal sash window), which can be very effective. Never underestimate the usefulness of thick curtains in reducing draughts.

The less successful methods of draught-proofing include a flexible metal strip nailed to the runners, and surface mounted rubber seals. These tend to be either unsightly, ineffectual, or both. There are however a growing number of specialist joinery firms who will take your sashes out, carry out any necessary repairs, and insert much more effective and long-lasting brushes or other seals around all the edges of the sashes. There are clear advantages to going down this route:

- Your repairs will be carried out professionally, which can include replacement of broken glass and missing putty.
- Air leakage will be reduced. This means less hot air going out and less cold coming in thus reducing uncomfortable draughts.
- Your windows will not rattle any more.
- There will be a noticeable reduction in noise from outside. Provided you are not in danger of destroying irreplaceable historic glass *(see page 126)*, you can have new thicker glass fitted at the same time, increasing the sound insulation of the window even further. (You will need to increase the size of the counterbalance weights accordingly.)
- The seals are nearly or totally invisible, depending on the system employed.
- This method is approved by English Heritage and local authority conservation officers.
- By using this system you will not have desecrated your home with plastic windows.
- Finally and perhaps most importantly **YOU WILL SAVE MONEY.**

There are a few points of detail to consider *(see illustrations)*:

Seals While brush seals (fine, nylon fibres) are generally used, there are other systems which employ seals made from urethane foam encapsulated in a polypropylene sheath; all types are made in varying sizes to fit different gaps.

Location of seals If the parting bead is used to carry the vertical seals, they will be visible and care must be taken not to paint over them, otherwise they will become clogged and lose their effectiveness.

Parting bead with seal These come in two forms, plastic or timber, the latter being the more expensive, especially if the original existing bead is to be re-used in this way. The disadvantage of plastic is that it usually comes in white only, and if painted, the paint will not stand up well to friction from the sash moving against it, leading to white streaks showing through, particularly if a dark colour has been used.

Gas fires and ventilation In order to function safely, gas-heating appliances need a supply of air. Larger appliances could malfunction if your draught-proofing is too effective – check with the manufacturer – requiring the introduction of a new vent.

Building regulations and 'trickle vents' With the intention of reducing the probability of condensation and consequent mould growth, the Building Regulations require a habitable room to be provided with a trickle vent ('background ventilation' of 800mm^2 in area). Interestingly, this is equivalent to the average gap around the average sash window. Fortunately the same regulation allows you to achieve this by opening the top sash of your fully weather-stripped window!

DAVID WRIGHTSON MCD BArch RIBA IHBC DipCons(IoAAS) is a partner in Acanthus Lawrence and Wrightson, Architects and Landscape Architects, Chiswick, London. For contact details see page 24.

METAL WINDOWS

▶ CRITTAL WINDOWS LIMITED

Springwood Drive, Braintree, Essex CM7 2YN
Tel 01376 324106 Fax 01376 349662

STEEL WINDOWS AND DOORS: Design, manufacture and installation of replica refurbishment steel windows and doors for all types of buildings. The modern steel windows include weather-stripping, corrosion protection, polyester powder finish and can incorporate double glazed units as well as leaded lights. *See also: display entry on page 134.*

▶ FABCO

Unit 23, Middleton Business Park, Yapton Road, Middleton-on-Sea, W Sussex PO22 6HS
Tel/Fax 01243 584289

METAL WINDOW CASEMENTS RESTORATION: Fabco offers a full conservation and restoration service for metal window casements etc – also the manufacture of casements and hoppers in ferrous and non-ferrous metals. *See also: display entry in Window Protection section, page 125.*

When contacting companies listed here, please let them know that you found them through *The Building Conservation Directory*

WINDOWS & DOORS

▶ C R CRANE & SON LTD

Manor Farm, Main Road, Nether Broughton, Leics LE14 3HB
Tel 01664 823366 Fax 01664 823534
Website www.crcrane.co.uk

SPECIALIST BUILDING AND JOINERY CONTRACTORS: *See also: profile entry in Building Contractors section, page 60.*

▶ CHILSTONE

Victoria Park, Fordcombe Road, Langton Green, Tunbridge Wells, Kent TN3 0RE
Tel 01892 740866 Fax 01892 740249

WINDOWS AND DOORS: Standard and special window surrounds, sills, heads and key stones. Also porticos and door surrounds. *See also: profile entry in Cast Stone section, page 116.*

▶ HADDONSTONE LIMITED

The Forge House, East Haddon, Northampton NN6 8DB
Tel 01604 770711 Fax 01604 770027
Website www.haddonstone.co.uk

STANDARD AND CUSTOM-MADE WINDOW SURROUNDS, CILLS, HEADS AND KEYSTONES. ALSO PORTICOS AND DOOR SURROUNDS: *See also: display entry in Cast Stone section, page 117.*

▶ HILL LEIGH LTD

Brue Way, Walrow Industrial Estate, Highbridge, Somerset TA9 4AW
Tel 01278 789156 Fax 01278 781768

TIMBER WINDOWS AND DOORS: Manufacturers of high performance windows, doors and conservatories. *See also: display entries in this section, page 136, and Conservatories section, page 156.*

▶ MARVIN ARCHITECTURAL

Gibbs House, Kennel Ride, Ascot, Berkshire SL5 7NT
Tel 01344 885995 Fax 01344 885455
E-mail uksales@marvin-architectural.com
Website www.marvin-architectural.com

SLIDING SASH WINDOWS: Marvin has been engineering and manufacturing made-to-order sliding sash windows for over 60 years. Marvin has over 11,000 windows and doors in its range and extensive custom design and size capabilities. Traditional craftsmanship and in-house technical support ensure solid wood windows of superior quality and performance. *See also: display entry on page 137.*

▶ REFURB-A-SASH

Queens Joinery, Queens Road, Twickenham, Middlesex TW1 4LJ
Tel 020 8892 1166 Fax 020 8892 0055

TRADITIONAL SASH WINDOW SPECIALISTS: Award winning company with services including refurbishment of existing windows (listed building specialists), full or part replacement in single or double-glazing and re-instatement of originals. French windows and casements also available. All products are hand made faithfully replicating original features using traditional joinery methods. All timber is preservative treated, hand prepared and spray paint finished. Corporate members of the Guild of Master Craftsmen and The Glass and Glazing Federation complying with their code of ethical practice.

▶ SASH WINDOW SPECIALISTS LIMITED

Unit 8 The Tannery Ind Est, The Midlands, Holt, Trowbridge BA14 6BB
Tel 0800 0838811/01225 783040 Fax 01225 783668

SASH WINDOWS AND DRAUGHTPROOFING: Your window problems are Sash Window Specialists' challenge. A company offering traditional craftsmanship coupled with today's technology, uniquely providing their invisible sash window draughtproofing system which stops draughts and rattles and also gives the traditional sliding sash an easy glide action. Built into the existing sashes – not nailed on – its invisibility makes it ideal and acceptable for listed buildings and sensitive conservation work. Sympathetic repairs and refurbishment including replica replacements using Georgian and Victorian moulding patterns in either soft or hardwood. Also available, bespoke timber double-glazed sashes with traditional mouldings, weights, cords and pulleys. *See also: display entry in this section, page 139.*

▶ STEEL WINDOW SERVICE AND SUPPLIES LTD

30 Oxford Road, Finsbury Park, London N4 3EY
Tel 020 7272 2294/6391 Fax 020 7281 2309
E-mail post@steelwindows.co.uk
Website www.steelwindows.co.uk

SUPPLY AND FIX, SERVICING AND REPAIR OF STEEL WINDOWS: *See also: display entry in Metal Windows section, page 134.*

▶ T & M GLASS

49 Leigh Road, Wimborne, Dorset BH21 1AE
Tel/Fax 01202 849619

WINDOWS AND DOORS: Misty, cracked or fogged up windows and doors. T & M specialise in double glazed sealed units into wooden and metal frames. Small repairs to complete refurbishments undertaken to any property. For a free, friendly and professional no obligation quote, telephone Trevor Pickering to discuss your requirements.

▶ THORVERTON STONE COMPANY
Seychelles Farm, Upton Pyne, Exeter, Devon EX5 5HY
Tel 01392 851822 Fax 01392 851833
E-mail caststone@thorvertonstone.co.uk
Website www.thorvertonstone.co.uk
RECONSTRUCTED STONE: *See also: display entry in Cast Stone section, page 116.*

▶ TIMBER TECH LTD
Le Catillon de Haut, Le Catillon, Grouville, Jersey JE3 9UR
Tel 01534 852837 Fax 01534 852847
TRADITIONAL TIMBER WINDOW AND EXTERIOR DOOR MANUFACTURERS: Michael O'Connor established Timber Tech in 1996 to produce high quality timber windows and doors for historic buildings. Products replicate the fine and unique detail of Jersey's and Guernsey's historic environment. Products are manufactured using the latest timber innovations and technology with a choice of double or single glazed insulated units. All products are made to the highest standards of build quality and design to enhance and sustain the appearance of character buildings. Competitive pricing guarantees good value for money. The team welcomes business opportunities from the UK and abroad and works closely with planning authorities ensuring that local requirements are met.

▶ WESTONCRAFT
George Streamer Building, White Hart Road, Slough, Berkshire SL1 2SF
Tel 01753 570980 Fax 01753 530198
BESPOKE JOINERY: Westoncraft is a company with a small, mature staff, specialising in windows, doors and conservatories. Their approach to their work, both on the shop floor and in the design of their products is to take personal pride in producing their very best. They have been accused of running a hobby rather than a business, and it is because of this that they feel that those involved in careful, sympathetic work, restoration or otherwise, would benefit from involving them in their projects. They are practised at combining traditional looks with contemporary insulation and security requirements.

STRUCTURE & FABRIC
Metal, Glass & Wood

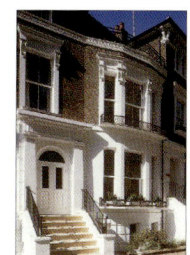

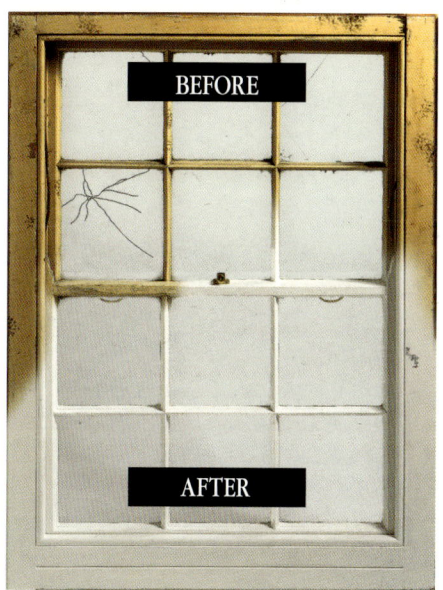

▶ ANELAYS
William Anelay Limited, Murton Way, Osbaldwick, York YO1 5UW
Tel 01904 412624 Fax 01904 413535
E-mail office@williamanelay.co.uk Website www.williamanelay.co.uk
BUILDING AND RESTORATION CONTRACTORS: *See also: profile entry in Building Contractors section, page 58.*

▶ ARTHUR E WOODWARD
1 Hall Mews, Leigh Hall Road, Leigh-on-Sea, Essex SS9 1QZ
Tel/Fax 01702 476345
SPECIALISTS IN GOOD QUALITY CARPENTRY AND BUILDING WORK: *See also: profile entry in Building Contractors section, page 58.*

▶ C R CRANE & SON LTD
Manor Farm, Main Road, Nether Broughton, Leics LE14 3HB
Tel 01664 823366 Fax 01664 823534 Website www.crcrane.co.uk
SPECIALIST BUILDING AND JOINERY CONTRACTORS: *See also: profile entry in Building Contractors section, page 60.*

▶ CARREK LTD
Mason's Yard, Wells Cathedral, Wells, Somerset BA5 2PA
Tel 01749 689000 Fax 01749 689089 Website www.carrek.co.uk
HISTORIC BUILDING REPAIR COMPANY: *See also: profile entry in Building Contractors section, page 60.*

▶ DEACON & SANDYS
Apple Pie Farm, Cranbrook Road, Benenden, Kent TN17 4EU
Tel 01580 243331 Fax 01580 243301
E-mail info@deaconandsandys.co.uk Website www.deaconandsandys.co.uk
INTERIOR WOODWORK: Specialists in 16th and 17th century interior woodwork in English oak as well as individual commissions, Deacon & Sandys make wall panelling with a hand-carved frieze, staircases with faceted newels and fretted splats, libraries of bookcases, exterior and interior doors and random width boarded floors. Antiquing and colouring by hand gives an authentic finish that is unrivalled. Important projects for the National Trust, the Honourable Society of Lincoln's Inn, English Heritage as well as Sanyo and ArtWood in Japan have been undertaken. Deacon & Sandys are also well known for new English oak furniture including hand-carved four-poster beds, refectory tables, chairs and dressers.

▶ F G GUEST JOINERY CONSULTANT
The Old Chapel, Park Street, Teddington, Middlesex TW11 0LT
Tel 020 8977 9907/020 8255 1004 Fax 020 8255 1004 Mobile 0467 826303
E-mail fguest@globalnet.co.uk
JOINERY CONSULTANT: Francis Guest and his team of mature joiners and cabinetmakers have been skilfully making traditional woodwork in London and the Home Counties for over 30 years. Most of their work comes through recommendation. Many listed buildings, churches and fine houses have examples of their work, and they have recently completed a major project at Ightham in Kent which involved dismantling some box pews dated 1640 and reassembling in another location in the church; other work included tracery work to a rood screen extension. In 1988 they were awarded a Design and Conservation Award in Richmond upon Thames.

▶ HEARNS (SPECIALISED) JOINERY LTD
101 Waldegrave Road, Teddington TW11 8LW
Tel 020 8977 1644/020 8977 0032 Fax 020 8977 1655 Website www.hearnsjoinery.co.uk
JOINERS, CABINET MAKERS AND WOODWORKERS

▶ KIRBY PEEL
9 Raper View, Aberford, Leeds LS25 3AF Tel 01132 813452
ARCHITECTURAL JOINERY AND TIMBER FRAME CONSERVATION: *See also: profile entry in Timber Frame Builders section, page 73.*

▶ L DANIELS & G ELDRIDGE
Windy Heights, High Cross, Froxfield, Petersfield, Hants GU32 1EH
Tel 01730 827472 Workshop 01730 828440
ARCHITECTURAL JOINERS AND CARPENTERS: L Daniels & G Eldridge is a family run business of three working partners, who have been in the wood working trade all their working lives. They make architectural joinery and curved work which they can fit if required, including purpose-made kitchens, stairs, doors, windows and mouldings. Their kitchens are constructed with morticed and tenoned front frames of solid timber, grooved into an 18mm melamine-faced carcase made of water-resistant board, with glued joints and doors hung on traditional brass butts. Stocks of air-dried English hardwoods are held and repair work is also undertaken.

FINE JOINERY

▶ MELCOMBE REGIS CONSTRUCTION

The Old Forge, Wyke Square, Weymouth DT4 9XP
Tel 01303 773239
E-mail info@buildersdorset.co.uk Website www.buildersdorset.co.uk

ARCHITECTURAL CARPENTRY AND FULL SERVICE BUILDING COMPANY: *See also: display entry in Building Contractors section, page 61.*

▶ MOTT GRAVES PROJECTS LIMITED

Sampleoak Lane, Chilworth, Guildford, Surrey GU4 8QW
Tel 01483 453326 Fax 01483 453325
E-mail james@mottgraves.fsbusiness.co.uk
Website www.mottgraves.co.uk

SPECIALIST JOINERY, CABINET WORK AND RESTORATION: This company provides a comprehensive woodworking service for private and professional clients, such as architects, project developers, building contractors and museums. The directors personally organise and manage each commission, carefully combining proven restoration methods with a sensitive approach to new work. They refurbished the oak panelled interior within County Hall, Westminster and have most recently restored the oak conservation area within Ingress Abbey, Ingress park, Dartford. Also, MGP regularly designs and manufactures high quality bespoke furniture and cabinets for fine interiors. Please contact James Mott for an information pack. *See also: display entries above, in Stone section, page 110 and in Building Contractors section, page 63.*

▶ N E J STEVENSON

Church Lawford Business Centre, Limestone Hall Lane, Church Lawford, Coventry CV23 9HD
Tel 024 765 44662 Fax 024 765 45345

DESIGNERS AND MAKERS OF DISTINCTIVE COMMISSIONED FURNITURE: *See also: profile entry in Cabinet Makers section, page 205.*

▶ ST BLAISE LTD

Westhill Barn, Evershot, Dorchester, Dorset DT2 0LD
Tel 01935 83662 Fax 01935 83017 E-mail stblaise@compuserve.com

NEW, REPAIR AND CONSERVATION JOINERY IN SOFTWOOD AND HOMEGROWN HARDWOODS: *See also: entry in Building Contractors section, page 65.*

FINE JOINERY

▶ TANKERDALE LTD

Johnson's Barns, Waterworks Road, Sheet, Petersfield, Hampshire GU32 2BY
Tel 01730 233792 Fax 01730 233922

JOINERY CONSERVATION AND RESTORATION: Tankerdale Ltd specialises in the conservation and restoration of historic joinery and furniture. Established in 1977 and covering all the British Isles, the company works frequently for the National Trust and are their official advisors on furniture and historic woodwork conservation. Tankerdale Ltd also works with English Heritage, museums, architects, contractors and private clients. Recent contracts include reinstating panelled rooms at Winfield House, Regent's Park and the conservation of fire-damaged doors from Abbeyleix House, Ireland. Contact Hugh Routh or John Hartley.

▶ TERNEX LTD

The Sawmill, 27 Ayot Green, Welwyn, Herts AL6 9BA
Tel 01707 324606 Fax 01707 334371

BESPOKE JOINERY IN ENGLISH AND IMPORTED HARDWOODS AND SOFTWOODS: *See also: profile entry in Timber Suppliers section, page 143.*

▶ TIPTREE JOINERY SERVICES

New Road, Tiptree, Colchester, Essex CO5 0HQ
Tel 01621 819220 Fax 01621 815499 E-mail office@tjevers.co.uk

SPECIALIST AND BESPOKE JOINERY MANUFACTURERS, RESTORATION AND CONSERVATION: Specialists in the manufacture of high quality traditional and architectural joinery. They offer a complete service from site survey, design to manufacture and installation. They have many years experience as joinery manufacturers. The high calibre of their craft skills and traditional techniques coupled with quality management skills, they believe enables them to offer a service that is second to none. They offer the practical experience and expertise required in providing a personal and professional service. For further information please contact Ian Garrod.

▶ WALDEN JOINERY

Empstead Works, Henley-on-Thames, Oxfordshire RG9 1UF
Tel 01491 572555 Fax 01491 410849 Website www.bwt.org.uk/walden

FINE JOINERY: Walden Joinery provides a high quality joinery service, combining traditional skills with modern techniques and plant. All joinery is purpose-made to clients' exact requirements, much is manufactured from detailed drawings, often to replace existing period pieces such as mouldings, windows, doors and panelling. Their ability to recreate period styles identically is of paramount importance when working in local conservation areas. Walden's recognises the importance of environmental considerations, and takes care to source its timber only from reputable suppliers, whose environmental policies are vetted.

▶ WALLIS JOINERY

Broadmead Works, Hart Street, Maidstone, Kent ME16 8RE
Tel 01622 690960 Fax 01622 693553

SPECIALIST JOINERY: Wallis Joinery has since it's formation in 1860 become one of the United Kingdom's leading specialist joinery companies offering an unparalleled standard of craftsmanship and expertise. Their high standard of joinery is exemplified by being twice awarded the prestigious Carpenters Award and receiving accreditation as a Quality Assured Company to BS 5750 Part 2. They are able to provide all components, from standard softwood joinery to complex restoration projects using hardwood, including purpose designed fixtures and fittings in veneers and laminates. While methods and materials constantly change in timber technology the constant factor with Wallis Joinery is the skills of its craftsmen to produce quality joinery. *See also: Wallis display entry in Building Contractors section, page 55.*

▶ WILDWOOD FINE FURNITURE AND JOINERY

Mint Cottage, Gilthwaiterigg Lane, Kendal, Cumbria LA9 6NT
Tel/Fax 01931 712804 (workshop) 01539 724707 (office)
E-mail wildwood@appleonline.net

PERIOD RESTORATION AND RENOVATION: Wildwood blends its clients' vision with the buildings' needs, reproducing, renovating, and installing all aspects of purpose made joinery. Wildwood can produce any period mouldings, including curved work. They are specialists in Georgian, Victorian and Regency periods, and are also commissioned to work on Tudor to present day. Special care is taken to match paintwork and glazing techniques. Working within its own environmental policy and Local Agenda 21 guidelines, Wildwood sources its timbers locally and from FSC managed woodlands.

TIMBER SUPPLIERS

▶ ALTHAM HARDWOOD CENTRE LTD

Altham Corn Mill, Burnley Road, Altham, Accrington, Lancs BB5 5UP
Tel 01282 771618 Fax 01282 777932 Website www.oak-beams.co.uk

HAND CUT OAK BEAMS: *See also: profile entry in Timber Frame Builders section, page 72.*

▶ ARTISAN OAK BUILDINGS LTD

80 London Road, Teynham, Kent ME9 9QH
Tel 01795 522121 Fax 01795 520744

SPECIALIST OAK SUPPLIERS: Stock holders of approximately 250 tonnes of new air-dried oak and 400 tonnes of reclaimed oak. Sections of timber include massive 12 metre long beams, joists, rafters and beautiful wide antique oak floorboards. Reclaimed pitch pine, elm and beech floorboards are also often available. Their flooring division also uses old timbers with original patina to produce decorative Versailles flooring panels. Reclaimed oak can be supplied 'as is' or cut, cleaned and treated to customers' requirements. Visit their two acre site, workshops and demonstration barn which houses a resource centre displaying many samples of interesting reclaimed building materials. Stockists of Rinaldi and Liberon products which can be seen and demonstrated.

▶ CARPENTER OAK & WOODLAND CO LTD

Hall Farm, Thickwood Lane, Colerne, Chippenham, Wiltshire SN14 8BE
Tel 01225 743089 01225 744100

SPECIALIST TIMBER SUPPLIES: Carpenter Oak & Woodland is a one-stop shop for your special timber needs: oak shingles, for historic church spires, traditionally cleft only from English oak: riven batten and oak tile pegs for historic roofs and riven laths for traditional plasterwork. Timber frames can be surveyed, repaired, and timber supplied: sawn, planed, or hewn timber for crucks, braces and all frame components, including hand-made oak framing pegs. Specialist fixings also available: rosehead nails and stainless steel products. Carpenter Oak & Woodland can be relied on for the finest quality materials, a rapid response and continuity of supply. *See also: display entry in Timber Frame Builders section, page 72.*

▶ ENGLISH WOODLANDS TIMBER LIMITED

Cocking Sawmills, Cocking, Near Midhurst, West Sussex GU29 0HS
Tel 01730 816941 Fax 01730 816874
E-mail sales@ewtimber.co.uk Website www.ewtimber.co.uk

TIMBER SUPPLIERS: English Woodlands Timber Limited supplies oak beams and air or kiln dried English hardwoods. Established in 1986, the company has an extensive range of hardwood stocks with all English species represented from ¾″ to 4″ thickness. A cut to size and planing service is offered alongside normal through and through plank sales. Oak beams are produced to order, supplied in all lengths and dimensions, including bends. The company recently supplied the oak laths used in construction of the innovative gridshell roof for the national centre at the Weald and Downland Open Air Museum in West Sussex.

▶ JAMESON JOINERY

Hook Farm, West Chiltington Lane, Billingshurst, West Sussex RH14 9DP
Tel 01403 782868 Fax 01403 786766

TIMBER AND JOINERY FOR OLD BUILDINGS: From their well equipped sawmill and joinery works in Sussex, Jameson Joinery produces traditionally jointed oak framed buildings, together with oak and elm flooring, oak beams, doors, windows, furniture, staircases, plasterer's laths and cabinetry. All joinery work is made to order for bespoke projects – beams up to 24″ wide can be tenoned and jointed in the traditional way and hand adze-finished. Traditional window and door fittings and nails are hand-forged on-site as are traditional square headed nuts. Deliveries nation-wide. *See also: display entry in Timber Frame Builders section, page 73.*

TIMBER SUPPLIERS

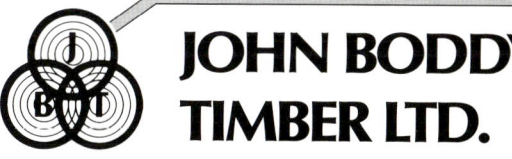

▶ MORGAN & CO (STROOD) LTD

Knight Road, Strood, Rochester, Kent ME2 2BA
Tel 01634 290909 Fax 01634 290800 E-mail info@morgantimber.co.uk

RESTORATION TIMBER: Morgan Timber, established in 1923, has one of the largest stocks of oak logs in the country. a versatile sawmill and a wealth of experience in cutting timber for restoration. Lengths up to 45 feet and shaped members can be cut from selected logs and structural oak can be strength graded to BS 5756. They also produce high quality kiln dried English oak and ash and have extensive machining facilities, including the production of solid hardwood flooring. Major projects supplied include the refurbishment of Windsor Castle, The Royal Courts of Justice, Rochester Castle, Ightham Mote, The Archbishop's Palace at Maidstone and the Parliamentary Buildings.

▶ ORIGINAL ARCHITECTURAL ANTIQUES CO LTD

22 Elliott Road, Love Lane Industrial Estate, Cirencester, Glos GL7 1YS
Tel 01285 653532 Fax 01285 644383
E-mail sales@originaluk.com Website www.originaluk.com

RECLAIMED TIMBER AND TIMBER FEATURES: *See also: profile entry in Architectural Salvage section, page 160.*

▶ ORIGINAL OAK

Ashlands, Burwash, East Sussex TN19 7HS
Tel 01435 882228

BUILDING AND RESTORATION OAK TIMBER/FLOORING SUPPLIERS: *See also: profile entry in Timber Flooring section, page 208.*

▶ TERNEX LTD

The Sawmill, 27 Ayot Green, Welwyn, Herts AL6 9BA
Tel 01707 324606 Fax 01707 334371

SAWMILL AND JOINERY COMPANY: Ternex is a long established sawmill and joinery company which produces English hardwoods, mainly oak, for beams and construction, flooring, furniture and joinery. Oak, strength-graded to BS 5756:1997 is available. Ternex also produces bespoke joinery, machined timber, flooring and mouldings in English and imported hardwoods and softwoods.

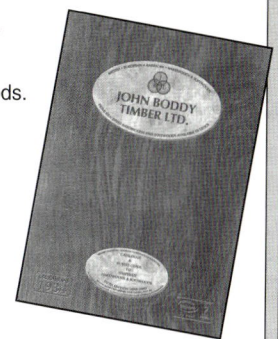

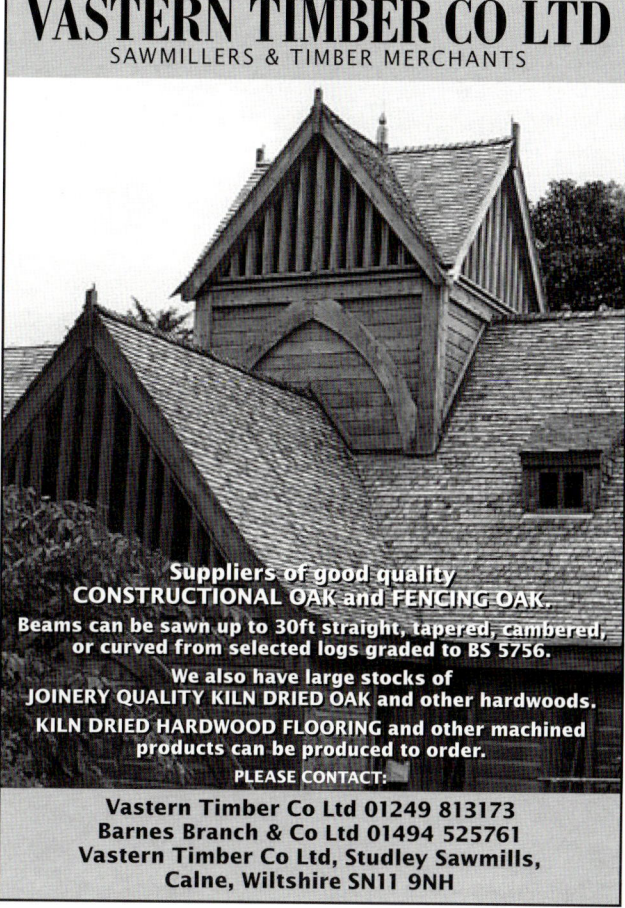

TIMBER SUPPLIERS

▶ TRADEWOOD FLOORING

22 Foxglove Close, Hesketh Bank, Preston, Lancs PR4 6TG
Tel 01772 811811 Fax 01772 812601
IMPORTERS AND DISTRIBUTORS OF QUALITY HARDWOOD FLOORING: *See also: display entry in Timber Flooring section, page 208.*

▶ WEALD AND DOWNLAND OPEN AIR MUSEUM

Singleton, Chichester, West Sussex PO18 0EU
Tel 01243 811363 Fax 01243 811475
E-mail wealddown@mistral.co.uk Website www.wealddown.co.uk
CONSERVATION SUPPLIES AND SERVICES: The Museum offers various supplies for use in timber frame building and the care of ancient buildings. *See also: profile entry in Courses & Training section, page 221.*

VENEERS

▶ CAPITAL VENEER CO LTD

12 Bow Industrial Park, Carpenters Road, Stratford, London E15 2DZ
Tel 020 8525 0300 Fax 020 8525 0070
SUPPLIERS OF OVER 100 SPECIES OF WOOD VENEERS: The Capital Veneer Co Ltd holds vast stocks of rare and precious varieties including many burrs and decorative veneers from all over the world. Alongside these, native and other temperate species are stocked, many in extra thicknesses for antique or constructional purposes. Whether your requirements are for an architectural project or a single piece of furniture, Capital Veneers welcomes all enquiries. Stock list sent on request.

TIMBER SUITABILITY

	EXTERNAL USE	INTERNAL USE	WORKABILITY	MOVEMENT	DENSITY	SUSTAINABILITY	GENERAL COMMENTS
SOFTWOODS							
Cedar *Cedrus spp*	A	A	A	B	B		
Cedar, Western Red *Thuja plicata*	A#	A	B	A	C		
Douglas Fir *Pseudotsuga menziesii*	A#	A	B	A	B		
Hemlock, Western *Tsuga heterophylla*	B	A	A	A	B		
Larch *Larix decidua*	B	A	B	A	B		
Parana Pine *Araucaria angustifolia*	X	A	A	B	B	2	
Pitch Pine *Pinus spp*	A/B	A	B	A	B		
Redwood, European *Pinus silvestris*	B	A	B	B	B		
Whitewood, European *Picea abies & Abies spp.*	B	A	A	B	C		
Yellow Pine *Pinus strobus*	X	A	A	A	C		
Yew *Taxus baccata*	A	A	B	A	B		
HARDWOODS							
Abura *Mytragyna ciliata*	X	A	B	A	B		
Afrormosia *Pericopsis ciliata*	A	A	B	A	A	1	
Afzelia *Afzelia spp*	A	A	C	A	A	2	
Agba *Gossweilerodendron balsamiferum*	A	A	B	A	B		
Ash *Fraxinus excelsior*	X	A	B	B	A		
Beech *Fagus sylvatica*	X	A	A/B	C	A		Prone to furniture beetle infestation
Birch *Betula alleghaniensis*	X	A	B	C	A		
Cherry (American and European) *Prunus serotina, Prunus Avium*	X	A	B	B	B		
Chestnut (sweet) *Castanea sativa*	A	A	B	A	B		
Elm *Ulmus spp*	X	A	B	B	B		
Idigbo *Terminalia ivorensis*	A#	A	B	A	B		
Iroko *Chlorophora excelca*	A#	A	B/C	A	A		
Keroing *Dipterocarpus spp*	A	X	B	C	A		
Lauan, Meranti & Seraya *Shorea spp*	B	A	A/B	A	A/B		
Lime *Tilia spp*	X	A	A	A	BC		
Mahogany (African) *Khaya spp*	A	A	B	A	B		
Mahogany (S America) *Swietenia macrophylla*	A	A	B	A	B	1*	
Maple (Rock) *Acer saccharum*	X	A	B	B	A		
Oak (American Red & Japanese) *Quercus spp*	X	A	B	B	A		
Oak (European & American white) *Quercus spp*	A#	A	B	B	A		
Obeche *Triplochiton scleroxylon*	X	A	A	A	C		
Opepe *Nauclea diderrichii*	A	X	C	A	A	2	
Plane *Platanus acerifoia*	X	A	A	B/C	B		See also Sycamore
Ramin *Gonystylus spp*	X	A	A	C	A		
Sapele *Entandrophragma*	A	A	B	B	B		
Sycamore *Acer pseudoplatanus*	X	A	A	B	B		Known in Scotland as 'plane'
Teak *Tectona grandis*	A	A	C	A	A		
Utile *Entandrophragma utile*	A	A	B	B	A		
Walnut (African) *Lavoa trichilioides*	A	A	B	A	B		
Walnut (American) *Juglans nigra*	A	A	B	A/B	A		
Walnut (European) *Juglans regia*	X	A	B	B	A		

KEY

	EXTERIOR USE	GENERALLY	DENSITY (kg/m^3)
A	Good	Good	640-800
B	Suitable with preservative	Acceptable	480-640
C	Not suitable	Use with caution	320-480
X	Not suitable	Not suitable	
#	*Avoid contact with ferrous metals in damp conditions*		

SUSTAINABILITY

1 "Endangered Species" – CITES (Convention on International Trade in Endangered species).

2 "Tropical species threatened in two or more countries in which they occur" – WCM (World Conservation Monitoring Centre) 1991.

* Three species of South American mahogany are listed under CITES – *Swietenia humulis* and *Swietenia mahogoni* under Appendix I & II (international) and *Swietenia macrophylla* under Appendix III (country specific; Brazil, Bolivia, Costa Rica and Mexico).

For further information on WCMC and CITES lists see the WCMC website: www.wcmc.org.uk.

▶ ALBION MANUFACTURING
The Granary, Silfield Road, Wymondham, Norfolk NR18 9AU
Tel 01953 605983 Fax 01953 606764
ARCHITECTURAL METALWORK AND WIREWORK SPECIALISTS: Albion Manufacturing has a wealth of experience in both new and restoration work from window guards to bronze doors. They have a comprehensive range of castings both old and new and in 2000 received a reward of distinction for gates and railings manufactured and installed for Messrs Willis Caroon in Ipswich.

▶ ARCHITECTURAL METALWORK CONSERVATION
Unit 19, Hoddesdon Industrial Centre, Pindar Road, Hoddesdon, Herts EN11 0DD
Tel /Fax 01992 443132
BLACKSMITHS AND FERROUS METALWORK CONSERVATORS: Understanding and valuing both the traditional methods of working iron and modern conservation practice, Architectural Metalwork Conservation are practitioners and consultants. Projects range from preventive maintenance and minimum intervention repairs through to carefully researched reproductions. The workshop specialises in the consolidation and indent repair of items such as cast iron windows and fine wrought ironwork. Advisories, reports, research and lectures are carried out for a number of registered charities and government funded bodies. Contact Keith Blackney for further information.

▶ BARR & GROSVENOR LTD
Jenner Street, Wolverhampton, West Midlands WV2 2AE
Tel 01902 352390 Fax 01902 871342
ARCHITECTURAL CAST IRON: Barr & Grosvenor Ltd is one of the few remaining foundries with the skills and patience to produce complex items as used in Victorian times. The company is involved at all stages in restoration projects requiring castings, whether still existing or long since removed. Their services can help stabilise, restore or recreate castings for all purposes. Most of their work is made to order so that essential attention to detail is guaranteed. Specification advice, pattern making, casting, machining and paint finishing are all available in house.

ARCHITECTURAL METALWORK

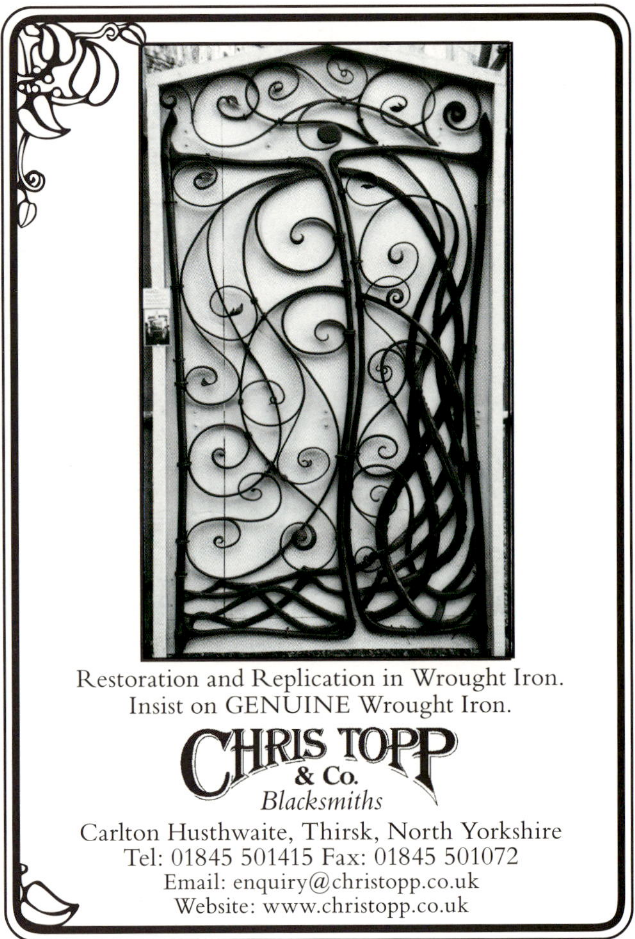

▶ C J BLACKSMITHS
The Smithy, Preston, Brockhurst, Shropshire SY4 5QA
Tel 01939 220357
E-mail sales@cjblacksmiths.co.uk
Website www.cjblacksmiths.co.uk
TRADITIONAL BLACKSMITHS

▶ CAMBRIAN CASTINGS (WALES) LTD
Unit 19, Crofty Industrial Estate, Penclawdd, Swansea SA4 3RS
Tel /Fax 01792 850912 Mobile 07966 157139
E-mail jon@cambriancastings.co.uk
Website www.cambriancastings.co.uk
ARCHITECTURAL METALWORK: Specialising in iron, bronze and aluminium casting, Cambrian Castings undertakes all types of architectural metalwork creation, renovation and repair. A full design and fabrication service for stairs, railings, grilles and screens; both decorative and structural, can be provided. The company specialises in bronze and cast iron grillework and complex cast iron components for park gates and interior/exterior stairs. Visit Cambrian Castings' website for examples, or contact Jon Barnett for more information.

▶ CARREK LTD
Mason's Yard, Wells Cathedral, Wells, Somerset BA5 2PA
Tel 01749 689000 Fax 01749 689089
Website www.carrek.co.uk
HISTORIC BUILDING REPAIR COMPANY: *See also: profile entry in Building Contractors section, page 60.*

▶ CASTAWAY CAST PRODUCTS & WOODWARE
Brocklesby Station, Brocklesby Road, Ulceby, North Lincolnshire DN39 6ST
Tel/Fax 01469 588995
CAST METAL PRODUCTS INCLUDING BESPOKE ITEMS: Supplying nation-wide, Castaway undertakes any cast metalwork project using aluminium, bronze and gun metal, grey and nodular (SG) iron, carbon steels and stainless steels. Castings for historic buildings include gutter sections and other drainage ware, airbricks, wall-retaining plates, brackets, gates and railings, signage, window frames and much more. Items can be made from drawings, photographs or from sight of originals – broken, corroded or intact. Items can be made from standard existing patterns or to your own designs in any quantities; from one upwards. In-house pattern making facilities are available if required. Contact John Wade to discuss your requirements in detail.

▶ CASTING REPAIRS LIMITED
Marine House, 18 Hipper Street South, Chesterfield, Derbyshire S40 1SS
Tel 01246 277656 Fax 01246 206519
RESTORATION OF IRONWORK: Casting Repairs Limited has over 40 years experience in the repair of decorative and structural ironwork. This covers work on buildings of major historical importance, large municipal buildings, public fountains, bandstands, statues, windows, gates and railings. The company can provide complete project management, where required, together with a combination of repair techniques including cold metal stitching, welding and brazing. With full fabrication facilities, new architectural ironwork can be designed and produced to specific requirements. Operating nationally, Casting Repairs brings an expert and cost effective approach to every project.

ARCHITECTURAL METALWORK

▶ CHRIS TOPP & COMPANY WROUGHT IRONWORKS

Lyndhurst, Carlton Husthwaite, Thirsk YO7 2BJ
Tel 01845 501415 Fax 01845 501072
E-mail enquiry@christopp.co.uk Website www.christopp.co.uk

BLACKSMITHS: Established in 1980, Chris Topp & Co has been one the pioneers in the return to the traditional art of the blacksmith. The company carries out a wide range of work throughout the UK in genuine wrought iron, cast iron and mild steel; including bespoke ironwork, restoration, renovation and repousee. A consultancy and design service is also available. Chris Topp & Co prides itself on using the highest craft standards and where appropriate the traditional material of the blacksmith, wrought iron. And as far as they know, through its sister company The Real Wrought Iron Company, Chris Topp & Co is the world's sole supplier of wrought iron.

▶ CROWNCAST LTD

The Foundry, Rushenden Road, Queenborough, Kent ME11 5HD
Tel 01795 662722 Fax 01795 666552

MANUFACTURERS OF IRON CASTINGS AND NON-FERROUS CASTINGS AND PATTERNS: Crowncast is a craft foundry which offers a bespoke service for special conservation and refurbishment projects. Crowncast's versatility means that they are often called on to cast replacement pieces using old castings as patterns. Projects have included lamp posts for Osborne House, gate posts for Kensington Palace Gardens and gates and fencing for Westminster and Greenwich. Recently they were heavily involved in the refurbishment of Brighton Railway Station.

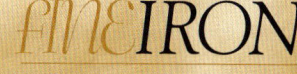

ARCHITECTURAL METALWORK

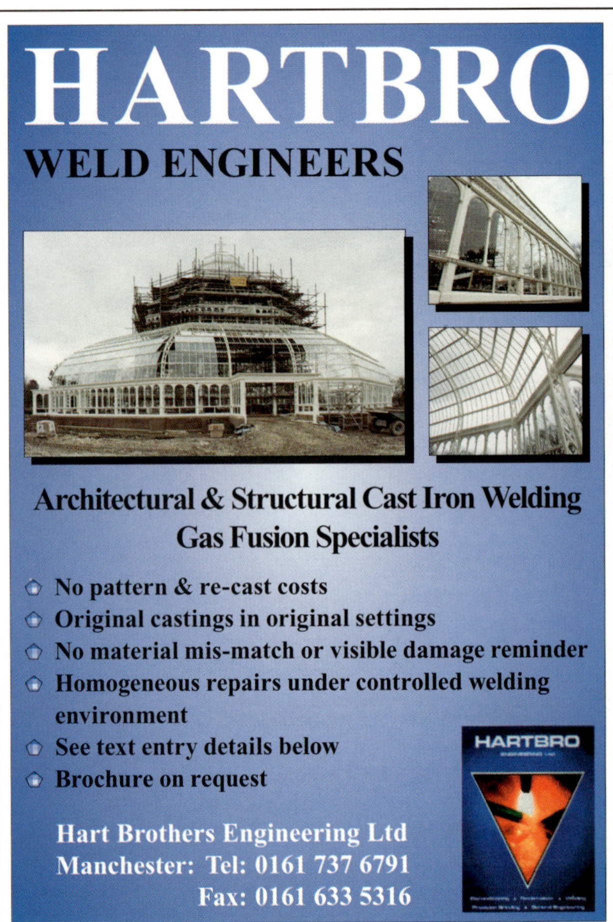

▶ DOROTHEA RESTORATIONS LTD

INCORPORATING ERNEST HOLE (ENGINEERS) LTD
Northern office and works: New Road, Whaley Bridge, via Stockport, Cheshire SK23 7JG
Tel 01663 733544 Fax 01663 734521
▸ Southern office and works: Riverside Business Park, St Anne's Road, St Anne's Park, Bristol BS4 4ED
Tel 0117 971 5337 Fax 0117 977 1677

RESTORATION OF ARCHITECTURAL METALWORK, TRADITIONAL MACHINERY AND MILLS: Celebrating 25 years as nation-wide leaders in traditional metalwork and engineering using authentic materials and methods, the company specialises in the repair and reinstatement of large scale wrought and cast iron structures. Free assistance offered to organisations preparing HLF submissions for municipal park restorations, public building and conservation area schemes. Recent projects include the 18th century main gates, Chirk Castle; Cookham Bridge, Maidenhead: Oswestry Park bandstand; Palace of Westminster gates; cast iron railings, British Museum; bronze leafwork, Albert Memorial; forged and leaded railings, Pavilion Gardens, Buxton. In-house services include condition surveys, historic records interpretation, repair of iron castings, engineering solutions to historic metalwork problems. *See also: display entry in this section, page 147.*

▶ EURA CONSERVATION LTD

Unit H3, Halesfield 19, Telford, Shropshire TF7 4QT
Tel 01952 680218 Fax 01952 585044 E-mail mail@eura.co.uk

CONSULTANTS AND CONSERVATORS FOR SCULPTURE, ARCHITECTURAL DECORATION AND GARDEN ORNAMENT: *See also: profile entry in Statuary section, page 115.*

▶ FABCO

Unit 23, Middleton Business Park, Yapton Road, Middleton-on-Sea, W Sussex PO22 6HS
Tel/Fax 01243 584289

WROUGHT AND CAST IRON SERVICE: Fabco offers a full service in the conservation and restoration of wrought and cast iron – also the manufacture of new decorative wrought iron. *See also: display entry in Window Protection section, page 125.*

▶ GEORGE JAMES & SONS

22 Cransley Hill, Broughton, Kettering, Northants NN14 1NB
Tel/Fax 01536 790295

BLACKSMITHS: George James & Sons were established in 1845 and are the fifth generation of a family blacksmithing in Northamptonshire. The firm has a reputation for producing specialist ironwork of a high quality. Commissions range from small hand forged items to large-scale architectural works. Services include the restoration and replication of existing work and new forged work in traditional or contemporary styles. Recent projects include restoration of mid 19th century ironwork at Harlaxton Manor, Lincolnshire and St Chad's Cathedral, Birmingham. Please ring Tim James to discuss your requirements.

▶ HART BROTHERS ENGINEERING LTD

Albion Works, Cobden Street, Pendleton, Salford M6 6LY
▸ Soho Works, Soho Street, Oldham OL4 2AD
Tel 0161 737 6791 Fax 0161 633 5316

GREY CAST IRON WELDING: Specialising in high quality casting repairs, Hartbro successfully undertakes remedial works to castings that require a high degree of skill to repair. The result being a reliable casting returned as close as possible to its original specification. Full cross-sectional welding of material damage ensures structural integrity throughout. Castings repaired from broken, worn, or parts-missing condition, of any weight, size or quantity. Hartbro casting repairs may be machined, filed, drilled or tapped to return your component to its original condition. Parts may be finished to a condition where their repair is virtually impossible to detect or identify that the item has been damaged. *See also: display entry above.*

► HODGSONS FORGE DECORATIVE METALWORK & RESTORATION

2 Wesley Road, Terrington St Clement, Kings Lynn, Norfolk PE34 4NG
Tel/Fax 01553 828637

HAND FORGED WROUGHT IRON SPECIALISTS: Master blacksmiths for five generations, now a partnership of Colin and Ian Hodgson, this family-run business is proud of the quality of their work and the service they provide. Their range of products covers all estate, domestic and industrial applications. All site work is undertaken including the dismantling and installation of ironwork. They can design or reproduce the most complex features including fine tracery work and wrought iron reproduction using traditional blacksmithing methods. Their most recent commissions include refurbishing ironwork at Knowle House, Sevenoaks, Kent and Waddesdon Manor, Bucks.

► J H PORTER & SON LTD

13 Cranleigh Mews, Cabul Road, London SW11 2QL
Tel 020 7978 5576 Fax 020 7924 7081

BLACKSMITHS: After 60 years in Kensington J H Porter & Son have moved their forge to larger premises. They manufacture, restore, reproduce and repair gates, railings, balustrades, balconies and staircases, together with security grilles, antiques, indoor/outdoor furniture etc. They work to their own or their clients' traditional or modern designs. Commissions include work for antique restorers, architects, artists, builders, churches, designers, English Heritage, estate managers, London Transport, the National Trust, photographers, shop-fitters and the general public. A work force of five blacksmiths produces all manner of ironwork to a very high standard.

ARCHITECTURAL METALWORK

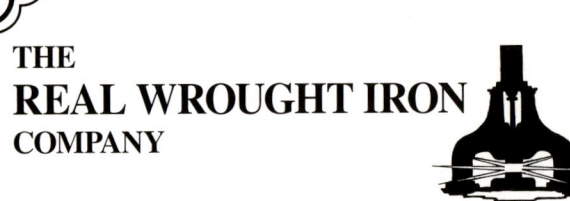

► JAMES HOYLE &SON

The Beehive Foundry, 48-50 Andrews Road, Cambridge Heath, London E8 4RL
Tel 020 7254 2335 Fax 020 7254 8811

ARCHITECTURAL METAL CASTING: The Hoyle family business has gained its experience in architectural casting work over five generations; one of the early projects using their castings was the building of the first underground railway in London. More recently they made and installed the ornate parapet lamp posts along the Thames Embankment to match the originals and other projects have included making the third arch fascia of Westminster Bridge and the castings for the refurbishment and new gates for the old Smithfield Market, balustrading within the Houses of Parliament and iron castings at the Natural History Museum. Alan Hoyle would be pleased to advise on and, or, make any casting from drawings, sketches, samples or just ideas.

► L C JAY LIMITED

19/21 Oak Street, Norwich, Norfolk NR3 3AU
Tel 01603 628798 Fax 01603 611910
E-mail enquiries@lcjay.co.uk
Website www.lcjay.co.uk

ENGINEERS AND METAL FOUNDERS: L C Jay Limited has over 50 years experience of casting in all ferrous and non-ferrous metals. The company uses standard drawings, 3D CAD files, photographs or the original item to manufacture castings ranging from columns, railings, gates, windows and staircases to decorative artefacts, plaques and unique metal sculptures. L C Jay's skilled pattern makers can produce new patterns for bespoke projects or use existing patterns from stock. A full graphic and structural design service is available in addition to on site refurbishment and installation works if required. Recent projects include the renovation of the Albert Memorial, The Queens gate in Kensington Palace Gardens and the Coalbrookdale gates in Hyde Park. *See also: display entry in this section, page 149.*

ARCHITECTURAL METALWORK

▶ LEANDER ARCHITECTURAL

Fletcher Foundry, Halstead Close, Dove Holes, Buxton, Derbyshire SK17 8BP
Tel 01298 814941 Fax 01298 814970
BESPOKE AND ALL TYPES OF STANDARD ARCHITECTURAL CASTINGS: *See also: display entry in Signage section, page 157.*

▶ MARSH BROS ENGINEERING SERVICES LTD

PO Box 3, Bakewell, Derbyshire DE45 1LT
Tel 01629 636532 Fax 01629 636003
E-mail info@marshbrothers.co.uk Website www.marshbrothers.co.uk
STRUCTURAL AND DECORATIVE CAST IRON: Marsh Bros provide a specialist service to restore decorative and structural cast iron components. The work undertaken includes staircases, railings, gates and crestings and cast iron roof structures for lantern and atrium roofs. Also, restoration and repair of cast iron bridge parapets, facades and arch castings. Marsh Brothers provides a bespoke supply and design service to restore and manufacture identical replacement parts where loss or damage has occurred. The company covers all areas of the UK and principal customers include county councils, Crown Estates and major civil contractors.

▶ MATHER & SMITH LTD

Hilton Road, Cobbswood Industrial Estate, Ashford, Kent TN23 1EW
Tel 01233 622214 Fax 01233 661565
E-mail sales@mjallen.co.uk Website www.mjallen.co.uk
IRON FOUNDER ARCHITECTURAL: Mather & Smith is probably the oldest established architectural iron founding and metalwork specialist in the UK, dating back to 1675, giving an unequalled depth of experience. Conservation and recent restoration works includes restoration of West Norwood Cemetery cast iron railings, and perimeter cast iron posts and railings at the Tower of London. Mather & Smith can offer a complete conservation and restoration service of all cast iron structures including railings, gates, stair balustrades, and balcony railings. They also offer a range of Victorian and modern cast iron wall railings and gates as well as a standard range of standard street furniture. A bespoke design and installation service is also offered. *See also: display entry on page 149.*

▶ PLOWDEN & SMITH

190 St Ann's Hill, London SW18 2RT
Tel 020 8874 4005 Fax 020 8874 7248
E-mail info@plowden-smith.com
Website www.plowden-smith.com
CONSERVATION AND RESTORATION: *See also: display entry in Fine Art Conservation section, page 202.*

▶ SHEPLEY ENGINEERS LIMITED

Westlakes Science Park, Ingwell Hall, Moor Row, Cumbria CA24 3JZ
Tel 01946 599022 Fax 01946 591933
E-mail engineers@shepley.vhe.co.uk
ARCHITECTURAL METALWORK RESTORERS AND RENOVATORS: Shepley Engineers have been responsible for engineering faithful and innovative solutions for many major architectural restoration projects. The company offers a comprehensive range of services acting as consultant, designer, contractor or as principal contractor depending on the size and specific requirements of the scheme. Major projects completed or in progress include: the Dorchester Hotel, London; Smithfield Market, London; the Curvilinear Range at the Botanical Gardens, Dublin; and the Palm House at Sefton Park, Liverpool.

▶ STRUCTURAL PERSPECTIVES

48 Holdsworth Road, Holmfield, Halifax HX2 9SZ
Tel/Fax 01422 240789
CAST AND WROUGHT IRON MATERIALS ASSESSMENTS: *See also: display entry and profile entry in Archaeologists section, page 16.*

ARCHITECTURAL METALWORK

GARDEN & STREET FURNITURE & LIGHTING

▶ BROXAP DOROTHEA

Rowhurst Industrial Estate, Chesterton, Newcastle under Lyme, Staffs ST5 6BD
Tel 01782 564411 Fax 01782 565357
E-mail sales@broxap.com
Website www.broxap.com

DESIGNERS, CONSULTANTS AND CONTRACTORS FOR NEW BUILD, REPRODUCTION AND RESTORATION WORK: *See also: display entry in Architectural Metalwork section, page 145.*

▶ THE CAST IRON COMPANY

8 Old Lodge Place, Twickenham TW1 1RQ
Tel 020 8744 9992 Fax 020 8744 1121
E-mail castiron@btconnect.com
Website www.castiron.co.uk

GARDEN AND STREET FEATURES: The Cast Iron Company design and produce traditional and modern street furniture. They also specialise in the reproduction and restoration of period metalwork. Their product range includes: signage; bollards and finger posts; lamps and lighting; litter bins and seats; posts and rails; tree grilles and guards; ornate gates and railings; bandstands and covered walkways. Recent contracts completed: the relighting of Battersea Bridge, cast railings for the Hon Artillery Company and pedestrian guard rail for the Royal Borough of Kensington and Chelsea. Catalogues detailing the full range are available from their sales office.

▶ CHILSTONE

Victoria Park, Fordcombe Road, Langton Green, Tunbridge Wells, Kent TN3 0RE
Tel 01892 740866 Fax 01892 740249

GARDEN AND STREET FEATURES: Reconstructed stone benches, bollards, etc. *See also: profile entry in Cast Stone section, page 116.*

▶ CRANBORNE STONE LIMITED

Cranborne Stone, West Orchard, Shaftesbury, Dorset SP7 0LJ
Tel 01258 472685 Fax 01258 471251
Website www.cranbornestone.com

ARCHITECTURAL AND GARDEN ORNAMENT: *See also: display entry in Cast Stone section, page 116.*

▶ HADDONSTONE LIMITED

The Forge House, East Haddon, Northampton NN6 8DB
Tel 01604 770711 Fax 01604 770027
Website www.haddonstone.co.uk

STANDARD RANGE INCLUDES URNS, TROUGHS, STATUARY, SUNDIALS, FOUNTAINS, POOL SURROUNDS, SEATS, COPINGS, TEMPLES, PAVILIONS AND BALUSTRADING: Custom-made architectural stonework a speciality. *See also: display entry in Cast Stone section, page 117.*

▶ SARABIAN LIMITED

Sarabian House, Kington St Michael, Chippenham, Wiltshire SN14 6JB
Tel/Fax 01249 750113
E-mail sarabian@fsbdial.co.uk

BUILT HERITAGE CLEANING SPECIALIST: *See also: profile entry in Masonry Cleaning section, page 174.*

GATES & RAILINGS

▶ EURA CONSERVATION LTD

Unit H3, Halesfield 19, Telford, Shropshire TF7 4QT
Tel 01952 680218 Fax 01952 585044
E-mail mail@eura.co.uk

CONSULTANTS AND CONSERVATORS FOR SCULPTURE, ARCHITECTURAL DECORATION AND GARDEN ORNAMENT: *See also: profile entry in Statuary section, page 115.*

▶ VICTORIAN CLASSIC STYLE

12 Brunswick Square, Herne Bay, Kent CT6 5QF
Tel/Fax 01227 364632

CAST IRON GATES AND RAILINGS: Victorian Classic Style is a small company dedicated to recreating cast iron gates and railings from the Victorian era for use in domestic situations. All their patterns are based on original English designs and have successfully been installed in local government conservation schemes involving English Heritage and the National Lottery. Delivery can be arranged nation-wide. They also have pattern work facilities available.

Derelict warehouses at the corner of Lower Duke Street and Henry Street, Liverpool (Roy Main)

THE TOWNSCAPE HERITAGE INITIATIVE

Renewing the heart of our historic towns and cities

JUDY CLIGMAN

The *Townscape Heritage Initiative* (THI) is the Heritage Lottery Fund's grant giving programme for the repair and regeneration of the historic environment in towns and cities throughout the UK. The scheme was born of HLF's desire to deliver sustainable conservation in historic urban areas by raising the standard of repair where the market has failed to do so, and by bringing new uses and new life back into areas which have lost their traditional economic base.

The THI was launched in 1998 with a £60 million budget for three annual bidding rounds and has now been extended for at least the rest of the life of HLF's current Strategic Plan (2002). Some £62 million has already been allocated to more than 80 schemes in the first three rounds. With highest priority given to applications from areas of social and economic deprivation, all demonstrate the key role historic buildings have to play in urban renaissance.

PARTNERSHIP

The THI offers funding for comprehensive and targeted programmes of repair, re-use and enhancement of historic urban areas. The main aim of the scheme is to make possible the continued viable use of the buildings which make up the special architectural character of historic urban areas, giving highest priority to the repair of historic buildings and to bringing derelict and under-used historic buildings back into use.

Grants are made to local partnerships which manage a common fund from which smaller grants are offered to property owners. An important feature of the THI is that it addresses the conservation needs of particular buildings within an overall strategy, rather than as self-contained restoration projects. The THI may only be one element of a much wider economic regeneration scheme. The partnerships are generally led by the local authority, which has the power to deliver the strategy, but they may also include community groups, building preservation trusts, development agencies, civic societies and other such organisations. These partnerships are responsible for putting together an integrated strategy for the social and economic regeneration of the area together with a pool of funding from a variety of sources to fund the strategy.

Applications are assessed in two stages, with 'stage one' applications submitted in May of each year and 'in principle' indications of support from HLF made by September. 'Stage two' submissions include the development of an action plan for the life of the scheme which sets out the proposed programme of work.

Under the THI individual building owners may apply for a grant from the pool of funding and the local authority and other major partners may carry out direct works to buildings in their ownership or to the public realm. We can contribute between £250,000 and £2 million towards each scheme.

The average size of grant is £745,265. In the first two rounds priority was given to schemes in Northern Ireland, Wales and Scotland which demonstrated problems of a severity rarely seen in England, where funding had been much more widely available for conservation areas, albeit on a modest scale, since the Town Schemes of the 1970's. Coastal and market towns featured highly in last year's bidding round.

THE APPLICATION

All the THI schemes should:

- involve a range of works to a number of buildings, structures or spaces within a defined area
- involve the local community
- benefit the wider community as well as those directly concerned with grant aided properties.

TYPES OF WORK FUNDED UNDER THE THI

The THI aims to contribute to the sustainability of local economies and to support the communities that live and work in each area. Projects might include bringing vacant floor space in historic buildings back into use or making an historic area more attractive as a location for businesses or as a tourist destination. We might support the repair and authentic reinstatements of elements lost from the 'public realm' or support the authentic reinstatement of architectural features to historic buildings and their settings.

The THI cannot support schemes which might fall under other HLF grant programmes such as active places of worship or urban parks and town squares; nor can it support schemes in which more than 25 per cent of the common fund is proposed for public realm works such as street paving and lighting, or which include routine maintenance that is the responsibility of the owner.

Two ongoing projects illustrate the kinds of scheme likely to win Heritage Lottery Fund support under the THI scheme, one concerning an industrial area of Liverpool and the other concerning two rural towns in Northern Ireland.

LIVERPOOL ROPE WALKS

In Liverpool the Duke Street/Bold Street Partnership had been in existence for two years before being awarded £1.5 million in the THI's first bidding round. They started work immediately on a four year programme. The partnership board consists of representatives from the public sector (mainly local authority), the private sector, the voluntary sector and four distinct local communities. The THI site is adjacent to a residential area and the project aims to create both new residential and commercial space, the second of these with the intention of creating extra employment for the local communities.

The Duke Street/Bold Street area was the site of many industries during the 18th century due to the proximity of the docks which formed the basis for the swift increase in population and wealth in the 18th and 19th centuries. Rope making in particular was important and many streets, Bold Street among them, owe their straightness and length to having originally been 'rope walks'. Bomb damage and the decline of the maritime industry both contributed to the area's decline.

The THI scheme is focusing on an important area of 18th and 19th century warehousing, surviving largely unaltered in a conservation area in the heart of Liverpool. Many of the buildings are listed and show characteristics unique to Liverpool such as combined merchants' residences and warehousing. Some original boundary walls still exist along with some railings and street signs. Public support for the scheme is high and economic and social deprivation is much in evidence.

▲ ***West Wemyss, Scotland**, one of the first THI schemes. Here, restored houses are being sold to first time buyers on low income and local people on local authority or housing association waiting lists. (Ewan Miles)*

◀ ***Creswell Model Village, Derbyshire** Although heritage provides the focus for THI schemes – in this case the conservation of a unique Victorian housing estate – environmental regeneration of an area invariably goes hand in hand with its social and economic regeneration (Zak Waters)*

Fine 18th and 19th century architecture in the High Street, Draperstown, Northern Ireland. *The town suffers from through-traffic and the decline of the rural economy. (James Morgan)*

The scheme includes repairing the structure and envelope of the targeted buildings and the repair and reinstatement of authentic historic surfaces. This will bring back into use vacant floor space to encourage the regeneration of the area as a commercial centre. The THI forms the basis of a wider programme within which roughly £20 million will be spent on the area over a four year period and which is intended to achieve the sympathetic re-development of gap sites and public realm works such as paving, street furniture and the planting of trees.

THE THI IN NORTHERN IRELAND

In Northern Ireland the THI has a strong community focus. This is largely because the planning service is centralised and is not the responsibility of the local authorities. In place of the local authority, the schemes are being led by local partnerships of community groups, regeneration bodies and building preservation trusts. The strength of this approach is the high level of community involvement and buy-in. The central planning service has signed up to supporting the THI and to implementing the planning controls needed to maintain the benefits of THI funding.

THE DRAPERS' TOWNS (MONEYMORE AND DRAPERSTOWN)

Supported with a grant of £1.2 million, this THI scheme is providing a comprehensive package to tackle the problems of disrepair and underuse in these towns, which originated as plantations of the Drapers' Company. The decline of the rural economy has left both towns with an architectural inheritance that they no longer have the resources to maintain, and they are also suffering from through traffic and a shift of trade. Behind the formal frontages lie charming simple buildings in very poor condition. Several courtyards of early, derelict outbuildings behind the main street frontages are also targeted.

The scheme involves a good mixture of major repairs to key buildings, projects to bring empty buildings back into use, reinstatement of architectural features and proposals to tackle gap sites in each town. Since the award was made, new funding has been coming forward from other sources - one local parish has been inspired to raise £50,000 through Sunday collections towards the repair of the corn store in Draperstown. The Drapers' Company has contributed almost £60,000 to the corn store and to the Manor House in Moneymore. Peace & Reconciliation funding of £175,000 has been confirmed and will go towards supporting various community projects.

SO FAR SO GOOD

The THI programme has been well received, particularly in Northern Ireland, Scotland and Wales (such as the repair of the run-down Lower Dock Street area of Newport in South Wales) where the needs are greatest. An independent evaluation of the programme by a team outside HLF will for the first time track a range of indicators of social and economic well-being in addition to the physical state of the urban fabric, to demonstrate the impact of conservation area funding over a ten year period.

THI schemes are seeing genuine community involvement and generation of local pride and interest in the fabric of declining historic urban areas, leading in turn to higher standards of conservation. The Townscape Heritage Initiative highlights the role historic buildings have to play in urban renaissance. We expect the benefits of these schemes to be wide reaching; town centre regeneration will reduce pressure on greenfield sites as well as attracting inward investment, tourism and local employment. We are supporting practical solutions, delivered by supporting local partnerships, which will bring people back into our towns and cities.

FURTHER INFORMATION

For further information on the *Townscape Heritage Initiative* please contact the HLF Information Team on 020 7591 6041/43/ 44/45.

Further information can also be found on the HLF website **www.hlf.org.uk**.

JUDY CLIGMAN is Deputy Director of Policy at the Heritage Lottery Fund, with particular responsibility for the Historic Buildings Environment.

CONSERVATORIES

WINDOWS · DOORS · **CONSERVATORIES** · STAIRS · **CRAFTED IN TIMBER**

Refinement, quality and a flexibility of design that ensures complete sympathy with any project of any period.

Hill Leigh, *established 1891*, creators of a range of traditionally-styled conservatories manufactured in Hardwood.

All these conservatories are bespoke, purpose-made products designed in conjunction with the client. They are suitable for commercial and residential purposes.

Please contact our technical sales department for further details.

Hill Leigh Ltd
Brue Way, Highbridge, Somerset TA9 4AW
Tel: 01278 789156 **Fax:** 01278 781768
email: enquiries@hill-leigh.co.uk
www.hill-leigh.co.uk

▶ HILL LEIGH LTD

Brue Way, Walrow Industrial Estate, Highbridge, Somerset TA9 4AW
Tel 01278 789156 Fax 01278 781768

TIMBER AND JOINERY SPECIALISTS: Manufacturers of high performance windows, doors, conservatories and also staircases. *See also: display entry above, and in Windows & Doors section, page 136.*

▶ MALBROOK CONSERVATORIES LTD

2 Crescent Stables, Upper Richmond Road, Putney, London SW15 2TN
Tel 020 8780 5522 Fax 020 8780 3344 Website www.malbrook.co.uk

BESPOKE CONSERVATORIES: Malbrook specialise in the design, production and installation of fine buildings in timber and glass. Particular attention is paid to the architectural details of the property, and the traditional joinery methods of construction are well suited to providing quality double-glazed structures for restoration and refurbishment projects. The experienced staff at Malbrook offer a full design service to assist clients and their architects in order to ensure the best possible solution. This attention to detail is maintained by close supervision throughout the installation process including all the finishing touches on site.

GARDEN BUILDINGS

▶ CHILSTONE

Victoria Park, Fordcombe Road, Langton Green, Tunbridge Wells, Kent TN3 0RE
Tel 01892 740866 Fax 01892 740249

GARDEN STRUCTURES: Reconstructed stone temples, follies, porticos. *See also: profile entry in Cast Stone section, page 116.*

CLOCKS

▶ THE CUMBRIA CLOCK COMPANY

Dacre, Penrith, Cumbria CA11 0HH
Tel/Fax 01768 486933

TOWER CLOCK RESTORATION: Their services include full restoration of mechanical movements, carillon and tune playing machines; the manufacture and installation of automatic-winding systems, and; the full restoration of dials. New clock systems and dials can be manufactured to the customer's requirements, complete with striking and chiming mechanisms. Their customer list includes: Salisbury and Hereford Cathedrals, the National Trust, Hampton Court Palace and many churches and local authorities throughout the country. Free quotations are given within the UK.

▶ SMITH OF DERBY

112 Alfreton Road, Derby DE21 4AU
Tel 01332 345569 Fax 01332 290642

CLOCKS AND ARCHITECTURAL FEATURES: Smith of Derby has since 1856 built, installed and maintained tower clocks on churches and public buildings. Regional centres provide clock maintenance, and the scheme provides year-round cover for site visits, faultfinding, replacement of hammer wires, weight lines and time motors. The company also has Architectural Metalwork and GRP fabrication divisions, which provide bespoke work for new developments and refurbishment schemes. Their work includes special projects for architects, interior designers and shopfitters. All work is done to commission, and examples include balustrades, hand beaten copper weathervanes, polished brass interior fittings and lighting units, and stainless steel shopfitting items. *See also: display entry above and in Architectural Metalwork section, page 152.*

SIGNAGE

STRUCTURE & FABRIC
Metal, Glass & Wood

GENERAL SUPPLIERS AND ARCHITECTURAL SALVAGE

Company	Type	Page
Ace Demolition – The Reclamation Centre	as	159
Ajeer Limited	as	159
Alpine Reclamation	as	159
Antique Buildings Limited	as	159
Architectural Heritage	as	189
Architectural Reclaim	as sp	159
Artisan Oak Buildings Ltd	as	142
Bursledon Brickworks Conservation Centre	sp	167
Cobar Services	as	160
Conservation Building Products Limited	as	160
Drummonds Architectural Antiques Limited	as	160
Ecomerchant Ltd	as	158
Glasgow Steel Nail Co Ltd	sp	179
H W Poulter & Son	as	189
Heritage Oak Buildings	as	76
Hirst Conservation Materials Ltd	sp	168
Jameson Joinery	sp	77
Jonathan Murray Fireplaces	as	160
Keyline East London	sp	115
LASSCO	as	160
MBL	sp	124
Mike Wye & Associates	sp	168
Minchinhampton Architectural Salvage Co	as	160
Nostalgia	as	189
Old House Store	sp	158
Original Architectural Antiques Company Ltd	as	160
Original Features	sp	205
Pew Corner Ltd	as	160
The Real Wrought Iron Company	as sp	150
Retrouvius Architectural Salvage	as	160
SALVO	as	159
Solopark Plc	as	161
South West Reclamation	as	162
Symonds Salvage	as	162
Walcot Reclamation	as	162
Weald & Downland Open Air Museum	sp	221
Westland	as	190

KEY
as architectural salvage
sp builders merchants and specialist suppliers

GENERAL BUILDING MATERIALS

GENERAL BUILDING MATERIALS

▶ CHALK DOWN LIME LTD

102 Fairlight Road, Hastings, East Sussex TN35 5EL
Tel 01424 443301 Fax 01580 830096 Mobile 0771 873 8708
E-mail chalkdownlime@supanet.com
MAINTENANCE AND REPAIR OF HISTORIC BUILDINGS: *See also: profile entry in the Mortars & Renders section, page 168.*

▶ G & N MARSHMAN

1 Nell Ball, Plaistow, Billingshurst, West Sussex RH14 0QB
Tel 01798 342427
MANUFACTURERS OF RIVEN OAK, CHESTNUT PLASTERERS' LATHS AND TILE AND STONE SLATE BATTENS: *See also: profile entry in Plasterwork section, page 214.*

▶ OLD HOUSE STORE

Hampstead Farm, Binfield Heath, Nr Henley-on-Thames, Oxfordshire RG9 4LG
Tel 0118 969 7711 Fax 0118 969 8822
E-mail info@oldhousestore.co.uk
Website www.oldhousestore.co.uk
SUPPLIERS OF TRADITIONAL BUILDING MATERIALS: *See also: display entry in this section, page 158.*

ARCHITECTURAL SALVAGE

ARCHITECTURAL SALVAGE

▶ ACE DEMOLITION – THE RECLAMATION CENTRE

Barrack Road, West Parley, Ferndown, Dorset BH22 8UB
Tel 01202 579222 Fax 01202 582043
ARCHITECTURAL SALVAGE AND TRADITIONAL BUILDING MATERIALS: Ace Demolition is a family run business that supplies quality reclaimed building materials and architectural artefacts at affordable prices with a prompt and personal service. Their comprehensive stock consists of mellow old bricks, slates, clay roof tiles and ridges, wood flooring, beams, joists, doors, windows, flagstones, railway sleepers, pavers, lamp posts, fireplaces and chimney pieces, period bathrooms and reconditioned taps, butlers' sinks, cast column radiators and many more architectural and period items. UK deliveries arranged. Open Monday – Friday 8.30 am – 5.30 pm and Saturday 9 am – 4 pm.

▶ AJEER LIMITED

Sugar Loaf Yard, Brightling Road, Woods Corner, Nr Heathfield, East Sussex TN21 9LJ
Tel 01424 838555 Fax 01424 838556
E-mail sales@ajeer.co.uk
Website www.ajeer.co.uk
RECLAIMED MATERIALS: Ajeer, possibly the largest reclaimed materials yard in Sussex is a leading source of building materials. The large site near Heathfield in East Sussex houses a vast range notably bricks, roof fittings and tiles, flagstones, reclaimed flooring, old oak and feature stones. Half a million bricks are stocked – new reclaimed, engineering, handmade, wirecut and Victorian pavers. Roofing materials include slates (new and old) replacement tiles (including concrete, machine and hand made peg tiles and pantiles) oak rafters and beams. Ajeer always has in stock an extensive range of reclaimed flooring from traditional pine and oak and parquet to the more modern beech. With over 15 years trading experience Ajeer has established a reputation particularly for its quality and range of flagstones from the colourful Indian flag to the more traditional York.

▶ ALPINE RECLAMATION

Old Goods Yard, Station Approach, Coulsdon North, Surrey CR3 2NR
Tel 020 8668 0123 Tel/Fax 01306 711711
RECLAIMED BUILDING AND LANDSCAPE MATERIALS: Established 1974, Alpine are specialist suppliers of a wide range of distinctive reclaimed weathered building and landscape materials including: handmade bricks - reds, dark and light multis, yellows, wirecuts, gaults, and pavers - granite cubes, random setts, York flagstones, knapped flints, oak beams and block flooring, weathered roof tiles and fittings and specials, Welsh slates, chimney pots, and odd architectural items. A brick and stone matching service is available. Alpine serves a wide delivery area with crane offload. Advice and personal attention provided; yard open by arrangement.

▶ ANTIQUE BUILDINGS LIMITED

Hunterswood Farm, Dunsfold, Surrey GU8 4NP
Tel 01483 200477 Fax 01483 200752
OAK BEAMS, BARN FRAMES, HANDMADE PEG TILES AND BRICKS: *See also: display entry in Reconstructed Buildings section, page 162.*

▶ ARCHITECTURAL RECLAIM

Theobalds Park Road, Enfield, Middx EN2 9BG
Tel 020 8367 1666 Fax 020 8367 6668
Website www.architecturalreclaim.com
ARCHITECTURAL SALVAGE: Architectural Reclaim buys and sells materials for the construction and furnishing of period properties. The company offers a brick matching service and stock tiles, sleepers, York stone, cobbles, Irish slate and paving slabs. Architectural features are also stocked, including stained glass, cast iron fireplaces, butler sinks, roll top baths, interior and exterior doors, block and strip flooring (mahoohoo, oak, pine). Contact Architectural Reclaim or visit their three-acre site, which is open seven days a week.

ARCHITECTURAL SALVAGE

▶ COBAR SERVICES

Howes Farm Cottage, Stowmarket Road, Ringshall, Stowmarket, Suffolk IP14 2JA
Tel 01473 658435 Fax 01473 658379

RECLAIMED BUILDING AND LANDSCAPING MATERIALS: Started in 1979, owned and run by a family with five generations of experience of brickmaking behind it, Cobar Services offers an extensive range of reclaimed materials with a distinct East Anglian flavour. Red and white bricks, oak beams, pan tiles, peg tiles, plain tiles, slates, pamments, and York stone flags are complemented by a carefully selected range of new materials with traditional character. All stock can be viewed at the yard in Ringshall village.

▶ CONSERVATION BUILDING PRODUCTS LIMITED

Forge Lane, Cradley Heath, Warley, West Midlands B64 5AL
Tel 01384 564219 Fax 01384 410625
E-mail simon@horsley.u-net.com
Website www.conservationbuildingproducts.co.uk

RECLAIMED BUILDING MATERIALS: Careful dismantling of old properties enables the company to reclaim bricks, tiles, slates, beams and quarry tiles used in period property restoration or the creation of new buildings with mellow-aged appearance. Interior staircases, floorboards, panelling, doors, stone and cast iron fireplaces, baths and basins are available renovated or 'as seen'. Exterior terrace flagstone, driveway cobble setts and paviors. Mill wheels, lamp posts, weathervanes, fountains, sundials, statues and urns for landscape design. Original stone carvings, repair work by 'in-house' stonemason. Phone or fax for advice, competitive quotes, samples, photographs or a brochure and map. Visitors welcome, crane offloading lorries make nation-wide deliveries. Inspirational stock – something for everyone.

▶ DRUMMONDS ARCHITECTURAL ANTIQUES LIMITED

The Kirkpatrick Buildings, 25 London Road (A3), Hindhead, Surrey GU26 6AB
Tel 01428 609444 Fax 01428 609445
Website www.drummonds-arch.co.uk

ARCHITECTURAL ANTIQUES, RESTORED BATHROOMS, RECLAIMED FLOORING, GARDEN DECORATION AND CONSERVATORIES: Specialist suppliers of fully restored antique bathrooms and fittings; reclaimed wood, tile and stone flooring; garden statuary, furniture, troughs and artefacts; wood, stone and iron fire surrounds; cast iron stoves and radiators; hardwood and pine doors; door and window furniture; gates and railings; antique furniture; lighting; windows; antique decorative items. Makers of the finest handmade cast iron baths and fittings; cast iron conservatories and brassware. Proper vitreous enamelling used on all baths.

▶ JONATHAN MURRAY FIREPLACES

358 Upper Richmond Road West, East Sheen, London SW14 7JT
Tel 020 8876 7934 Fax 020 8876 0869
E-mail jmf358@aol.com
Website www.fireplaces-uk.com

ANTIQUE AND REPRODUCTION FIREPLACES: *See also: profile entry in Fireplaces section, page 189.*

▶ THE LONDON ARCHITECTURAL SALVAGE AND SUPPLY CO LTD

Saint Michael's, Mark Street (off Paul Street), London EC2A 4ER
Tel 020 7749 9944 Fax 020 7749 9941
E-mail st.michaels@lassco.co.uk
Website www.lassco.co.uk

ARCHITECTUR AL SALVAGE: LASSCO is London's largest, longest established and best known specialist in architectural salvage. Divided between five companies, the LASSCO group sells an unrivalled range of decorative features and details; chimneypieces, panelled rooms and garden ornament; ecclesiastical furniture, stained glass, mirrors and all types of lighting; pub and museum interiors; doors, radiators, bathrooms, kitchens; reclaimed and new hardwood flooring, as well as a range of high quality replicas provided by the new LASSCO House & Garden. All are constantly sourced, often from prestigious London sites, restored and on display at Shoreditch, Islington and Bermondsey, six days a week 10am–5pm.

▶ MINCHINHAMPTON ARCHITECTURAL SALVAGE CO

New Catbrain, Cirencester Road, Nr Minchinhampton, Glos GL6 8PE
Tel 01285 760886 Fax 01285 760838
E-mail masco@catbrain.com
Website www.catbrain.com

ARCHITECTURAL ANTIQUES, GARDEN STATUARY, CHIMNEYPIECES AND TRADITIONAL FLOORING: Minchinhampton Architectural specialises in large architectural features and garden ornament. The company also carries extensive stocks of both hardwood and softwood flooring. Situated in the heart of the Cotswolds, the company offers a comprehensive sourcing service, friendly advice and a new garden design/consultancy department. Opening hours are Monday to Friday 9am–5pm and Saturday 9am–3pm. Telephone or e-mail for further details. Subscribers to the Salvo Code.

▶ ORIGINAL ARCHITECTURAL ANTIQUES CO LTD

22 Elliott Road, Love Lane Industrial Estate, Cirencester, Glos GL7 1YS
Tel 01285 653532 Fax 01285 644383
E-mail sales@originaluk.com
Website www.originaluk.com

ARCHITECTURAL SALVAGE: The Original is a leading source for quality architectural items in the Cotswolds with over 30 years experience and an established reputation for fairness and good value with both trade and private clients. The company specialises in fully restored antique oak doors, oak beams, period and reproduction fireplaces, stone troughs, well-heads, gates and fine garden ornament of all descriptions. Photographs by e-mail or post on request. Delivery arranged. Credit cards accepted. Monday to Saturday 9am-5pm, Sunday 10am-4pm. Subscribers to the Salvo Code.

▶ PEW CORNER LTD

Artington Manor Farm, Old Portsmouth Road, Guildford, Surrey GU3 1LP
Tel 01483 533337 Fax 01483 535554
E-mail pewcorner@pewcorner.co.uk
Website www.pewcorner.co.uk

RECLAIMED CHURCH INTERIORS AND HANDMADE FURNITURE: *See also: entry on Antique & Furniture Restorers map, page 207.*

▶ RETROUVIUS ARCHITECTURAL RECLAMATION & DESIGN

Office – 32 York House, Upper Montagu Street, London W1H 1FR
Tel/Fax 020 7724 3387
▶ Warehouse – 2A Ravensworth Road, London NW10 5NR
Tel 020 8960 6060 Mobile 0378 210855
E-mail mail@retrouvius.com
Website www.salvo.co.uk/dealers/retrouvius
Contact Adam Hills or Maria Speake

ARCHITECTRUAL ANTIQUES AND BUILDING COMPONENTS: Retrouvius seeks out architectural antiques and building components nation-wide. They work with architects, builders, and demolition companies to purchase unwanted fittings and fixtures directly from site. Retrouvius abides by the SALVO Code. Both partners are architecturally trained and have worked in the conservation world. The company offers a unique and personal design service helping with the integration of salvaged materials into interior and landscape projects. Their office is located in the heart of London's West End, with a warehouse 15 minutes away. Please phone for further information.

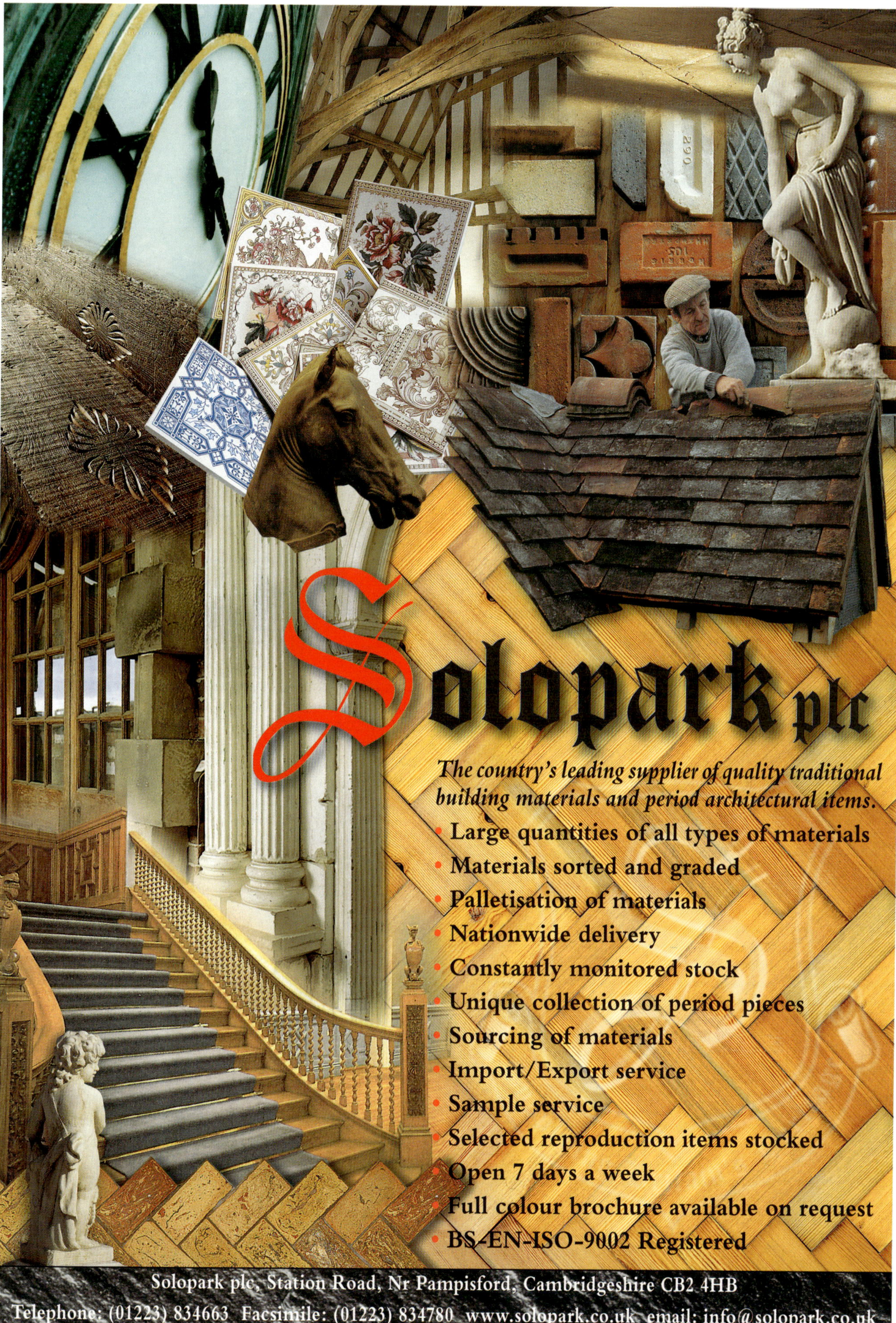
Solopark plc
The country's leading supplier of quality traditional building materials and period architectural items.
• Large quantities of all types of materials
• Materials sorted and graded
• Palletisation of materials
• Nationwide delivery
• Constantly monitored stock
• Unique collection of period pieces
• Sourcing of materials
• Import/Export service
• Sample service
• Selected reproduction items stocked
• Open 7 days a week
• Full colour brochure available on request
• BS-EN-ISO-9002 Registered
Solopark plc, Station Road, Nr Pampisford, Cambridgeshire CB2 4HB
Telephone: (01223) 834663 Facsimile: (01223) 834780 www.solopark.co.uk email: info@solopark.co.uk

ARCHITECTURAL SALVAGE

▶ SOUTH WEST RECLAMATION

Gwilliams Yard, Edington, Bridgwater, Somerset TA7 9JN
Tel 01278 723173 Fax 01278 722800
Website www.southwest-rec.co.uk

RECLAIMED BUILDING MATERIALS AND ARCHITECTURAL ANTIQUES: A friendly, well organised company offering a comprehensive range of quality traditional renovation materials. Bridgwater clay roofing tiles, Welsh slates, ridge, bricks and natural flagstones. A good range of character oak and elm beams, joists and floorboards is always available. Period fireplaces, doors, church pews, stone troughs, staddle stones and gates feature among the varied stock on offer, also an extensive range of slate and terracotta flooring. Subscribers to the Salvo Code.

▶ SYMONDS SALVAGE

Colts Yard, Pluckley Road, Bethersden, Kent TN26 3DD
Tel 01233 820724/ 01233 850677 Mobile 0585 306055

SUPPLIERS OF RECLAIMED BUILDING MATERIALS: Symonds Salvage supply bricks, tiles, slates, stone, large quantities of oak beams, and items of architectural interest including, sanitary ware etc. They also supply barns for re-erection suitable for garaging and granaries. Open Mon-Fri 8.30am to 5.30pm. Sat 8.30am to 1.00pm.

▶ WALCOT RECLAMATION

108 Walcot Street, Bath BA1 5BG
Tel 01225 444404 Fax 01225 448163
E-mail rick@walcot.com/jane@reproshop.com
Website www.walcot.com/www.reproshop.com

ARCHITECTURAL ANTIQUES AND TRADITIONAL BUILDING MATERIALS: One of the country's leading dealers in affordable architectural antiques and salvage, Walcot has two large city centre sites. (1) The Walcot Street yard specialises in: chimney pieces, fireplaces, panelling, doors, wrought iron, bathroom fittings, brassware, etc. Stock is recorded on database with photographic support. A shipping and export service is available as is on site parking. Contact details above. (2) The Riverside Depot supplies all types of reclaimed flooring, roofing, paving, oak beams and stonework. Patinated oak and elm flooring a speciality.

▶ WESTLAND LONDON

St Michael's Church, Leonard Street, London EC2A 4ER
Tel 020 7739 8094 Fax 020 7729 3620
E-mail westland@westland.co.uk
Website www.westland.co.uk

PERIOD AND PRESTIGIOUS CHIMNEYPIECES, ARCHITECTURAL ELEMENTS, PANELLING, FOUNTAINS, STATUARY, PAINTINGS AND FURNITURE: *See also: display entry in Fireplaces section, page 190.*

RECONSTRUCTED BUILDINGS

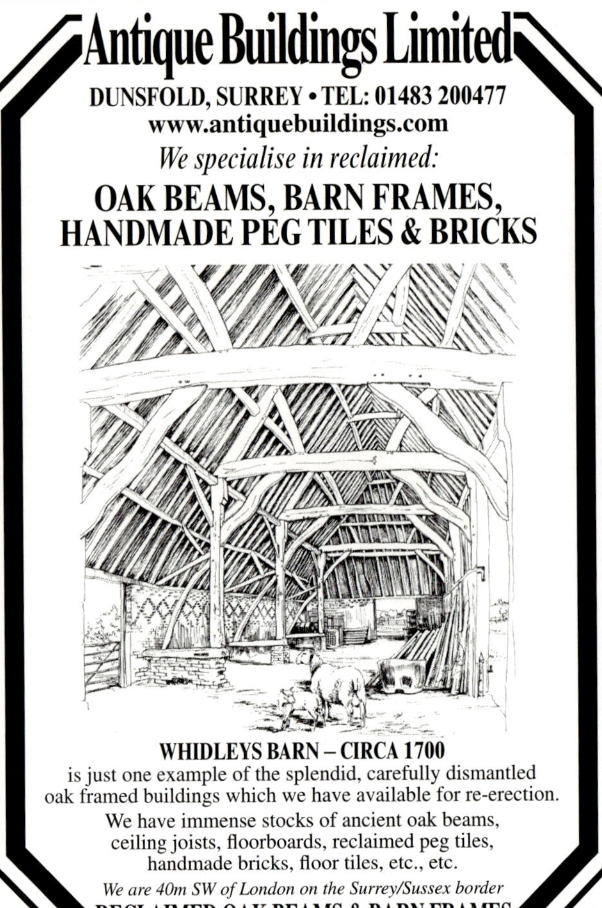

Antique Buildings Limited
DUNSFOLD, SURREY • TEL: 01483 200477
www.antiquebuildings.com
We specialise in reclaimed:
OAK BEAMS, BARN FRAMES,
HANDMADE PEG TILES & BRICKS
WHIDLEYS BARN – CIRCA 1700
is just one example of the splendid, carefully dismantled oak framed buildings which we have available for re-erection.
We have immense stocks of ancient oak beams, ceiling joists, floorboards, reclaimed peg tiles, handmade bricks, floor tiles, etc., etc.
We are 40m SW of London on the Surrey/Sussex border
RECLAIMED OAK BEAMS & BARN FRAMES
SUPPLIED FOR RE-ERECTION

▶ PERIOD PROJECTS LIMITED

Park Street, Stoke-by-Nayland, Suffolk CO6 4SE
Tel 01206 262322 Fax 01206 262588

OAK FRAMED BUILDINGS: Established 1980, Period Projects Limited has become well known in the South East and the Home Counties for its work in oak framed construction whether new-build or supply of quality ancient oak framed buildings for reconstruction. PPL also offers a range of part oak, part soft wood framed cart lodges and studios in a standardised design format. Recent projects include the supply and reconstruction of a 17th century oak framed thatched barn in a country park for a local authority, a new oak framed wing for a Grade II* farmhouse, and a complete design and build for an oak framed barn music room.

the SALVO CODE

Architectural Salvage Dealers subscribing to the Salvo Code agree to:

1. Not buy any item if there is the slightest suspicion that it may be stolen.
2. Not to knowingly buy any item removed from listed buildings without listed building consent or from sites of scheduled monuments without scheduled monument consent.
3. To record the registration numbers of vehicles belonging to persons unknown who offer items for sale.
4. Where possible to keep a record of the provenance of an item, including the date of manufacture, where it was removed from, and any previous owners.
5. To agree to their business details being held on a list of businesses which subscribe to the Code.
6. To display a copy of this Code in a public position within their premises.

Reprinted with kind permission of Salvo Magazine.

At present **102** dealers have agreed to abide by the Code. The list is available from Salvo, see page 159.

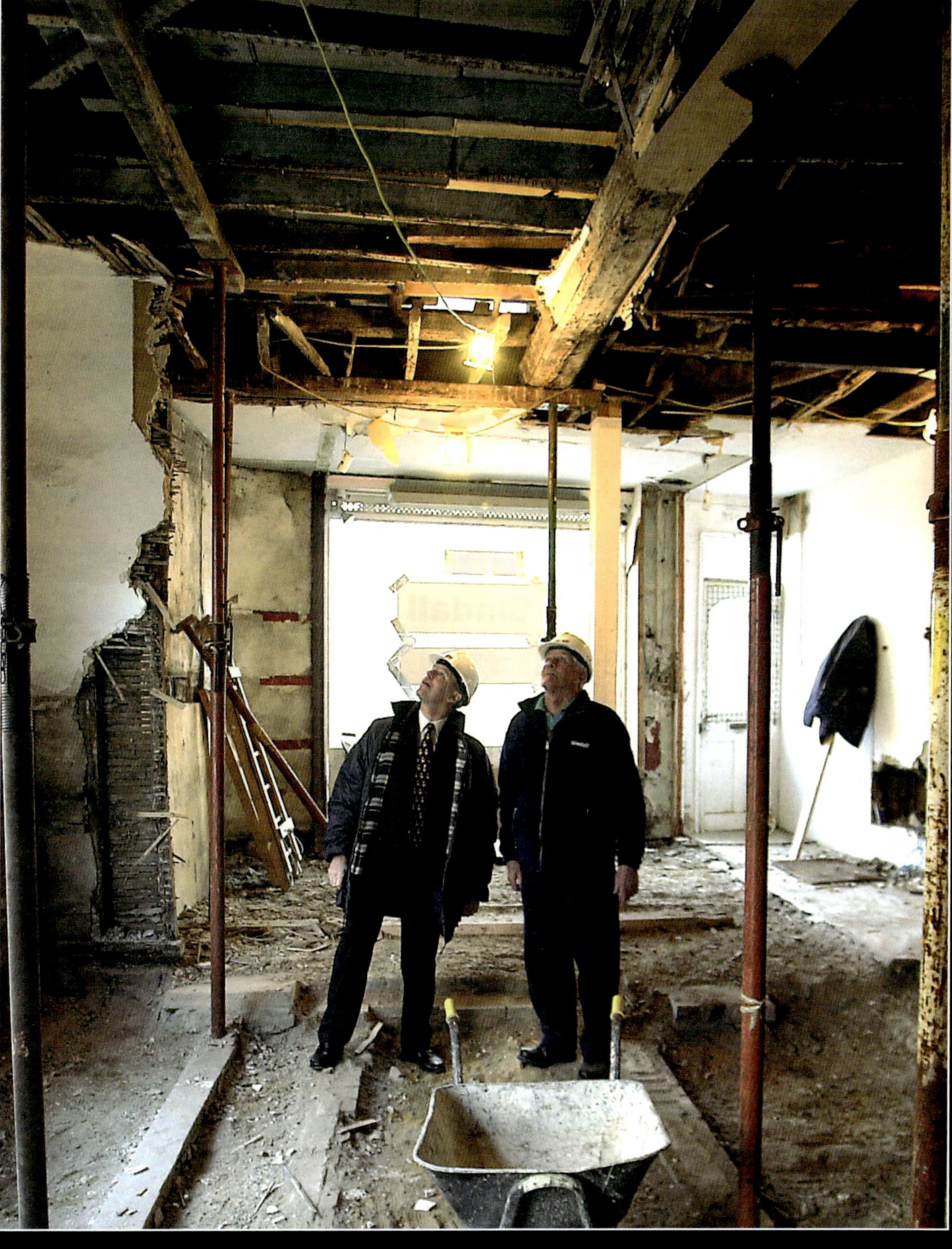

MOISERS HALL MUSEUM, BURY ST EDMUNDS
The refurbishment of this grand Victorian building in Bury St Edmunds, which was originally a corn exchange, is being carried out by Sindall Ltd for St Edmundsbury District Council

4.1 REMEDIAL

DAMP AND DECAY

		Page
Abbey Heritage Ltd	er	104
Balmoral Stone Ltd	dd	176
Bedford Timber Preservation Company	dd er nd	177
Blackfriars Paints	dd	183
Boniface Associates	nd	45
Calder Preservation Services	dd	176
David Ball Restoration (London) Limited	nd	109
Demaus Building Diagnostics Ltd	dd nd tt	176
G B Geotechnics Ltd	dd nd	50
Gifford and Partners	ec	188
Hanwell Instruments Limited	dd ec nd	178
Helifix Limited	er	179
The Hertfordshire Roofing & Renovation Company	dd	59
Hutton+Rostron Environmental Investigations Limited	dd ec nd	176
I J P Building Conservation	nd	61
I W Payne & Company Ltd	dd	61
Ian Russell Engineering	nd	41
Lightwright Associates	ec	188
Maysand	dd	176
Mowlem Rattee & Kett	er	62
Munters Property Damage Restoration Ltd	dd ec nd	177
Oscar Faber	nd	188
Plowden & Smith Ltd	ec	202
Reddiseals	er	124
Renlon Ltd	dd	178
Ridout Associates	dd nd	177
Robinsons Preservation Limited	dd er nd tt	177
Rotafix Ltd	er	178
Ryder & Dutton	nd	48
Taywood Engineering Limited	dd	68
Terminix Limited	dd er	178
Thermo Lignum UK Ltd	dd	178
W T Specialist Contracts	er	69
Ward & Dale Smith, Chartered Building Surveyors	tt	178

KEY
- dd damp and decay treatment
- ec environmental control
- er epoxy resin repairs
- nd non destructive investigations
- tt structural timber testing

LIME MORTARS AND RENDERS – SUPPLIERS

		Page
Acanthus Plain & Decorative Plastering	hy lm	167
Bleaklow Industries	hy lm	167
The Bulmer Brick and Tile Co Ltd	hy lm pz	167
Bursledon Brickworks Conservation Centre	hy lm pz re	167
Calder Traditional Building Services	lm	168
Castle Cement Limited	hy lm pz	168
Cathedral Works Organisation (Chichester) Limited	hy lm pz re	107
Chalk Down Lime Ltd	lm re wd	168
Cornish Lime Company	hy lm pz re	169
H J Chard & Sons	hy lm pz re	168
Hirst Conservation Materials Ltd	hy lm pz re	168
Historic Buildings Conservation Limited	wd	39
Keim Mineral Paints Ltd	cp	184
Keyline	hy lm re wd	115
The Lime Centre	hy lm pz re	169
Mike Wye & Associates	hy lm pz re	168
Milestone Lime	hy lm	168
Nimbus Conservation Limited	pz	110
Old House Store	hy lm pz wd	169
Rose of Jericho	hy lm	170
St Astier	hy	170
The Traditional Lime Co	hy lm re	170
Twyford Lime Products	lm	169
Ty-Mawr Lime Ltd	hy lm pz re	170
Yorkshire Decorative Plasterers Ltd	re	216

KEY
- cp stone consolidants
- hy hydraulic lime
- lm non-hydraulic lime (lime putty)
- pz pozzolanic additives
- re hair and fibre reinforcement
- wd wattle and daub

FIXINGS

		Page
Avon Stainless Fasteners	fx zw	179
Fred Brodnax Architectural Blacksmiths	fx	148
Glasgow Steel Nail Co Ltd	fx zt	179
Helifix Limited	zr zw	179
Old House Store	zt	179
Protovale	zw	179
Tamworth Scaffolding Company	te	83
Taunton Fabrications	zr zw	179

KEY
- fx nuts, bolts, screws and nails
- te temporary protection
- zr roof ties and plates
- zt traditional nails
- zw wall ties and plates

MASONRY CLEANING

CONTRACTORS

		Page
SCOTLAND		
Abbey Heritage Ltd	ac cc gf ls mc pr sc wt	171
Flirok UK Limited	ac gf mc mp wt	173
Holden Conservation Ltd	ac cc ls pr sc wt	115
Nicholas Boyes Stone Conservation	mc	173
YORKSHIRE		
Abbey Heritage Ltd	ac cc gf ls mc pr sc wt	171
Anelays	ac cc gf pr sc wt	58
Flirok UK Limited	ac gf mc mp wt	173
NORTH-WEST		
Burleigh Stone Cleaning & Restoration Co Ltd	mc	172
Burnaby Cleaning Co Ltd	ac cc gf wt	171
Maysand Limited	ac cc gf sc wt	173
Stoneguard	ac cc gf mc pr sc wt	174
WALES		
Abbey Masonry and Restoration	ac cc gf pr sc wt	104
Taylor Dalton, Heritage Building Contractors	mc pr	67
William Taylor Stonemasons	sc wt	112
WEST MIDLANDS		
Cameron (UK) Plc	ac cc gf pr sc wt	171
Centreline Architectural Sculpture	mc	107
Russcott Conservation and Masonry	ac cc gf ls pr sc wt	111
Wells Masonry Services Ltd	mc	114
William Sapcote & Sons Ltd	ac	68
EAST MIDLANDS		
Flirok UK Limited	ac gf mc mp wt	173
Hirst Conservation	mc	173
SOUTH-EAST – NORTH OF THE THAMES		
A F Jones (Stonemasons)	ac cc gf mc pr sc wt	171
Abbey Heritage Ltd	ac cc gf ls mc pr sc wt	171
Ashby Stone Masonry Limited	mc	105
Cliveden Conservation Workshop Ltd	ac cc gf ls pr sc wt	108
J Joslin (Contractors) Ltd	ac pr sc wt	108
Kaizen Stone Cleaning	mc	172
Stoneguard	ac cc gf mc pr sc wt	174
Suffolk Brick & Stone Cleaning Co Ltd	ac mc pr	174
LONDON		
Cleaning & Restoration of Historic Buildings (London)	mc	171
David Ball Restoration (London) Limited	ac cc gf mc pr sc wt	173
H W Poulter & Son	mc	189
Holden Conservation Ltd	ac cc ls pr sc wt	115
Ken Negus Ltd	ac cc gf pr wt	173
O'Reilly Period Cornice Restoration & Cleaning	pr sc	215
PAYE Stonework & Restoration	ac cc gf pr sc wt	173
Priest Restoration	ac cc pr sc wt	111
Stonewest Limited	ac cc gf pr sc wt	174
SOUTH-EAST – SOUTH OF THE THAMES		
C Ginn Building Restoration	mc	172
Cathedral Works Organisation (Chichester) Limited	ac cc mc pr sc wt	107
SOUTH-WEST		
Carrek Limited	ac cc gf ls pr sc wt	60
Flowplant Group Limited	cc gf mc mp	172
Nimbus Conservation Limited	mc wt	110
Sarabian Limited	ac mc pr	174
St Blaise Ltd	cc sc mc pr	174
Stonehealth Limited	ac cc mc mp pr sc	173
Wells Cathedral Stonemasons Ltd	mc	113
Western Stone	ac cc gf mc sc wt	113

SUPPLIERS

		Page
Blackfriars Paints	pp	185
Flirok UK Limited	mp	173
Flowplant Group Limited	mp	172
Keim Mineral Paints Ltd	mp pp	184
Stonehealth Limited	mp pp	174
Strippers Paint Removers Ltd	pp	174

KEY
- ac air/water abrasive cleaning
- cc chemical cleaning
- gf graffiti protection and removal
- ls laser cleaning
- mc masonry cleaning
- mp masonry cleaning products and materials
- pp poultices
- pr paint removal and poulticing
- sc steam cleaning
- wt low pressure water cleaning

Regional designation is according to office location. Many firms operate nationally.

BIRD CONTROL

		Page
Balmoral Stone Ltd	bc	178
Microbee Bird Control Limited	bc	178
Splitlath Limited	bc	64
Terminix Limited	bc	178

KEY
- bc bird control

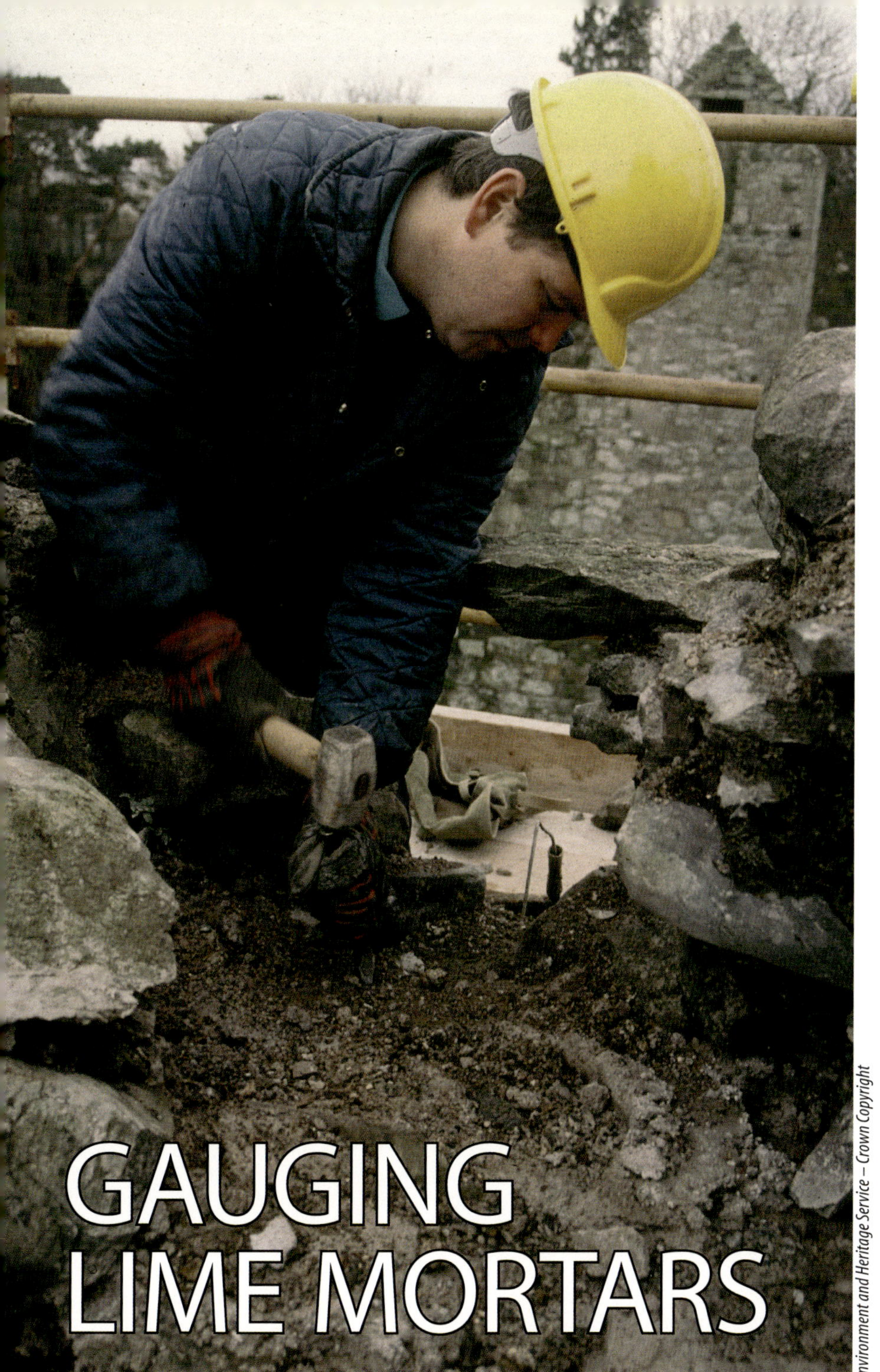

Stone mason Steven Savage working on Grey Abbey Church, Northern Ireland.
After a major research programme by mouldings specialist Stuart Harrison, EHS is conserving the ruin.

GAUGING LIME MORTARS

PETER ELLIS

For as long as lime mortars have been used, they have been mixed with reactive materials known as 'pozzolans'. These additives, which include such common materials as fragments of pottery and certain types of brick, may have been introduced into the mix for various reasons, but their effect was to produce a mortar that had some hydraulic properties, was less permeable and generally more durable than an ordinary lime mortar. The addition of small quantities of pozzolan or any other material that achieves this effect is known as gauging.

A non-hydraulic lime mortar sets by a simple chemical reaction between lime (calcium hydroxide) and airborne carbon dioxide alone, whereas a mortar with some hydraulicity also sets by a reaction between calcium and the reactive components, principally silica. (Hydraulicity is actually defined as the ability to set under water, without any carbon dioxide.)

The current debate is focused on the blending of non-hydraulic lime putty and hydraulic hydrate.

It is the author's view that gauging with hydraulic hydrate is poor practice. Although it is important to make clear that the principal proponents of this practice can provide many examples of the successful use of these 'complex mixes', long-term durability of these mortars has yet to be assessed.

The combination of ordinary lime and hydraulic lime in the same mortar causes concern for the following reasons:

- **There is little historical precedent.** This practice is likely to be a development of the Portland cement / lime hybrid mortars typified by the 1:1:6 and 1:2:9 mixes used in the 20th century. Lime in these mixes was added as a plasticiser to improve workability, although in a 1:2:9 it was presumably expected that the lime would carbonate and assist the setting process. Analysis of these hybrids regularly finds lime still present after 50 years. The setting cement impedes the carbonation of lime.
- **It is a dangerous assumption** that the addition of, for example, lime putty to an eminently hydraulic hydrate will produce a moderately hydraulic lime. The chemistry of hydraulic lime is complex and the setting processes are delicate. Data available on this[1] is preliminary and suggests significantly reduced compressive strength, increased water vapour permeability, and far worse performance in salt crystallisation tests. It is extremely important to define this practice. The blending of non-hydraulic and hydraulic materials can range from a hydraulic hydrate with five per cent putty added to a putty with five per cent hydraulic hydrate added, and every variable between. The end-product will have varying properties.
- **This blending is now not necessary.** We are fortunate to have a wide range of hydraulic hydrates, lime putty and pozzolanic additives available in the UK, and there is no application in historic building repair where a blend is likely to outperform the correct grade of hydraulic hydrate, or pozzolanic lime. The only possible exception to this is the addition of less than eight per cent by volume of lime putty to a hydraulic hydrate to improve plasticity. This will improve workability, hopefully not at the expense of durability, although thorough mixing of the hydrate mortar often makes the addition of putty unnecessary. In the UK, specialists first became aware of this practice when Jura-kalk, an eminently hydraulic hydrate from Switzerland became available. It was advised that this material had been routinely blended with lime putty in equal proportions in Denmark and elsewhere in Europe. Jura-kalk is a binder containing very little lime[2,3] and is a complex blend of compounds of calcium and silica (principally C2S), calcium and alumina (principally C3A), calcium, alumina and silica (C2AS), calcium, alumina and iron (C4AF) and calcium carbonate. The addition of lime to this is likely to be different to the addition of lime to a material that does contain lime. Many hydraulic hydrates notably from England and France have significant lime content.

A BRIEF HISTORY OF LIME

Lime was produced by burning locally available limestone in a coal or wood fired kiln at a temperature rarely in excess of 900°C. The properties of the lime produced were largely influenced by the chemical composition of the limestone burnt, and limestones containing clay minerals produced a lime with weak hydraulic properties. The hydraulicity of these limes was likely to be weak because of the low firing temperatures, and weakened further if the lime produced was stored as putty. Certain compounds, notably the calcium aluminates, but also any di-calcium silicates present will hydrate or begin hydration in the lime pit. This explains why the 'hot mixes' where quicklime is mixed with water and aggregate on site have different properties.

The belief that most historic limes were non-hydraulic or only very weakly hydraulic is supported by the fact that certain additives have been added historically to alter the performance characteristics. These 'heated' materials contained silica, alumina and iron which became reactive towards alkalis including lime.

The earliest mortars analysed from Jericho in the Jordan Valley, and Tel-Ramad, Syria[4], dating from 7000 BC contain stable end-products of pozzolanic reactions although it is not possible to conclude whether the pozzolanic material was deliberately added, or naturally present in the aggregate.

Pozzolanic materials in the form of crushed brick and tile were deliberately added in large quantity in mortars of Minoan Crete of *c* 1000 BC, ancient Greece, and the Roman period. Evidence suggests that the Romans used crushed brick and tile before they discovered the naturally occurring pozzolanic aggregates from around Vesuvius.

The practice, not so much 'gauging with pozzolans' but more the deliberate inclusion of pozzolanic materials as aggregate, was lost in Britain after 400 AD but continued in Europe as demonstrated by the analysis of samples from the Byzantine Empire[5], Venetian renders[6], Sistine Chapel plasters[6], and recently analysed samples from the 13th century Moorish castle in Gibraltar which have the ingredients of ancient mortars – carbonated lime, calcium silicate hydrate, brick particles, quartz sand and limestone particles.

Further evidence that limes were generally non- or only very weakly hydraulic is demonstrated by Vicat[7] in early 19th century France, and his frustration with the limes available, and his exhaustive trials to find a binder that would prove durable for hydraulic engineering works. His work, and others', notably John Smeaton in late 18th century England, led to the recognition of natural hydraulic lime, the manufacture of artificial hydraulic limes, the invention of Parker's 'Roman Cement', and in 1824 to the first Portland cement patent.

THE SPECIFICATION OF REPAIR MORTARS

This is a complex issue and each building and its particular condition and problems must be considered individually. Two considerations are of paramount importance:

1 **The 'Like-for-Like' philosophy.** In most cases it is technically and aesthetically appropriate to carry out repairs using a mortar to match the existing or original material, replacing like for like. This requires proper analysis to ascertain exactly what material was used, and demands a detailed understanding of materials currently available. This does not imply that poor mortars should be matched, particularly where their use might be harmful to original fabric.

2 **Mortars should be durable yet sacrificial to the building fabric.** This normally entails preparing the mortar from the constituents required and in the right proportions to ensure that the result is both more porous and more permeable than the stone or brick. This is so that the mortar age, decay and ultimately fail before the masonry – hence the term 'sacrificial'.

THE CHARACTERISTICS OF THE VARIOUS LIME TYPES

1 **Non-hydraulic Lime Putty (Fat Limes)**
Many traditional limes were non-hydraulic, as is most modern lime putty. They set by the reaction with atmospheric carbon dioxide in the presence of moisture alone. A non-hydraulic lime mortar is soft, porous, permeable and plastic. They are used for bedding mortars, for internal and external pointing mortars, and for internal plasters. Internal putty plasters have been and still are on occasion gauged with gypsum to accelerate the set and reduce shrinkage. This was commonly done from *c* 1760 to ceilings and especially run and cast work, but never to walls where there is a risk of damp as gypsum is slightly soluble in water and sulphate salts migrate and crystallise on the plaster surface.

2 **Impure Lime Putty (Lean Limes)**
Most traditional limes, but sadly few (if any) modern limes fall into this category. They contain impurities such as coal or wood ash, unburnt or partially burnt limestone and a small proportion of reactive silica produced by the de-hydroxylation of clay minerals in the limestone. Some contained a small proportion of di-calcium silicate (C2S). The setting process was principally carbonation augmented by a very weak hydraulic reaction as the C2S hydrates and the silica reacts with lime. The un-converted calcium carbonate and fuel ash also played a positive role. These limes were not as hydraulic as today's 'feebly hydraulic' classification. These limes, supplied as both putty and quicklime, were used to build most things in Britain, including much of London. They were also used for base-coat plasters and external renders.

3 **Traditional Hydraulic Limes**
Certain limestones with high clay mineral content produced hydraulic limes that would be today classified as 'feebly hydraulic' (or sometimes possibly moderately hydraulic). The principal examples of these are the Lias limestones from Somerset, Devon, and Aberthaw in South Wales. Arden lime in Scotland is another example. These were invariably supplied as quicklime to be mixed with water and sand on site and used immediately. Putty made from these limes would set quickly, and usage advice for Totternhoe lime, a hydraulic chalk lime from Bedfordshire which was slightly less hydraulic than Lias limes, was to slake only enough on a Friday necessary for the following week's work.

These traditional hydraulic limes produced durable mortars which were used widely. For example, Lias lime from Devon was used at the Tower of London from the 15th century.
The hydraulicity of the lime produced from these complex raw materials is determined by kiln temperature, and indeed Blue Circle cement is now made from Aberthaw limestone. Traditional kilns rarely got hot enough for complete combination, and the hydraulicity of these limes was largely due to a pozzolanic silica/ lime reaction together with the hydration of limited C2S and C2F (di-calcium ferrite).

4 **Modern Hydraulic Hydrates**
These are produced from clay mineral rich limestones similar to those used to make the traditional hydraulic limes, but once burnt, the material is passed through a hydrating plant, where sufficient water is added to convert the quicklime to calcium hydroxide but not to hydrate the C2S. However, any calcium aluminates are likely to be hydrated by this process. These range in hydraulicity from feebly to eminently hydraulic depending on factors such as kiln temperature and length of time in the kiln, as well as the chemical composition of the limestone. Some of these materials are subsequently blended with pozzolanic additives and in some cases white cement. The Foresight project[2] has identified C3A (tri-calcium aluminate) and C4AF (tetra calcium alumino ferrite) in most types tested indicating kiln temperatures in excess of 1,000°C, hotter than traditional lime kilns. These modern hydrates are therefore more hydraulic than the earlier materials. Available data also suggests an inversely proportional relationship between hydraulicity and permeability.[1,8]
These hydrates, in particular the less hydraulic grades, have a part to play in historic building repair in applications where reduced porosity and increased strength are advantageous and where reduced vapour permeability is acceptable. These applications include external mortars and renders especially in exposed or aggressive environments. There is less risk of failure when work must proceed in winter as they set more quickly and are thus vulnerable to frost for a shorter period. They are clearly appropriate for repairs to hard mortars such as 'Roman Cement', and a better option than cement

based mortars. They are rarely appropriate for pointing mortars or internal plasters.

5 **Pozzolanic Lime Mortars**
The durability of pozzolanic lime mortars of correct mix design is proven beyond doubt. A pozzolan is defined as a material that is capable of reacting with lime in the presence of water at ordinary temperatures to produce cementitious compounds. The essential difference between these and modern hydraulic hydrates is that the reaction takes place in solution. The pozzolanic reaction products and the compounds produced on ageing will differ and depend on the calcium to silica ratio in solution. Modern hydraulic hydrates derive their hydraulic properties from the subsequent hydration of compounds of principally calcium and silica produced by solid state reaction in the kiln. The chemistry is of the same chemical nature, but it is not the same.

Pozzolans vary in reactivity, and historically include naturally occurring volcanic Italian pozzolana and Santorini earth as well as artificial forms including brick and tile powder. The varieties most used in the UK are the metakaolin Metastar 501, certain brick dusts of known reactivity and Trass from Germany, but other forms include HTI (ceramic 'high temperature insulation') and PFA ('pulverised fuel ash'). The ground slags are not true pozzolans as they may themselves be cementitious; these are classed as latent hydraulic binders. The addition of ten per cent pozzolan improves durability and strength and slightly reduces porosity and permeability. The pozzolan reacts with the lime and does not set in isolation as occurs when hydraulic lime or cement is added. Pozzolans are added to lime putty mortars where there is doubt about durability and the reduced porosity is not disadvantageous.

To quote Vitruvius from *De Architectura* in the first century BC: 'If to river or sea sand, potsherds ground and passed through a sieve, in the proportion of one third part, be added, the mortar will be the better for use.'

REFERENCES

1 Jeanne Marie Teutonico, Geoff Ashall et al, *International RILEM Workshop Proceedings PRO 12*. English Heritage and BRE UK (NB Full results to be published by English Heritage by end of 2001)

2 Paul Livesey et al, *Foresight Project*. University of Bristol (To be published 2001)

3 Dave Hughes and Simon Swann, *Hydraulic Limes – a preliminary investigation*. Lime News, Volume 6, 1998

4 Joseph Davidovits (Institute of Applied Archaeological Sciences. Miami, Florida) *Ancient and Modern Concretes: What is the real difference?* Concrete International, December 1987

5 L Binda, G Baronio, C Tedeschi (Politecnico of Milan, Italy), *Experimental Study on the Mechanical Role of Thick Mortar Joints in Reproduced Byzantine Masonry*. International RILEM Workshop Proceedings PRO 12

6 E Charola (USA) and F Henriques (Universidade Nova de Lisboa, Portugal), *Hydraulicity in Lime Mortars Revisited*. International RILEM Workshop PRO 12

7 LJ Vicat, *A Treatise on Calcareous Mortars and Cements, Artificial and Natural*. 1837, Re-printed Donhead Publishing Ltd, 1997

8 P Banfill & A Forster (Heriot Watt University, Scotland), *A Relationship Between Hydraulicity and Permeability of Hydraulic Lime*. International RILEM Workshop, PRO 12

PETER ELLIS originally trained as a conservator of paintings in London but has worked with older buildings for the majority of his career. For the last few years he has been manager of Rose of Jericho, analysts and manufacturers of materials for the conservation and repair of historic buildings. He has had many papers published including *Analysis of Mortars (to include historic mortars) by differential thermal analysis* in the International RILEM Workshop proceedings PRO 12.

MORTARS & RENDERS

▶ ACANTHUS PLAIN & DECORATIVE PLASTERING
5 Hansford Square, Combe Down, Bath BA2 5LQ
Tel/Fax 01225 837223
SUPPLIERS OF MORTARS AND RENDERS AND TRADITIONAL LIME FINISHES: *See also: profile entry in Plasterwork section, page 214.*

▶ BALMORAL STONE LTD
31 Bankhead Drive, Sighthill, Edinburgh EH11 4DN
Tel 0131 453 4777 Fax 0131 453 6077
STONEMASONRY AND RESTORATION CONTRACTORS: *See also: display entry in Stone section, page 106.*

▶ THE BULMER BRICK & TILE CO LTD
Bulmer, Nr Sudbury, Suffolk CO10 7EF
Tel 01787 269232 Fax 01787 269040
LIME PRODUCTS: Stockists of lime putty, hydraulic lime and associated products. Also producers of ground brick pozzolans. *See also: profile entry in Brick Suppliers section, page 99.*

▶ BURSLEDON BRICKWORKS CONSERVATION CENTRE
Bursledon Brickworks, Coal Park Lane, Swanwick, Southampton SO31 7GW
Tel/Fax 01489 576248
E-mail bursledon@ndirect.co.uk
TRADITIONAL BUILDING PRODUCTS AND EDUCATIONAL COURSES: The Centre, administered by an educational trust, is based at the 19th century Bursledon Brickworks near Southampton. It has a museum and educational function and also supplies traditional building materials including lime putty, hydraulic lime, ready-mixed lime mortars, limewash, hair, stone dust, pozzolanic additives, ochres and pigments, pointing irons, riven laths, battens, air-dried oak, traditional drawn glass and cast iron door furniture. Other products can be sourced. The Centre runs courses on traditional materials, techniques and the conservation of the built environment at Introductory, CPD and Post Graduate level. The Centre also provides consultancy advice on building conservation, and fully serviced facilities for seminars, exhibitions and conferences.

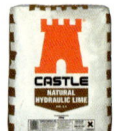

▶ CALDER TRADITIONAL BUILDING SERVICES

'Woodhurst' Cattlegate Road, Crews Hill, Enfield, Middx EN2 8AU
Tel 01707 876515 Fax 01707 872413
LIME MORTARS, PLASTERING AND RENDERING AND LIMEWASH: *See also: Calder Group display entry in Building Contractors section, page 59.*

▶ CARREK LTD

Mason's Yard, Wells Cathedral, Wells, Somerset BA5 2PA
Tel 01749 689000 Fax 01749 689089
Website www.carrek.co.uk
HISTORIC BUILDING REPAIR COMPANY: *See also: profile entry in Building Contractors section, page 60.*

▶ CHALK DOWN LIME LTD

102 Fairlight Road, Hastings, East Sussex TN35 5EL
Tel 01424 443301 Fax 01580 830096 Mobile 0771 873 8708
E-mail chalkdownlime@supanet.com
SPECIALISTS IN MORTARS AND RENDERS: Chalk Down Lime stocks traditional building materials and offers a mortar analysis service. Suppliers of matured slaked lime putty, ready mixed mortars and renders, mortars made to individual requirements, limewash, natural pigments, laths, coal tar etc. Conservation projects undertaken by a team of dedicated craftsmen and professionals. They aim to provide a comprehensive service in the maintenance and repair of historic buildings.

▶ H J CHARD & SONS

1 Cole Road, Bristol BS2 0UG
▶ Builders' Merchants 01179 777681
LIME PUTTY AND RELATED PRODUCTS: H J Chard & Sons manufacture for sale direct to the trade and retail users lime putty from best quality Buxton non-hydraulic white quicklime. Lime putty is used in the production of a wide range of sand-lime mortars, including hair mortars. Other products available include pigments, tallow, linseed oil, PFA, HTI powder and French hydraulic lime.

▶ HIRST CONSERVATION MATERIALS LTD

Laughton, Sleaford, Lincolnshire NG34 0HE Tel 01529 497517 Fax 01529 497518
E-mail materials@hirst-conservation.com
Website www.hirst-conservation.com
CONSERVATION MATERIALS: The company produces lime putty, mortars, plasters, renders, daubs, grouts, limewashes, coatings and paints for historic building repair. A range of other conservation materials is also supplied. All products are supported by technical and practical guidance. Analysis and research services are available enabling historic materials to be closely matched where required. Supplies can be delivered anywhere in Europe. Short courses are available on application. This company is a subsidiary of Hirst Conservation who can provide a team of specialists to apply the materials, giving a single source of responsibility. *See also: display entry on the inside front cover.*

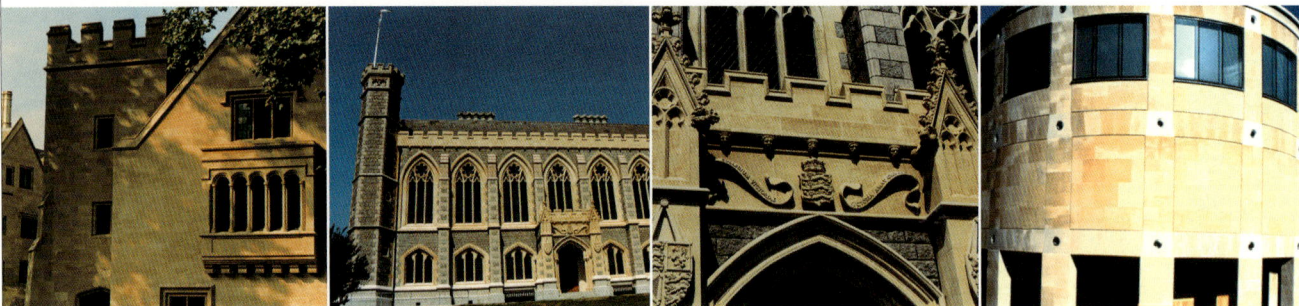

▶ MIKE WYE & ASSOCIATES

Buckland Filleigh Sawmills, Buckland Filleigh, Beaworthy, Devon EX21 5RN
Tel/Fax 01409 281644
E-mail sales@mikewye.co.uk
Website www.mikewye.co.uk
LIME PUTTY, LIME MORTARS, PLASTERS AND LIMEWASH: Mike Wye & Associates supplies the finest quality traditional lime products at unbeatable prices, together with a comprehensive range of natural building materials, paints, varnishes and waxes. Visit their website for latest offers, limewash colour charts, guidesheets and practical courses programmes.

▶ MILESTONE LIME (subsidiary of Whippletree Hardwoods)

Milestone Farm, Barley Road, Flint Cross, nr Royston, Herts SG8 7QD
Tel 01763 208966
TRADITIONAL LIME AND RELATED PRODUCTS: Suppliers of quality slaked lime putty and hydraulic limes for conservation and restoration of ancient buildings. Also stockists and producers of riven and sawn oak, sweet chestnut, hazel and larch lath. *See also: Whippletree Hardwoods display entry in Timber Suppliers section, page 144.*

▶ ROSE OF JERICHO

at St Blaise Ltd, Westhill Barn, Evershot, Dorchester DT2 0LD
Tel 01935 83676 Fax 01935 83017
E-mail stblaise@compuserve.com
Website www.rose-of-jericho.demon.co.uk

SUPPLIERS OF TRADITIONAL MORTARS AND PAINTS: Rose of Jericho manufactures and supplies feebly and non-hydraulic lime putties and a wide range of pozzolanic additives. Non and feeble hydraulic mortars, plasters and renders are made together with lime based conservation materials. A wide range of aggregates are stocked allowing accurate historic, geological, geographical, and technically appropriate repair or matching mixes. Rose of Jericho manufactures limewashes, distempers, permeable emulsions, flat oils, eggshells in historic and contemporary colours. Sophisticated expert mortar analysis using DTA. *See also: display entry in this section, page 170, and in Interiors Consultants & Conservators section, page 199.*

▶ STATS CONSULTANCY (STATS Limited, founded 1974)

Porterswood House, Porters Wood, St Albans, Herts AL3 6PQ
Tel 01727 833261 Fax 01727 835682
E-mail ian.sims@stats.co.uk
Contact Dr Ian Sims

SPECIALIST ENGINEERING, MATERIALS AND ENVIRONMENTAL CONSULTANTS: *See also: profile entry in Materials Analysis section, page 45.*

▶ TWYFORD LIME PRODUCTS

1 Twyford Place, Tiverton, Devon EX16 6AP
Tel 01884 255407 Fax 01884 242446
E-mail arhunt@1twyford.fsnet.co.uk

MORTARS AND RENDERS: Manufacturer of lime putty, lime mortars, plasters and lime washes, skills for the repair and maintenance of historic buildings, conservation and repair of cob buildings.

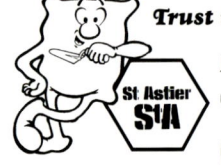

▶ TY-MAWR LIME LTD
Ty-Mawr, Llangasty, Brecon, Powys, Wales LD3 7PJ
Tel 01874 658249 Fax 01874 658502
E-mail tymawr@lime.org.uk
Website www.lime.org.uk

PRODUCERS AND SUPPLIERS OF NATURAL AND TRADITIONAL BUILDING MATERIALS: Ty-Mawr's extensive list of building materials includes lime, earth, natural paint and insulation materials for use in restoration, refurbishment and construction. Lime, paint/finishes and building technologies courses are run regularly and paint and mortar analysis/matching services are provided. A colour brochure and over 40 technical sheets are available.

When contacting companies listed here, please let them know that you found them through *The Building Conservation Directory*

▶ A F JONES (STONEMASONS)

33 Bedford Road, Reading, Berkshire RG1 7EX
Tel 0118 957 3537 Fax 0118 957 4334

MASTER STONEMASONS: Established in 1865 by Arthur F Jones and continued today by G A and A G Jones, this firm has accumulated experience and expertise gained by five generations of stone masons. There are currently 20 highly skilled craftsmen many of whom have spent their working lives with A F Jones, restoring, carving, cleaning and conserving stone facades. A F Jones offers a complete service from consultancy and specification to production and site fixing. Their style of management is non-confrontational and above all, fair and honest. A F Jones is an approved contractor for The Churches Conservation Trust, English Heritage the National Trust and Historic Royal Palaces. *See also: display entry in Stone section, page 105.*

▶ ABBEY HERITAGE LTD

Dartford House, Two Rivers, Station Lane, Witney, Oxon OX8 6BH
Tel 01993 709699 Fax 01993 709959
E-mail stone@abbeyh.co.uk
▶ York Office Tel 01904 567332
▶ Edinburgh Office Tel 0131 228 2281

FACADE CLEANING, LASER CLEANING, RESTORATION AND MASONRY: Leaders in the field of laser cleaning for buildings, artefacts and artworks, Abbey specialises in facade cleaning – stone, terracotta, faience, granite, marble and brickwork – incorporating Jos/Torc and Doff systems, also anti-graffiti treatments. As part of their complete package they can implement masonry restoration and new stone projects. Abbey Heritage works closely with leading architects, surveyors, local authorities and consultancies throughout the UK. Dedication to detail and a high level of managerial input has earned them a reputation for integrity and reliability, generating consistent repeat business. As principal contractor Abbey provides associated leadwork, roofing, decoration, joinery, bronze and ironwork services. They offer full national and international consultancy services.

▶ CAMERON (UK) PLC

Cockshades, Wybunbury, Nantwich, Cheshire CW5 7HA
Tel 01270 841122 Fax 01270 841520
E-mail cameron@dial.pipex.com

STONE/BRICK CLEANING AND RELATED SERVICES: Cameron's wealth of experience and vast knowledge of available cleaning techniques enable their trained personnel to carry out their skills on all building fabrics. *See also: display entry in Stone section, page 107.*

▶ CLEANING & RESTORATION OF HISTORIC BUILDINGS (LONDON)

6 Greenways, Abbots Langley, Watford, Herts WD5 0EU
Tel/Fax 01923 270350 Mobile 07831 483974
E-mail djfrost_London@virginbiz.com
Website www.historicbuildingslondon.co.uk

MASONRY CLEANING: David Frost and his internationally recognised team are passionate about stone cleaning. Mr Frost is a consultant, scientist and craftsman applying the right technique for every material, in even the most awkward places, normally using a variety of approaches to suit the varied and often delicate surfaces encountered; including decorative glass and wood. The team uses Flirok and new SoftClean micro-abrasive systems and lasers (25 and 500 models). Tensid anti-graffiti and Neolith chemicals and Jos are used, also cone jet nebulous sprays and other mildly abrasive systems and poultices. Staff training is of paramount importance; techniques are constantly monitored and upgraded. Most work is with historic buildings, both interior and exterior. Please ring for a brochure, or for some straightforward advice.

MASONRY CLEANING

BURLEIGH

STONE CLEANING & RESTORATION CO. LTD.
ESTABLISHED 1975

FOR A COMPLETE PROFESSIONAL SERVICE

Technical Advice, Quotations, Surveys, Drawings, Main Contractors, Specialist Sub-Contractor, Direct Works. A first class service in all aspects of Building Restoration for over 25 years. Our skilled craftsmen have extensive experience of work on Historic, Listed and Modern Buildings of every size.

Recently Completed Restoration Projects
Trinity Presbyterian Church, Wrexham
St Francis Xavier School, Liverpool
Westin Hotel, Dublin
St Catherines Church, Whitchurch, Shropshire
St Thomas Church, Lydiate, Merseyside
Littlewoods Store, Liverpool
Port Sunlight Village – Planned Maintenance

Judge our history of service for yourself, send for our Brochure, Contracts Completed List and References

The Old Stables, 56 Balliol Road,
Bootle, Merseyside L20 7EJ

Telephone 0151 922 3366
Fax 0151 922 3377
email: info@burleighstone.co.uk
http://www.burleighstone.co.uk

Neolith is a division of the Flowplant Group Ltd

Watt Road Churchfields Industrial Estate
Salisbury Wiltshire SP2 7UD
Tel 01722 325 424 Fax 01722 411 329
E-mail: info@flowplant.co.uk
Internet: www.flowplant.co.uk

Stone Federation Great Britain

MASONRY CLEANING

▶ DAVID BALL RESTORATION (LONDON) LIMITED

104A Consort Road, London SE15 2PR
Tel 020 7277 7775 Fax 020 7635 0556 E-mail mail@dbr.uk.com

STONE CLEANING SPECIALISTS: *See also: display entry and profile entry in Stone section, page 109.*

▶ FLIROK UK LTD

16 Foxdale, Stamford, Lincs PE9 2UZ
Tel 01780 763420 Fax 01780 753450 Mobile 0860 525322
E-mail sjsservices@compuserve.com Website www.flirok.co.uk
▸ Yorkshire office Tel 01924 480090 Fax 01924 498830
▸ Scottish office Tel 0141 6331647 Fax 0141 5776762

MASONRY CLEANING EQUIPMENT: Flirok UK Ltd is a new company formed in the United Kingdom by its Belgian parent to introduce a new range of quality specialist cleaning machines for stone and surfaces, including graffiti removal. Over seven years Flirok Belgium has researched and developed these high quality machines employing a wet and dry air assisted system capable of using very low volumes of granulates operating at low pressures, from 0.1 bar (1.5lbs/square inch). The system offers major environmental benefits being chemically-free and discharging low volumes of granulates at source. The machine's gentleness will retain the character and originality of the surface, without creating a totally new appearance, vital for conservation and renovation work. Latest information and pictures on Website www.flirok.co.uk

▶ HIRST CONSERVATION

Laughton, Sleaford, Lincolnshire NG34 0HE
Tel 01529 497449 Fax 01529 497518

STONE CLEANING AND RELATED SERVICES: *See also: display entry on the inside front cover and profile entry in Building Contractors section, page 62.*

▶ MAYSAND LIMITED

109-111 Windsor Road, Oldham, Lancs OL8 1RH
Tel 0161 628 8888 Fax 0161 627 0996 E-mail sales@maysand.co.uk

MASONRY RESTORATION AND CONSERVATION: Maysand's expert team of surveyors and craftsmen provides the full range of masonry and building conservation services for historic, ecclesiastical and commercial projects. The company's range of services includes masonry stabilisation, stone and brickwork cleaning, repair and conservation/restoration, re-pointing, traditional stone masonry and new-build cladding. Members of Stone Federation Great Britain.
See also: display entry in Stone section, page 110.

▶ NEGUS

(Ken Negus Ltd), 90 Garfield Road, London SW19 8SB
Tel 020 8543 9266 Fax 020 8543 9100
E-mail enquiries@kennegus.co.uk Website www.kennegus.co.uk

A family run company established in 1978 by Ken Negus and now managed by Graham Negus. They are able to offer a comprehensive service in the cleaning and conservation of stonework, brickwork, terracotta, granite, marble etc, and also paint and graffiti removal. Among the contracts they have been involved in are: Big Ben, Royal Courts of Justice (internal cleaning), City University, RIBA and many other well known public buildings, together with many churches, banks, building societies, private houses etc. Brochure available on request.

▶ NICOLAS BOYES STONE CONSERVATION

9 Palmerston Place Lane, Edinburgh EH12 5AE
Tel 0131 225 4438 Fax 0131 226 3673 Website www.nb-sc.co.uk

STONE CONSERVATION SERVICES: Throughout Scotland and Northern England Nicolas Boyes Stone Conservation specialises in the conservation of carved and decorative stonework – sculpture, statuary, architectural decoration and historic building fabric in traditional materials – and decorative plasterwork. Replication of subjects can be undertaken. The Phoenix Conservation Laser system is often employed to remove soiling, and environmental monitoring and preventive conservation services can be provided as part of the full service.
The company conducts important conservation work from recording through to report with sensitivity and imagination, on time, within budget and with good conservation rigour.

▶ PAYE STONEWORK & RESTORATION LTD

44-46 Borough Road, London SE1 0AJ
Tel 020 7928 4000 Fax 020 7928 4004 Website www.payestone.co.uk

MASONRY CLEANING SPECIALISTS: *See also: profile and display entries in Stone section, page 111.*

MASONRY CLEANING

▶ SARABIAN LIMITED

Sarabian House, Kington St Michael, Chippenham, Wiltshire SN14 6JB
Tel/Fax 01249 750113
E-mail sarabian@fsbdial.co.uk

BUILT HERITAGE CLEANING SPECIALIST: Establishing a reputation for consistently producing high quality results in cleaning building fabrics, wrought iron, lead, bronze and brass architecture with sympathetic cleaning methods. Sarabian Limited can provide DOFF, powered granulate system, poulticing and mild chemical treatments for fabric cleaning and paint removal.

▶ ST BLAISE LTD

Westhill Barn, Evershot, Dorchester, Dorset DT2 0LD
Tel 01935 83662 Fax 01935 83017
E-mail stblaise@compuserve.com

JOS, DOFF, MORA, NEBULOUS, MICRO, DRY AND WET AIR ABRASIVE AND POULTICING: *See also: entries in Building Contractors section, page 65.*

▶ STONEGUARD

St Martins House, High Street, Ruislip, Middlesex HA4 7AU
Tel 01895 675577 Fax 01895 679125
Website www.stoneguard.co.uk
▸ with offices also in Bath, Birmingham, Bradford, Manchester and Stirling

MASONRY CLEANING AND PROTECTION: *See also: profile entry in Building Contractors section, page 66.*

▶ STONEHEALTH LIMITED

Bowers Court, Broadwell, Dursley, Glos GL11 4JE
Tel 01453 540600 Fax 01453 540609
E-mail info@stonehealth.com
Website www.stonehealth.com

STONE CLEANING SYSTEMS: Best known for the supply of leading specialist equipment, such as JOS/Torc and DOFF, the company also offers other cleaning methods for restoration and conservation. Stonehealth was responsible for establishing the 'Register' of trained operators, seen as important in helping to maintain standards, by allowing authenticity of a contractor to be checked prior to contract commencement. Stonehealth carries a wide up-to-date range of safer chemical products; from prevention, eg anti-graffiti, floor treatments, through to restoration, eg soot, metal oxide, graffiti removal etc. Capable of supplying 'specifically formulated' products and even extremely innovative ones, such as the company's new product, applied as cream/paste, but peeled off 24 hours later as a latex film containing removed dirt and pollutants. *See also: display entry in this section, page 173.*

▶ STONEWEST LTD

Lamberts Place, St James's Road, Croydon CR9 2HX
Tel 020 8684 6646 Fax 020 8684 9323

BUILDING CONTRACTORS AND STONE MASONS: *See also: display entry in Building Contractors section, page 64.*

▶ SUFFOLK BRICK & STONE CLEANING COMPANY LIMITED

Dickens House, Old Stowmarket Road, Woolpit, Bury St Edmunds, Suffolk IP30 9QS
Tel 01359 242650 Fax 01359 242621

BRICK AND STONE CLEANING: Established in 1984, Suffolk Brick & Stone Cleaning Company has been involved in many major projects in the Eastern Counties using sympathetic methods to clean and restore brick and stone work. These projects include: Old Addenbrookes, Cambridge; Norwich Castle; Bedford Town Hall; St Dunstans, Fleet Street. The company is a registered Jos system contractor and uses other sympathetic cleaning methods as appropriate. Masonry repairs, pointing, waterproofing and paint removal are also undertaken as part of the overall service or as separate assignments. Please contact Greg Simonds, General Manager, for more information and helpful advice.

PAINT REMOVAL

▶ HIRST CONSERVATION

Laughton, Sleaford, Lincolnshire NG34 0HE
Tel 01529 497449 Fax 01529 497518

REMOVAL OF SPECIFIC PAINT LAYERS TO INTERIOR AND EXTERIOR SURFACES: *See also: display entry on the inside front cover and profile entry in Building Contractors section, page 62.*

▶ STONEHEALTH LIMITED

Bowers Court, Broadwell, Dursley, Glos GL11 4JE
Tel 01453 540600 Fax 01453 540609
E-mail info@stonehealth.com
Website www.stonehealth.com

PAINT REMOVAL AND ANTI GRAFFITI PRODUCTS: *See also: display entry on page 173 and profile entry on this page.*

ENVIRONMENTAL MONITORING

Holistic and Sustainable Solutions to Conservation

JAGJIT SINGH

Deterioration of historical building materials is attributed to changes in their environment. The majority of environmental problems are associated with those defects in the fabric that lead to water penetration, condensation and dampness in the building fabric. Severe salt efflorescence, damp staining, blistering of finishes and timber decay in buildings are mainly the result of water penetration.

However the causes of deterioration are also influenced by the building's internal environment. Humidity, temperature and ventilation all contribute to this microclimate, which will vary depending upon the building structure and the envelope of the internal building fabric.

ENVIRONMENTAL MONITORING

There is little point in dealing with decay if the causes of the decay are not dealt with first. Indeed, it is often necessary to treat the cause alone. When dealing with historic building fabric the historic value of the original material often justifies retaining partially decayed material, provided that neither its integrity, nor that of the building of which it is part, is jeopardised in any way.

Where the causes of decay are not obvious it is necessary to carry out a thorough study of the environmental conditions to identify the cause of decay. This is done by employing a range of hand held instrumentation, physical sampling and sensor technology to monitor various parameters within the fabric of the building.

Environmental monitoring may also be justified where the recurrence of a defect is unlikely to be detected before extensive damage has been caused, for example in the roof space above an auditorium. In this case long-term environmental monitoring will be required.

***Monitoring St George** A probe installed by English Heritage monitors variations in relative humidity and ambient temperature of a magnificent but highly vulnerable wall painting in the castle chapel at Farleigh Hungerford near Bath.*

ENVIRONMENTAL MONITORING, METHODOLOGY AND OVERVIEW OF APPROACH

The first step in the investigation of a problem building is to carry out a thorough inspection of the building for defects. Then:

- Establish moisture contents in affected materials, such as timber, plaster, masonry, insulation materials and textiles.
- Establish the humidity, temperature and dew point in the environment, both internally and externally. (The dew point is the point at which air-borne moisture condenses due to a fall in temperature, for example in a porous masonry wall which is cold on one side and warm on the other.)
- Investigate in greater detail as necessary the moisture profiles in large dimension timbers and across masonry masses.

This information can be determined by:

- Measuring moisture contents of timber with resistance based moisture meters. Probes can also be used to measure moisture contents at depth in large section timbers and those built into masonry.
- Surface moisture readings in plaster and masonry using moisture meters. These will indicate if a wall is dry but can give false readings of dampness (see below).
- Where possible, mortar samples should be taken of the areas affected to determine accurately the moisture and salt content of the masonry. This does, however, have the disadvantage of not being non-destructive.
- Data loggers used to measure the environmental parameters (temperature, humidity and dew point in particular) both internally and externally.
- Specialist probes used to measure moisture across masonry walls.

The results of all or some of the above tests will establish the cause and enable a solution to the problem to be put forward.

Mortar sample analysis

Mortar sample analysis is one of the most important tools in establishing accurately the moisture levels in masonry and plasters. Where moisture levels are high it is also possible to determine how long there has been a damp problem from the salt content, a high salt content indicating a long-established problem. Mortar sample analysis can also be useful to determine the type of salt when trying to establish whether there is a genuine problem with rising dampness. However taking samples of mortar or plaster for analysis has the disadvantage of causing some damage, and might not be appropriate where, for example, ornate plasterwork is concerned.

Timber moisture contents

Timber moisture contents above 20 per cent indicate unacceptably high moisture levels in the building. If this is a general moisture level rather than localised then this is likely to be associated with high humidity in the building. Localised high readings are more likely to be associated with a building defect. For instance, high readings in a built-in end of a timber would indicate that the wall was damp, posing the threat of future timber decay. The options are to isolate the timber from the wall, provide an air gap around the timber to allow the timber to breathe, or to eradicate the source of damp and monitor the timber as the wall dries out. Which option is chosen will be determined according to each situation.

Masonry moisture monitoring profiles across walls

Measurement of the moisture across the thickness of a wall is a specialised task as there are no instruments available off the shelf for carrying this out. Tailor made probes are used containing hygroscopic materials (materials which absorb moisture). These are placed in the wall at varying depths and sealed off from the outside environment. After some time the probes are removed and their moisture content analysed. This method will give an indication of moisture levels across the thickness of the

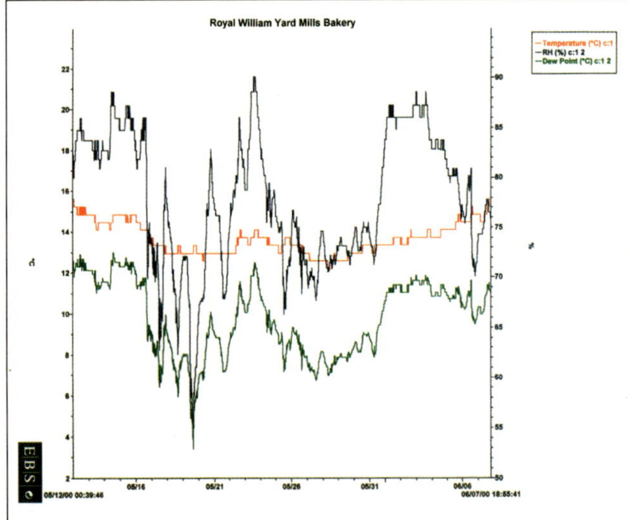

An example of a resistograph measurement of moisture content in timber

wall and combined with temperature and humidity readings both internally and externally will give an indication of the moisture source (see below). However, it must be pointed out that the use of hygroscopic material to measure moisture is inaccurate at higher moisture levels.

Environmental data loggers

Data loggers measuring temperature and humidity are useful to determine whether there is, for instance, an abnormally high humidity or whether there is a risk of condensation in a building.

If readings are taken on both the interiors and exteriors of the building, dew points can be calculated within materials such as masonry masses.

STABILISING THE HISTORIC ENVIRONMENT

For the holistic and sustainable conservation and preservation of historic buildings, stable environmental conditions are important.

Once investigations have been completed, a strategy can be devised to stabilise the building's environment. Various building works may be required to prevent further water penetration and to maximise ventilation to damp affected materials. Correction of these building defects, combined with measures to dry out the wet areas and to protect any decorative interior finishes by allowing ventilation of the wet areas, will prevent further deterioration. If thoughtfully and competently carried out, such work may extend the life of the building indefinitely, with dignity.

Until the drying out of the building fabric and its associated timber elements is completed, any other actions to remedy the deterioration problems will be ineffective and a waste of time and resources.

In some situations it may well be necessary to introduce both continuous long-term monitoring and preventative maintenance. Long term monitoring may be necessary for the following reasons

- To provide information on the state of moisture equilibrium and balance (moisture sources, reservoirs and sinks) in the building's environment, its fabric and its structural elements as it dries out.
- To allow co-ordination and scheduling of work stages to prioritise remedial work to achieve acceptable levels of moisture in the masonry and timber and to prevent future deterioration problems.
- To allow a cost-effective, long-term holistic approach to environmental stabilisation of the historic environment.

RECOMMENDED READING

Singh J, *Building Mycology, Management of Health and Decay in Buildings*, Spon, London 1994

Singh J, *Dry rot and other wood-destroying fungi: their occurrence, biology, pathology and control.* Indoor + Built Environment 1999; 8: 3-20

Dr JAGJIT SINGH, Director of Environmental Building Solutions Ltd, is an independent consultant specialising in building health problems, heritage conservation and environmental issues. His current research focuses on interrelationships of building structures and materials with their environments and occupants.

DAMP & TIMBER DECAY

▶ BALMORAL STONE LTD

31 Bankhead Drive, Sighthill, Edinburgh EH11 4DN
Tel 0131 453 4777 Fax 0131 453 6077
WOODWORM AND DRY ROT: *See also: display entry in Stone section, page 106.*

▶ CALDER PRESERVATION SERVICES

'Woodhurst' Cattlegate Road, Crews Hill, Enfield, Middx EN2 8AU
Tel 01707 876515 Fax 01707 872413
SOLUTIONS TO DAMP AND DECAY: Working exclusively in London north of the Thames up to Hertford town, Calder solves period properties' damp and decay problems. A Sovereign Approved contractor, Calder uses innovative and sympathetic solutions to eradicate wet and dry rot and timber infestations. Structural timber engineering *in situ* restoration and repairs, damp proofing and tanking treatments form a large part of their work. Established 1966, this family business' portfolio of projects includes many historic buildings, mostly in residential but also in commercial use. Fully insured, 30 year protected guarantees are available. *See also: Calder Group display entry in Building Contractors section, page 59.*

▶ DELTA MEMBRANE SYSTEMS LTD

Unit 7, Bassett Business Centre, Hurricane Way, North Weald, Essex CM16 7AA
Tel 01992 523811 Fax 01992 524046 Website www.deltamembranes.co.uk
DAMP TREATMENT: Delta Membrane Systems Ltd provides a range of products which are designed to provide lasting and cost effective solutions for damp contaminated and degraded buildings. From basements to vaults and from floors to walls, Delta has systems available which can completely transform almost any damp area. The systems are British Board of Agre'ment approved and also come with peace of mind in the form of a 30 year guarantee. A free site visit and assessment is available by contacting Delta on 01992 523811, alternatively visit the company website, or request a brochure.

▶ DEMAUS BUILDING DIAGNOSTICS LTD

Stagbatch Farm, Leominster, Herefordshire HR6 9DA
Tel 01568 615662 E-mail info@demaus.co.uk Website www.demaus.co.uk
STRUCTURAL TIMBER TESTING AND BUILDING DIAGNOSTICS: Demaus Building Diagnostics specialises in the detection and assessment of decay, weakness and fire damage in structural timber using non-destructive techniques. *See also: profile entry in Non-destructive Investigations section, page 50.*

▶ HUTTON+ROSTRON ENVIRONMENTAL INVESTIGATIONS LIMITED

Netley House, Gomshall, Surrey GU5 9QA
Tel 01483 203221 Fax 01483 202911
E-mail ei@handr.co.uk Website www.handr.co.uk
Contact Tim Hutton MA MSc MRCVS
CONSULTANTS ON TIMBER DECAY, BUILDING FAILURES AND ENVIRONMENTS: Simple solutions to common problems and expertise covering biodeterioration, structural decay, timber strength grading, damp, environmental health, non-destructive surveying and building monitoring systems. H+R carry out independent site and laboratory investigations providing specifications for remedial work or conservation. Expert witness work is also undertaken. They operate the Rothound® dry rot search dogs and install Curator® electronic environmental and structural monitoring systems. Resurgam®, a division of H+R, specialises in building conservation. Clients include The Royal Household, National Trust, English Heritage, national and local government, engineers, surveyors and property owners. *See also: Resurgam profile entry in Heritage Consultants section, page 39.*

▶ MAYSAND PRESERVATION CO LTD

109-111 Windsor Road, Oldham, Lancs OL8 1RH
Tel 0161 628 8888 Fax 0161 627 0996 E-mail sales@maysand.co.uk
MASONRY AND TIMBER CONSERVATION CONTRACTORS: Maysand's preservation team includes highly skilled technical staff to offer solutions required to preserve and conserve the fabric of buildings. These include preventative treatment against dampness, fungal and insect infestations within buildings. Other services include chemical damp-proofing, underground waterproofing tanking systems, timber engineering, specialist timber preserving treatments and building contracting services for historic, ecclesiastical and commercial projects. Members of BWPDC. *See also: display entry in Stone section, page 110.*

DAMP & TIMBER DECAY

▶ RENLON LTD

Richardson House, Boundary Business Court, Church Road, Mitcham, Surrey CR4 3TD
Tel 020 8687 4000 Fax 020 8687 4040
E-mail surrey@renlon.com Website www.renlon.com

DAMP AND TIMBER DECAY PROTECTION: Renlon provide a full range of services to protect buildings, including basement waterproofing and condensation control. Qualified and experienced surveyors work together with the client's professional team to correctly diagnose the cause of problems and design proposals to suit the property. Methods proposed offer the least disturbance to historic fabric and appearance whilst providing full protection with Lloyds insurance backed guarantees. Now available without the use of toxic chemicals. Many of the South East's oldest and most valuable buildings are protected by Renlon guarantees.

▶ TERMINIX LIMITED

Heritage House, 234 High Street, Sutton, Surrey SM1 1NX
Tel 020 8661 6600 Fax 020 8642 0677 Branches 0800 789500

SPECIALISTS IN SOLVING DAMP AND TIMBER DECAY PROBLEMS: Founded over 50 years ago Terminix (formerly Peter Cox) has extensive experience in the repair of historic and listed buildings specialising in damp proofing, waterproofing, treatments for woodworm and fungal decay, epoxy resin repairs, wall stabilisation and condensation control. Specialist insurance to cover these problems is also available. Its unique Transfusion system for installing a remedial dpc has been proven in use in countless properties. A nation-wide service is provided and most work carries a 30 year guarantee. A member of the BWPDA, the company is also registered under ISO 9002.

▶ WARD & DALE SMITH, CHARTERED BUILDING SURVEYORS

The Walker Hall, Market Square, Evesham, Worcs WR11 4RW
Tel 01386 446623 Fax 01386 48215 E-mail wds@ricsonline.org

INDEPENDENT SURVEYS, REPORTS, ADVICE ON TIMBER AND DAMP RELATED PROBLEMS

TANKING

▶ CALDER PRESERVATION SERVICES

'Woodhurst', Cattlegate Road, Crews Hill, Enfield, Middx EN2 8AU
Tel 01707 876515 Fax 01707 872413

SOLUTIONS TO DAMP AND DECAY: *See also: profile entry in Damp & Timber Decay section, page 176.*

ENVIRONMENTAL MONITORING

▶ HANWELL INSTRUMENTS LTD

Unit 12, Mead Business Centre, Mead Lane, Hertford, Herts SG13 7BJ
Tel 01992 550078 Fax 01992 589496
E-mail sales@hanwell.com
Website www.hanwell.com

SPECIALISTS IN ENVIRONMENTAL MONITORING AND CONTROL SOLUTIONS: Hanwell have built a solid reputation as true specialists in environmental monitoring and control, providing simple solutions to common problems. Currently available hardware includes miniature temperature and humidity loggers, Lux loggers, Lux and UV loggers, Shock and Tilt transportation loggers, wire free radio telemetry monitoring and control systems. Other products include dehumidifiers, humidifiers and a combined humidifier and dehumidifier unit. Their client list includes many of the World's major heritage institutions. Hanwell's policy is one of continual development and improvement. Please contact Hanwell Instruments to discuss your requirements.

▶ MUNTERS

Blackstone Road, Huntingdon, Cambs PE29 6EE
Tel 01480 442327 Fax 01480 458333

ENVIRONMENTAL MONITORING AND CONTROL: *See also: display entry in Damp & Timber Decay section, page 177.*

EPOXY RESIN REPAIRS

▶ ROBINSONS PRESERVATION LIMITED

38 Kansas Avenue, Salford M5 2GL
Tel 0161 872 3133 Fax 0161 872 6167

STRUCTURAL TIMBER REPAIRS: *See also: display entry in Damp & Timber Decay section, page 177.*

▶ ROTAFIX

Rotafix House, Abercraf, Swansea SA9 1UX
Tel 01639 730481 Fax 01639 730858
E-mail rotafixltd@aol.com Website www.rotafix.co.uk

TIMBER ENGINEERING USING THE RESIWOOD™ SYSTEM: Rotafix is a leading European organisation in the development, formulation and manufacture of polymer systems for use in the repair, restoration and upgrading of ancient and modern timber structures – water mills, windmills, wind turbines, castles, cottages, museums, bridges, barns and boats. Contact Rotafix for literature and invaluable information on structural timber repair systems. Rotafix provides two 1-day CITB approved certificated training courses. The first covers the Principles of Timber Engineering with particular reference to restoration and repair and the second covers Repair and Restoration of Brickwork, Stone, Masonry and Concrete.

▶ TERMINIX LIMITED

Heritage House, 234 High Street, Sutton, Surrey SM1 1NX
Tel 020 8661 6600 Fax 020 8642 0677 Branches 0800 789500

EPOXY RESIN REPAIRS: *See also: profile entry in Damp & Timber Decay section on this page.*

BIRD CONTROL

▶ BALMORAL STONE LTD

31 Bankhead Drive, Sighthill, Edinburgh EH11 4DN
Tel 0131 453 4777 Fax 0131 453 6077

BIRD CONTROL: *See also: display entry in Stone section, page 106.*

▶ MICROBEE BIRD CONTROL LIMITED

Unit 1, Windsor Park, 50 Windsor Avenue, London SW19 2TJ
Tel 020 8540 9968 Fax 020 8540 7477
E-mail microbee.co.uk
Website www.microbee.co.uk

PEST BIRD CONTROL: Established in 1984 the company develops, manufactures and installs a wide range of devices to exclude and prevent feral pigeons from landing on buildings. These include Micropoint anti-perching pins, and Microwire stainless steel anti-perching wires as well as anti-pigeon enclosure netting to protect light wells. These products can be seen (with binoculars) on Canada House, Trafalgar Square, The Royal Opera House and thousands of other locations ranging from Royal Palaces to rubbish dumps. The company also fits exclusion netting to protect light wells and can sterilise and remove pigeon fouling.

INSECT ERADICATION

▶ THERMO LIGNUM UK LTD

19 Grand Union Centre, West Row, London W10 5AS
Tel 020 8964 3964 Fax 020 8964 2969 Mobile 07946 830013
E-mail thermolignum@btinternet.com
Website www.thermolignum.com

CHEMICAL FREE ERADICATION OF INSECT PESTS: Eradication of insect pests from all types of organic material including timber, textiles, leather, paper, etc using the unique 'Warmair' and 'Noxia' processes. Thermo Lignum® eco-friendly techniques use only naturally occurring elements, sound biological principles and established laws of physics to eradicate damage-causing insects (woodworm, moth, carpet beetle etc) in objects ranging from art to architecture: furniture, frames, sculptures, upholstery, structural timber, architectural components. A mobile service is available for large collections, immoveable objects and timber structures within buildings. Clients include the national museums and galleries, major auction houses, fine art dealers, leading restorers and conservators, architects and surveyors and owners of historic houses and buildings.

STRUCTURAL METAL TIES & FIXINGS

▶ HELIFIX LIMITED

21 Warple Way, London W3 0RX Tel 020 8735 5222 Fax 020 8735 5223
E-mail info@helifix.co.uk
Website www.helifix.co.uk

STRUCTURAL REPAIR AND STABILISATION: For all remedial applications, Helifix has developed a comprehensive, cost-effective range of high performance, grade 304 and 316 stainless steel ties, fixings and non-disruptive masonry repair and reinforcement techniques. They are all independently tested, fully proven, widely used, manufactured to ISO9002 quality assured standards and supported by full technical data. These specially engineered stress free products are fully concealed for a sympathetic repair and provide a secure, reliable connection in bricks, blocks, stone, concrete and timber in buildings and masonry structures of all types and ages. Helifix provides a fully technical support service with advice, site visits, design specifications and quality installation by their national network of trained approved installers.

▶ PROTOVALE OXFORD LTD

Rectory Lane Trading Estate, Kingston Bagpuize, Abingdon, Oxfordshire OX13 5AS
Tel 01865 821277 Fax 01865 820573
E-mail info@protovale.co.uk
Website www.protovale.co.uk

WALL TIE LOCATORS: The Imp® wall tie locator rapidly locates mild and galvanized steel wall ties with pinpoint accuracy. Due to its unique technology the Imp does not require constant retuning. This combined with the loud audio tone, makes the Imp fast, accurate and easy to use. The Imp is also used by structural conservators and structural investigators throughout the world to assist in locating and mapping ferrous inclusions, from simple dog cramps to major structural steel. Protovale's research and development department, working in conjunction with GB Geotechnics Limited of Cambridge, has built a down hole cover meter for the precise location of cramps and cathodic protection.

▶ TAUNTON FABRICATIONS

Cornishway East, Galmington Trading Estate, Taunton, Somerset TA1 5LZ
Tel 01823 324266 Fax 01823 351862
E-mail mail@tauntonfab.co.uk
Website www.tauntonfab.co.uk

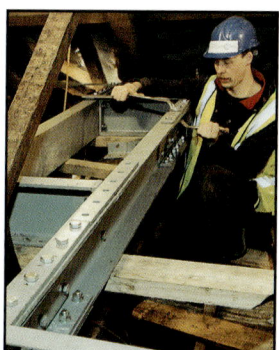

STRUCTURAL METAL FIXINGS: A unique and specialised division within Taunton Fabrications, requiring a great deal of attention to detail, is the repair and renovation of steelwork and steel reinforcing of masonry/timber to listed and architecturally protected buildings. Normally working in conjunction with conservation architects, engineers and builders, consideration to the type of materials to be used with an unobtrusive approach to the final visual effect is of paramount importance. Access to some sites can often be extremely demanding and the company's experienced and specialist staff, with over 30 years experience, are able to liaise closely with the client to offer a well planned and thought out approach to practical issues. *See also: display entry on this page.*

TRADITIONAL NAILS

▶ GLASGOW STEEL NAIL CO LTD

Lowmoss, Bishopbriggs, Glasgow G64 2HX
Tel 0141 762 3355 Fax 0141 762 0914
E-mail 101453.3140@compuserve.com
Website ourworld.compuserve.com/homepages/glasgowsteelnail

MANUFACTURERS OF TRADITIONAL CUT NAILS

▶ OLD HOUSE STORE

Hampstead Farm, Binfield Heath, Nr Henley-on-Thames, Oxfordshire RG9 4LG
Tel 0118 969 7711 Fax 0118 969 8822
E-mail info@oldhousestore.co.uk
Website www.oldhousestore.co.uk

SUPPLIERS OF TRADITIONAL ROSEHEAD NAILS: *See also: display entry in General Building Materials section, page 158.*

GENERAL FASTENINGS

▶ AVON STAINLESS FASTENERS

Avondale Business Centre, Woodland Way, Kingswood, Bristol BS15 1AW
Tel 0117 960 6665 Fax 0117 960 6668
E-mail sales@asfast.sagehost.co.uk

STAINLESS FASTENINGS: Full range of stainless steel fasteners (nuts, bolts, screws, woodscrews, coach screws, coach bolts, studding, plate washers, nails, tying wire, plain bar etc) for the timber building and conservation industries. Product guide available on request.

STRUCTURAL TIMBER TESTING

▶ DEMAUS BUILDING DIAGNOSTICS LTD

Stagbatch Farm, Leominster, Herefordshire HR6 9DA
Tel 01568 615662 E-mail info@demaus.co.uk
Website www.demaus.co.uk

STRUCTURAL TIMBER TESTING AND BUILDING DIAGNOSTICS: Demaus Building Diagnostics specialises in the detection and assessment of decay, weakness and fire damage in structural timber using non-destructive techniques. *See also: profile entry in Non-destructive Investigations section, page 50.*

One side-wall sprinkler located above the capital provides protection for this room and its contents. The house remained in continuous use throughout the installation, and by using special dust-extraction drills, this table was laid just one hour after installation.

The dome of the US Library of Congress – spot the sprinkler!

FIRE SUPPRESSION IN HISTORIC BUILDINGS

DAVID GIBBON and IAIN FORBES

For many years fire suppression systems have been distrusted. There is a mental connection between sprinklers, the principal form of fire suppression used today, and fire alarms; at a basic level some people may suppose that if one sprinkler in a building goes off, they all go off. But in reality sprinklers and alarms perform quite different roles. Sprinklers are purely mechanical devices that deliver water to the point where it is needed most, not electronic devices like smoke and fire detectors. They are a simple and robust form of technology that has been in use for more than 100 years. In recent years much work has gone into reducing the size and number of sprinkler heads.

The objective of a fire suppression system is to 'knock down' and 'hold down' a fire so that it can be finally put out and prevented from re-kindling. The final extinguishment and 'break up' of a fire is usually the task of the fire service.

Where a fire cannot be 'seen' or reached because, for instance, the fire is seated within a bookcase or in a void, a properly designed sprinkler system will, nevertheless, control the fire by its wetting action, preventing it from growing outwards and upwards. Such a controlled fire will burn out, usually in 20 minutes, or be extinguished by steam.

By contrast with a suppression (sprinkler) system, a fire extinguishing system such as a gas or mist system, has a far more difficult task, particularly where it depends on a single release of gas or mist. If the fire is not fully extinguished, it can easily re-kindle, so such systems must be designed to carefully match results on test fires in test rooms identical to the actual rooms in the building. Furthermore, they may not work if, for instance, a door or a window is left open.

CONSIDERATIONS FOR INSTALLATION

A sprinkler system consists of three components: a water supply, distribution pipework and sprinkler heads. There are no detectors other than the sprinkler heads themselves and there is no necessity for electrical linkages except to provide automatic notification – for example, to the fire brigade and building managers.

WATER SUPPLY

The mains water supply may be sufficient in itself but if not, it will be necessary to install a storage tank and, if an adequate head of water is not available, a pump. For many historic buildings, the tank size required by the Loss Prevention Council (LPC) Code may, in practice, be unnecessarily large. This is one of the areas where expert engineering design may be called for to find a non-standard design solution.

Typically a fire in a residential building will be controlled by a few sprinkler heads requiring only a small supply of water to which the fire brigade can, if necessary, connect up when they arrive on the scene. In some cases a lake or a pond may be conveniently situated to provide an assured supply. Pumps can be powered by a diesel engine as a precaution against a failure of the electricity supply, perhaps due to the fire. However, diesel engines require a fair amount of maintenance. In a recent project on a historic building and its particularly valuable contents, the insurers have agreed to accept a sprinkler supply which relies on mains water backed up by a small electric pump and tank alone.

PIPEWORK

In the past, distribution pipework was laid using steel pipes, but now copper, stainless steel and, most recently, plastic pipes suitable for potable water are also used. The pipe jointing system is a critical issue. In the case of plastic pipes, solvent welded joints eliminate all hotwork and avoid the problems of using large tools in confined spaces. Plastic pipes can also be cut to size on site and fitted into awkward spaces, so avoiding the need for costly prefabrication.

It is essential to design plastic pipework to maintain water flow in a fire; usually this means that each sprinkler head must be fed from two directions. The pipework is isolated from the supply by robust main valves and provision is made to drain down the system and to simulate the discharge of a sprinkler head for test purposes.

SPRINKLER HEADS

Sprinkler heads come in various designs. Modern miniature sprinkler heads respond rapidly when exposed to heat from a fire. The linkage in them is commonly a glass bulb containing alcohol that causes the bulb to shatter at a precise temperature. There are two main types: one sprays water equally in all directions and is usually set in a ceiling;

The mechanical works for this sprinkler system are hidden in cupboards beneath two public telephones

CATHEDRAL COMMUNICATIONS LIMITED

the other is a sidewall sprinkler which has a deflector fitted to it which enables it to throw the water from the wall. It is the latter type of sprinkler that has a particular application in historic buildings. Both types have considerably enhanced coverage by comparison with their predecessors.

So-called 'concealed' sprinkler heads are also available which are hidden behind a flat plate that drops off in a fire. Unfortunately the plate has to stand slightly proud of the adjoining surface making it visible and the performance of a 'concealed' head in a fire is slower so the whole system must be correspondingly beefed up. These considerations usually negate any benefit in historic buildings.

PROS AND CONS

Effective fire prevention measures such as staff training and smoke detection systems are often relied on as an alternative to fire suppression. Yet early warning by smoke detection is very far from being a guarantee against fire loss, and it can be completely ineffective, even with automatic dialling. Disasters such as Uppark and Windsor Castle happened despite the prompt attendance of the fire service.

Most fires in sprinklered buildings are controlled by the action of one or two sprinkler heads at most. Sprinkler heads are basically valves that are held shut by a temperature sensitive linkage. They are rated to respond to the sort of temperature that can only be the result of a flaming fire. By the time a sprinkler head goes off the fire that set it off has, by definition, reached the stage where only water will stop it. By acting directly onto the seat of a fire, sprinklers reduce the level of water damage compared with the only alternative, the fireman's hose or hydraulic monitor. By acting early in the development of a fire, both fire and water damage are minimised. The often repeated assertion that sprinklers are synonymous with water damage is therefore nonsense.

Properly designed, a sprinkler system provides the highest possible assurance against significant fire damage to buildings and their contents. When, a few years ago, the IRA firebombed Oxford Street, those shops that were unprotected by sprinklers were gutted. Those that had sprinklers were trading again immediately. Life safety is also enhanced in a sprinklered building. The only blemishes on the near perfect safety record of sprinklers have resulted from failures that have occurred where either the system was not correctly designed or when the main valves were closed off at the time of the fire.

THE EXPERTS' VERDICT

The benefits of sprinklers have long been recognised by the insurance industry. Discounts on premiums are given for properties protected by sprinklers. The Loss Prevention Council (LPC) in the UK sets very exacting standards for the installation of sprinklers on behalf of the industry. These are set out in the *LPC Rules for Automatic Sprinkler Installations (incorporating BS 5306: Part 2)*.

The benefits of sprinklers are less well recognised in the Building Regulations, largely because sprinklers have been thought of as providing protection to property rather than life safety. However, this situation is changing and, increasingly, regulatory authorities are prepared to take carefully designed fire engineering solutions into consideration.

Although English Heritage has given no clear lead on the value of sprinklers in historic buildings, the case for fitting them has gained authoritative support from Historic Scotland by way of Technical Advice Note 14 *Installation of Sprinkler Systems in Historic Buildings*. In America the debate is virtually over. For instance, the National Parks Authority, which is responsible for many of the USA's most precious historic buildings, makes widespread use of fire engineering techniques and has installed sprinkler systems in many of the buildings in its care. The fire codes in most of the individual states are based on the NFPA (National Fire Protection Association of America) Codes. Of particular relevance are NFPA 914 *Recommended Practice for Fire Protection of Historic Structures* and NFPA 909 1997 *Standards for the Protection of Cultural Resources Including Museums, Libraries, Places of Worship and Historic Projects.*

INSTALLATIONS IN HISTORIC BUILDINGS

The case for installing fire suppression systems in historic buildings is a strong one. Old buildings are particularly susceptible to fire damage, a large number of listed buildings are damaged by fire every year, many of them being burnt to the ground. If a modern warehouse is destroyed by fire, the building and its contents can usually be replaced if the money is there to do so, whereas historic buildings and their contents are, by definition, irreplaceable except by replica, which is never quite the same thing. It is therefore important to take measures to prevent fires from getting out of control.

Those who are familiar with sprinkler installations in commercial buildings will be understandably horrified at the thought of serried ranks of sprinkler heads popping through drawing room ceilings and the nightmare of threading bulky pipes through the delicate fabric of an old building. It does not have to be like that. The design of sprinkler systems for historic buildings requires a completely fresh approach. For a start, the level of protection required can be drastically reduced by comparison with a warehouse building. The average sized bedroom in a country house can usually be protected by no more than a single sprinkler head. In a relatively elaborate interior they are easy to hide particularly by using the sidewall type of sprinkler head which give remarkably good coverage. In plainer interiors much can be done to minimise their visual impact, bearing in mind that a modern sprinkler head is a fraction of the size of a smoke detector and can be finished to match the surrounding decorations.

The use of sprinklers may enable existing doors to be retained without alteration, as a carefully positioned sprinkler head may enable even a modestly constructed door to withstand the effects of fire, not for half an hour or an hour but, in effect, indefinitely. In the same way glazing can be similarly 'upgraded' and local authority building control officers acknowledge these 'trade-offs'.

A plastic sprinkler pipe with solvent-welded joints being installed in a convenient ceiling void below the floor joists

One of the first sprinkler systems installed in a historic building in the UK was at Duff House, a William Adam masterpiece in the north of Scotland. This project, which was managed by Historic Scotland in the early 1990s, was designed to provide comprehensive protection for the building. The system was remarkably successful in discreetly placing the sprinkler heads so that they go virtually unnoticed. In the mid 1990s a grand Georgian country house in Northern Ireland was the next major project of its kind. Here the sprinkler system was run in LPC approved plastic pipework, which enabled it to be installed with virtually no damage to the fabric of the building, whilst the building remained in use. Not only is the installation almost invisible, but much of the clutter of earlier fire safety installations was removed at the same time. Significant projects followed at the ancient Parliament House and at the National Library of Scotland.

One of the most interesting of the more recent sprinkler systems is at Newhailes on the outskirts of Edinburgh. This delightful house of the Scottish Enlightenment by James Smith, 1686, was refitted and extended by William Adam from about 1720. Here the National Trust for Scotland has embarked on a fascinating conservation project. The philosophy of this project, 'to conserve as found', is at variance with the requirements for visitor access. The fire officer's requirements started out being, frankly, horrendous but they virtually melted away when the intention of providing a sprinkler system was introduced. The challenge at Newhailes is to minimise visual or physical intervention. Newhailes is an extraordinary survival with a very simple yet fragile interior. All involved in the project are agreed on one thing: if it can be done here, it can be done anywhere.

Achieving such objectives often means that tortuous pipe routes have to be found through the fabric. In some cases pipe sizes far below those normally associated with commercial installations may be the only answer. The starting point must be the availability of accessible voids. The system must be designed to fit the building. To guarantee the effectiveness of such a bespoke system, the hydraulic design of every part of the system needs to be carefully worked out. This dictates a very different approach to the normal industry standard for sprinkler installations. By contrast, in a warehouse or a large modern retail unit the obvious arrangement for sprinkler pipework is in a grid. Such sprinkler systems are invariably designed and installed by specialist contractors in accordance with a system of rules. This approach is not suited to historic buildings where a more fundamental level of engineering design is required and, inevitably, the detail is not known in advance of the work.

Other issues that have been addressed in recent projects include the use of pipework materials that ensure that the sprinkler water comes out clean, the elimination of 'hot work' and, in one case, the use of sprinkler pipes to actually heat the building, so greatly reducing the total quantity of services installed.

Alternative forms of fire suppression which offer the prospect of even less water damage are being developed. These work on the principle that the hot smoke rising could draw in a mist of fine water droplets, smothering the fire. There are, however, a number of drawbacks. As with gas extinguishing systems, the enclosure must be near perfect. The technology and its underlying engineering principles are nothing like as tried and tested, and the systems are not simply mechanical like sprinklers so there is a risk of accidental discharge. As yet the reliability of such systems is not proven, so the extent to which regulatory authorities will accept their use to trade off more basic life-safety measures will be, at best, limited.

If there is a weakness in the case for installing sprinklers in historic buildings it is a lack of experience and expertise among building conservationists, consulting engineers and contractors. Nevertheless, the fundamental case for the application of fire engineering principles to the protection of historic buildings, as opposed to 'cook book' and regulations-based approaches, is overwhelming. Already, those in the field of building conservation who have grasped this particular nettle, look at projects like the restoration of Windsor Castle (where no fire suppression was installed) with some amazement, knowing that a single sprinkler head would have prevented the original tragedy.

DAVID GIBBON is a chartered building surveyor and project manager with Gibbon Lawson McKee Ltd in Edinburgh. He has been involved in various fire safety projects including the National Library of Scotland major refurbishment project and he is a member of the Buildings Committee of the National Trust for Scotland.
IAIN FORBES is a chartered engineer with Forbes Leslie Network in Glasgow. Fire suppression projects include national museums, libraries and galleries in Scotland and Ireland and the National Trust for Scotland historic buildings and collections.

PROTECTIVE

FIRE PROTECTION AND SECURITY

	CONTRACTORS	CONSULTANTS	PRODUCTS	Page
A C Wallbridge & Co Ltd	lp			84
B A J System Design Ltd			fp	182
Bramah			se	123
Carrek Ltd	lp			60
The Chapman Bathurst Partnership Ltd	lp			187
Church Conservation Limited	lp			84
Donald Insall Associates Ltd		fs		29
Eura Conservation Ltd	se	se		183
Gibbon, Lawson, McKee Limited		fs		182
Gifford and Partners	fs se	fs se		188
International Fire Consultants Limited		fs		182
Keim Mineral Paints Ltd			fb	184
Leaderflush & Shapland			fb	138
Lightwright Associates		fs se		188
MBL	se			124
Oscar Faber		fs		188
Redpath Buchanan	lp			84

KEY
ch chimney linings
fb fire resistant coatings
fp fire protection systems
fs fire safety consultants
lp lightning protection
se security products and services

PAINTS AND DECORATIVE FINISHES

		Page
Between Time	lw	58
Bursledon Brickworks Conservation Centre	lw pg	167
Chalk Down Lime Ltd	lw	168
Classidur/Blackfriar Paints	pd	183
Cornish Lime Company	lw pg	169
Ecomerchant	pd	183
Farrow & Ball	pd pg	184
Hirst Conservation Materials Ltd	lw pd	184
Keim Mineral Paints Ltd	pd	184
Keyline	lw pg	115
Knowles & Son (Oxford) Ltd	pd	62
The Lime Centre	lw pg	169
Mike Wye & Associates	lw pg	168
Old House Store	lw pd	184
Rose of Jericho	lw pd pg	184
Sanderson	pd	201
The Traditional Lime Co	lw pg	170
Ty-Mawr Lime Ltd	lw pg pd	184

KEY
lw limewash pd paints and decorative finishes pg pigments

FIRE ESCAPE LADDERS

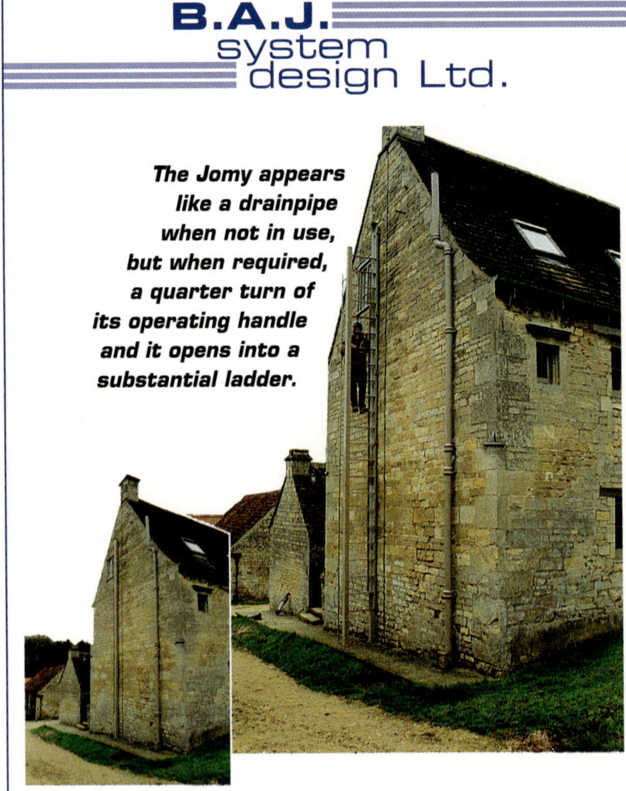

Unit 45 A & B Hartlebury Trading Estate, Hartlebury, Nr Kidderminster, Worcestershire DY10 4JB
Telephone: 01299 250052 Fax: 01299 251153

FIRE PROTECTION

► GIBBON, LAWSON, McKEE LIMITED

41A Thistle Street Lane South West
Edinburgh EH2 1EW
Tel 0131 225 4235 Fax 0131 220 0499

FIRE PROTECTION: It has been recognised for some time that fire suppression (particularly sprinklers) holds the key to the protection of historic building fabric and life safety, but until recently the problems of how to install a system in a historic building appeared to be insurmountable. GLM are now able to provide a cost effective service based on practical experience. *See also: profile entries in Architects section, page 29 and Surveyors section, page 46.*

► INTERNATIONAL FIRE CONSULTANTS LIMITED

20 Park Street, Princes Risborough, Bucks HP27 9AH
Tel 01844 275500 Fax 01844 274002

FIRE SAFETY ENGINEERS: IFC is a team of 16 full time staff offering a timely and professional service to its clients. IFC has extensive knowledge and experience in dealing with fire safety and building conservation issues in a sympathetic manner. The firm's engineers have detailed structural knowledge and are able to offer clients full architectural standard outputs. Services available include; building surveys, fire risk assessment, design and specification advice, co-ordination and design of fire tests, fire safety management, emergency plans and staff training. IFC has worked for English Heritage, The Historic Royal Palaces Agency, the National Trust as well as many private owners of historic and listed properties.

CATHEDRAL COMMUNICATIONS LIMITED

CHIMNEYS

▶ THE NATIONAL ASSOCIATION OF CHIMNEY ENGINEERS (NACE)

PO Box 5666, Belper, Derbyshire DE56 0YX
Tel 01773 599095 Fax 01773 599195

ALL TYPES OF CHIMNEY CONSTRUCTION, REPAIR AND MAINTENANCE: Previously known as the National Association of Chimney Lining Engineers, (NACLE) the National Association of Chimney Engineers (NACE) is now an association of independent companies specialising in all types of chimney works. The association runs a register of competent companies with categories of specialisation for each company. In addition to this NACE runs training schemes for operatives leading to an NVQ in chimney engineering and information seminars for professionals (surveyors, architects clerk of works etc) who require background information on chimneys, their design and correct construction.

CHIMNEY FLUE LINERS

▶ CICO CHIMNEY LININGS LIMITED

Freepost, Westleton, Saxmundham, Suffolk IP17 3EF
Freephone 0500 833787 Fax 01728 648428
E-mail cico@chimney-problems.co.uk
Website www.chimney-problems.co.uk

NATIONAL CHIMNEY LINING SERVICE: Established for over 15 years, CICO Chimney Linings provides a choice of either CICO's Refractory Cast-In-Situ linings (independently approved by the BBA) or CICO Seal – the answer where a thinner parging coat is required. CICO normally offers surveys without obligation, nation-wide.

INSURANCE

▶ LA PLAYA

PO Box 992, Waterbeach, Cambridge CB5 9SQ
Tel 01223 522411 Fax 01223 864124
E-mail property@laplaya.co.uk Website www.laplaya.co.uk
Contact Matthew Mullee

INSURANCE: La Playa offers specialist insurance and advice for listed and historic buildings, and for household contents including antiques, fine art, works of art and jewellery. Private Client Director Matthew Mullee has a wealth of experience in the field. Contact him for advice or a second opinion on risk management and security issues; insurance during extension, conversion and repair; use of craftsmen and specialist repairers/suppliers; liabilities for domestic staff and public (for open days etc); valuations for art and antiques. *See also: editorial article on page 56.*

SECURITY

▶ EURA CONSERVATION LTD

Unit H3, Halesfield 19, Telford, Shropshire TF7 4QT
Tel 01952 680218 Fax 01952 585044
E-mail mail@eura.co.uk

SECURITY FIXINGS: *See also: profile entry in Statuary section, page 115.*

PAINTS & FINISHES

▶ CLASSIDUR

Blackfriar Paints & Varnishes, Blackfriar Road, Nailsea, Bristol BS48 4DJ
Tel 01275 854911 Fax 01275 858108

MICROPOROUS PAINTS: *See also: display entry on this page.*

▶ ECOMERCHANT LTD

The Old Filling, Head Hill Road, Goodnestone, Nr Faversham, Kent ME13 9BY
Tel 01795 530130 Fax 01795 530430
Website www.ecomerchant.co.uk

ENVIRONMENTALLY FRIENDLY PAINTS: *See also: display entry in General Building Materials section, page 158.*

▶ FARROW & BALL

Showroom: 249 Fulham Road, London
Tel 01202 876141 Fax 01202 873793

MANUFACTURERS OF TRADITIONAL PAPERS AND PAINT: *See also: display entry on page 184.*

PAINTS & FINISHES

SERVICES & TREATMENT
Protective

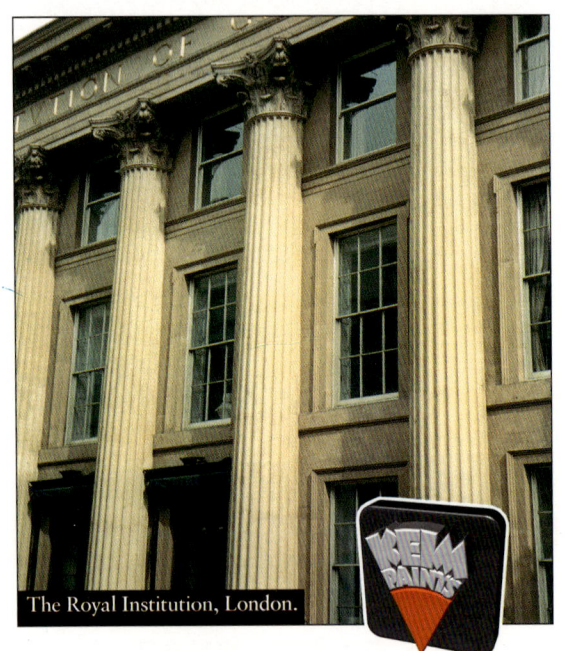

▶ HIRST CONSERVATION MATERIALS LTD

Laughton, Sleaford, Lincolnshire NG34 0HE
Tel 01529 497517 Fax 01529 497518

PAINTS AND RELATED PRODUCTS: *See also: display entry on the inside front cover and profile entry in Mortars & Renders section, page 168.*

▶ LISA OESTREICHER

Jubilee House, High Street, Tisbury, Wiltshire SP3 6HA
Tel 01747 871717 Fax 01747 871718

ARCHITECTURAL PAINT ANALYSIS: Lisa Oestreicher provides a full range of analytical skills and techniques for the study of paint and interior finishes within historic buildings. These include the identification of pigments and media as well as archival research. Full reports are prepared to provide a detailed insight into the historical development of interior and exterior decorative schemes, for documentation purposes, conservation or accurate restoration. Assistance can also be given in the design and implementation of programmes of conservation and decoration. Recent clients include the National Trust, The Victoria & Albert Museum, architects, conservators and owners.

▶ OLD HOUSE STORE

Hampstead Farm, Binfield Heath, Nr Henley-on-Thames, Oxfordshire RG9 4LG
Tel 0118 969 7711 Fax 0118 969 8822
E-mail info@oldhousestore.co.uk
Website www.oldhousestore.co.uk

TRADITIONAL AND ORGANIC PAINTS AND FINISHES: *See also: display entry in General Building Materials section, page 158.*

▶ ROSE OF JERICHO

at St Blaise Ltd, Westhill Barn, Evershot, Dorchester DT2 0LD
Tel 01935 83676 Fax 01935 83017
E-mail stblaise@compuserve.com

TRADITIONAL PAINTS, LIMEWASHES, DISTEMPERS, HISTORIC AND NEW COLOURS: *See also: entries in Mortars & Renders section, page 170, and St Blaise Conservation entries on pages 65 and 199.*

▶ TY-MAWR LIME LTD

Ty-Mawr, Llangasty, Brecon, Powys, Wales LD3 7PJ
Tel 01874 658249 Fax 01874 658502
E-mail tymawr@lime.org.uk
Website www.lime.org.uk

PRODUCERS AND SUPPLIERS OF NATURAL AND TRADITIONAL BUILDING MATERIALS: *See also: profile entry in Mortars & Renders section, page 170.*

STONE CONSOLIDATION

Halts decay and prolongs life

ELIZABETH GARROD

Natural weathering of stone is inevitable, but some types have a structure that makes them more durable than others.

There are three major causes of deterioration in natural stone; pollutants, frost and crystallisation of soluble salts. Water penetration is the main instigator of decay, and structure is the most important factor influencing the ability of stone to resist decay processes.

THE PROBLEM

To illustrate some of the problems a stonework wall may be subjected to throughout its lifetime, imagine the following common scenario:

There is constant wetting and drying of the surface of a limestone wall over the years. Gradually more of the exposed surfaces begin to erode a little. This is natural weathering. A decade of extremely wet weather interspersed with unusually low temperatures heightens the natural weathering. The atmospheric pollutant, sulphur dioxide, combines with the calcium carbonate in the stone and creates a hard gypsum layer covering the surface of the stone. This means that moisture inside the stone cannot escape and salts may crystallise behind this hard layer and eventually cause spalling, which leaves a weak exposed surface, more vulnerable to natural weathering. The owners of the building notice the problem but choose the wrong type of treatment, perhaps a water repellent treatment that does not breathe. It doesn't penetrate any further than the surface and the result is a hard, impervious outer surface like the gypsum. Ultimately the same problem arises as before; spalling occurs and weathering is accelerated.

What, if anything, can be done for this stonework? If it is not possible to shelter or isolate the material from the weather in some way, and continual erosion will lead to the loss of the original, then conservation through consolidation is the answer.

PRINCIPLES

The fundamental principle of conservation is to alleviate the problems affecting a building or monument in a way that doesn't detract from its history and doesn't endanger but promotes its future. This should be done, wherever possible, with the absolute minimum of intervention and in a way that is reversible should a better way come along in the future.

When talking about stone consolidation it is important to be clear on the difference between a consolidant and a preservative. The aim of a preservative is to totally preserve the stone in whatever state of weathering it has reached and stop all future decay; this

Testing samples of different consolidants at Howden Abbey, North Yorkshire. The door surround and the pilaster provide a good example of severe stone decay.

generally means applying a coating to the surface of the stone which totally protects it from the effects of the atmosphere around it. Consolidation on the other hand should aim to stabilise the friable material whilst still allowing weathering to take place as a result of natural processes and at a natural rate. When considered in this context, consolidation appears to agree with the principles of conservation. However, the application of a chemical to a stone is usually considered a more significant intervention than most. Simple mechanical repairs and alterations such as erecting guttering to protect the stone for example are clearly more readily reversible, and 'traditional' coatings such as lime wash are considered to be sacrificial – that is to say that they provide protection as they decay. Furthermore, stone can cope with getting wet, as long as it can also get dry; it can also cope with minimal loss around the edges: indeed some stones could have been put in certain places purely to be sacrificial. When is it necessary to introduce such a significant intervention and apply a chemical to stabilise the effect of weathering?

A thorough assessment of the damage to the stone and the examination of all possible options, bearing conservation principles in mind, should always precede a decision on whether to simply carry out repairs or whether to use some form of consolidant.

THE IDEAL

The only requirement of a consolidant is to reduce the rate of decay of the stone surface and the most successful treatments are those which least alter the characteristics of treated stone leaving it similar to the underlying sound stone.

The ideal polymer for use in stone consolidation would be one that can reverse the degradation of a stone, returning it as nearly as possible to its original condition. In order to achieve this the treated stone should mimic sound stone in as many characteristics as possible. Some characteristics are, however, more important than others. The most important of these are strength, porosity, permeability, thermal dilation and colour. Of all the polymers, silanes seem to hold out the most promise although they may not be suitable in every situation. The theoretical end-product of polymerisation of the simplest silanes is silica, which is present as a cementing mineral in many sandstones and may mimic the behaviour of a natural cement more closely than many other polymers.

DIFFICULTIES

Traditionally waxes and linseed oil were used as water repellents but they had several flaws:

- there was no possibility of deep penetration by these viscous substances;
- they tended to discolour the surface of the stone;
- the surface picked up dirt very easily;
- after a longer period of exposure there could be a breakthrough of salt efflorescence;
- generally the water repellency of these substances caused problems.

For modern consolidant treatments these issues still remain but the most important are depth of penetration and water repellency.

DEPTH OF PENETRATION

There is agreement that treatments confined to the outer surface of stonework are dangerous since they can result in spalling. However, there is no agreement on what would be an appropriate depth of treatment beyond the fact that it is obviously necessary to treat the stone deeply enough to consolidate the full thickness of the decayed zone. It is necessary for the consolidant to penetrate at least 25mm into the stone to reach all areas of friable material.

WATER REPELLENCY

Since water plays such an important role in the decay of stonework, research into consolidants has often focused on their ability to make stone water repellent. However, this may prevent the harmless effects of natural weathering as well as the more damaging ones, causing the appearance of the stone to change.

More seriously, if a consolidant or preservative is water repellent, not only will the ingress of water be lessened but also water already inside the stone will be hampered from making its way to the surface to evaporate. The result of this can be salt crystallisation and eventually spalling. To prevent this, the stone would have to be totally dry throughout the period when the consolidant is being applied (almost an impossibility in our climate) to be certain that no moisture is trapped.

Water repellents are usually marketed as being vapour permeable, suggesting that moisture will be able to escape to the surface of a wall and evaporate. Unfortunately water ingress may continue as vapour at the surface, as water escapes through cracks at mortar joints, as rising damp, or by transfer from surrounding stonework, often entering the structure faster than it can escape. Water repellents can reduce the amount of moisture in stone but they cannot be guaranteed to exclude all moisture.

Although opinion is divided on whether the use of water repellents can reduce stone decay, it is clear that any material that completely blocks surface porosity will lead to accelerated stone decay. Such materials should never be used.

SOLUTIONS

An ambient low humidity, a dry building and a high penetration depth are all very important in the application of a consolidant. So during the decision-making process and before application it is necessary to investigate the needs and condition of the building, monitor its environment and assess the properties of the consolidant.

There has been lots of research into consolidants over the last century to determine colour changes, strength changes, their effect on properties of different stone and even biological growth tests. *But* as yet, no one has whole-heartedly recommended one product or indeed one type of product as ideal for use in stone conservation.

There are many products on the market, all with different brand names, but their main constituents are similar and can therefore be classified into categories.

SILANE-BASED MATERIALS

Silane-based materials are generally organosilicon compounds which polymerise inside the stone. Some water is needed to aid the reaction, but the amount is critical; a high humidity means the reaction may take place too quickly and too much water leaves no space for the polymer to form. The end product of polymerisation is silica, similar to the natural silica deposits which bind many sandstones. Penetration can be quite deep but this depends greatly on the product used and the conditions in which it is applied.

By the production of silica there is a definite consolidating effect and many silane-based products seem to increase the strength (flexural, compressive, tensile etc) of damaged stone. Unfortunately, there is some colour change with most types of silanes, although studies show that this usually lessens after about 18 months. Porosity, water absorption and pore size distribution have shown to be affected by the treatment, a little in some cases and a lot in others. This influences resistance to salt crystallisation and freeze/thaw action. Where there is a new area of stabilised decayed material, moisture evaporation has to take place within the stone and this may lead to salt crystallisation at the boundary between treated and non-treated stone.

These are the main silane-based products and their main features:

- tetraalkoxysilanes – have little water repellency
- alkyl trialkoxysilanes (such as brethane) – less consolidation, but good water repellency
- polysiloxanes – flexibility and more water repellency
- silicon hydrides – use presents many health and safety problems
- halogen bearing silanes – generate damaging acids, so thought to be too dangerous to use in conservation.

ORGANIC-BASED MATERIALS

These products can be applied by themselves or dissolved in an appropriate solvent. They generally have good adhesion to the substrate and are good at taking up dimensional changes in stone (such as thermal expansion and contraction). The disadvantage of using organic-based materials is that they can be vulnerable to heat or ultra-violet (UV) light and generally the penetration depth depends greatly on the ability of the solvent to carry the consolidant into the stone and the percentage of moisture in the stone. Many products have a very low penetration depth.

These are the main organic-based products and their main features:

- acrylic consolidants – there can be some colour changes
- vinyl consolidants – very unstable in heat and light and tend to pick up dirt
- epoxies – the treated stone will be prone to yellowing and the appearance of a white powder
- polyurethanes – can alter many of the properties of the stone it is applied to, including strength, porosity, and brittleness; sometimes having quite a detrimental effect
- polyesters – have an extremely poor resistance to UV radiation and acid rain; not ideal for stone conservation
- perfluoropolyethers – good water repellents, their advantage in stone conservation is the fact that they are reversible and stable in UV light, but they have limited cohesive properties
- fluorinated elastomers – water absorption, vapour permeability and porosity can be altered by them, but cohesion is good.

INORGANIC TREATMENTS

- fluorosilicates – cannot be used on limestones since they react badly with calcium carbonate, and they are not very effective on sandstones
- barium-hydroxide – there is sometimes a colour change, but it can be a very good consolidant if applied correctly and kept wet for an accurate amount of time
- limewater – an old product that is still in use, it is reversible and simple to use.

SURFACE COATINGS

Lime treatments and shelter coatings can be applied to a surface to act as a sacrificial layer after repairs and treatment have taken place. They should be breathable and can be colour tinted to match the weathered stone.

Shelter coats are ideal in extreme environments since they provide a sacrificial layer designed to weather and protect the underlying stone, therefore halting decay and prolonging life. It is essential to keep shelter coatings in good repair with regular maintenance.

THE FUTURE

Current research, mainly centred on the properties of each type of consolidant, is largely being done in the laboratory, although these tests often give different results to those performed on actual buildings. Nevertheless, most non-proprietary research concludes that there is no ideal consolidant for use on historic stonework, confirming comments made 80 years ago; either some discoloration of the stone is caused or the altered properties of the stone cause additional problems. There is also little data on the effective lifespan of treatments beyond one or two years.

For new buildings there have been tentative suggestions of dipping building materials into a silane-based product before building. For existing buildings, there are different methods of application to consider. For example, if a water repellent consolidant is applied to each stone individually and not to the mortar in between, a good quality lime mortar could be used sacrificially to help any water escape from the building. Alternatively several carefully chosen stones could become sacrificial by not applying a water repellent to them, so that any water in the structure could make its way to them and escape. These stones may have to be monitored and replaced on a regular basis but it may aid the rest of the building.

There is some new research into dispersed hydrated lime in the recent RILEM publication *Historic Mortars: Characteristics and tests* and more extensive tests could certainly be very interesting (Strotmann, 1999).

Whatever treatment is chosen, it is important not to forget the source of damp. Most old stone buildings were built with complicated systems of water removal in place. Over time stone gullies acting as guttering may have eroded away and the system will begin to fail, resulting in stone decay. With the principles of conservation uppermost in mind the best way to consolidate the stonework of a building is to find those places where water is causing decay and make minor repairs to the building to re-institute its own system of water removal. Minor repairs and regular maintenance can in themselves halt decay and prolong life.

ELIZABETH GARROD has a BSc in Heritage Conservation and has worked at BRE for the last three years. Recently she has been involved in projects investigating lime mortars for English Heritage and extensively testing the properties of limestones and sandstones from quarries all over the UK.

HEATING AND LIGHTING SERVICES

	PRODUCTS	SERVICES	PAGE
Adrian Cox Associates		sv	40
Alpine Reclamation	li		159
Architectural Reclaim	ra		159
Best & Lloyd Ltd	li		187
The Cast Iron Company	lf li		152
The Chapman Bathurst Partnership Ltd		ht lc sv	188
Chelsom Limited	lc lf li		187
Chris Reading & Associates	lc li		188
Cico Chimney Linings Ltd	ch		183
E G Swingler & Sons	ch		82
Emberheat		sv	188
Exodus Electronic Limited	lf		187
Gifford and Partners		ec sv	188
H W Poulter & Son	ra		189
Hanwell Instruments Limited		ec	178
Hart Brothers Engineering Ltd		sv	148
Hutton+Rostron Environmental Investigations Limited		ec	176
I J P Building Conservation		sv	61
Illumin Glass Studio	li		129
James Hoyle & Son	lf li		149
Jonathan Murray Fireplaces	ra		189
Julian Harrap Architects		sv	31
LASSCO	li		160
Light & Design Associates		lc	187
Lighting Design & Consultancy		lc sv	188
Lightwright Associates		ec ht lc sv	188
Mather & Smith Ltd / M J Allen Group	lf		149
Munters Property Damage Restoration Ltd		ec sv	177
National Association of Chimney Engineers	ch		183
Original Architectural Antiques Company Ltd	li		160
Oscar Faber Consulting Engineers		lc sv	188
Peter Stephen & Partners		lc sv	41
Plowden & Smith Ltd		ec	202
R Hamilton & Company Limited	lf li	lc sv	187
Robert Bloxham-Jones Associates		sv	188
Stoneworks of Bath Ltd	ra		189
Taywood Engineering Limited		sv	68
Walcot Reclamation Ltd	li ra		162
Westland & Company	li		190
Wilkinson Plc	li		207

KEY

ch	chimney linings	lf	light fittings: display lighting
ec	environmental control	lc	lighting consultants
ht	heating engineers	ra	radiators and stoves
li	light fittings: antique and decorative	sv	services engineers

LIGHTING CONTROLS

ANTIQUE & DECORATIVE LIGHTING

▶ BEST & LLOYD LIMITED

Cambray Works, William St West, Smethwick, West Midlands B66 2NX
Tel 0121 558 1191 Fax 0121 565 3547 E-mail sales@bestandlloyd.co.uk

DESIGNERS AND MANUFACTURERS OF FINE LIGHTING: Extensive cast pattern library covering all periods including Georgian, Rococo, Adam and Hepplewhite, Victorian etc. Range includes hand finished picture lights in a range of finishes, traditional cast chandeliers and the BestLite Collection. Specialists in design, manufacture and restoration of lighting for ecclesiastical and listed buildings.

▶ CHELSOM LIMITED

Head Office: Heritage House, Clifton Road, Blackpool, Lancashire FY4 4QA
Tel 01253 831400 Fax 01253 698098

DECORATIVE LIGHTING AND REFURBISHMENT OF PERIOD LUMINAIRES: Chelsom have been supplying high quality lighting to the contract market for over 50 years. The range covers all types of luminaires in both traditional and modern styles. Chelsom specialise in the restoration of period light fittings and their conversion to low energy light sources when required. They also design and manufacture bespoke lighting to client specification. Restoration work takes place at their factory in Blackpool where there is also an experienced technical design department. Chelsom have a trade showroom in Fulham, London.

▶ R HAMILTON & COMPANY LIMITED

Quarry Industrial Estate, Mere, Wiltshire BA12 6LA
Tel 01747 860088 Fax 01747 861032

LIGHTING SPECIALISTS: Hamiltons manufactures to order wiring accessories and lighting control systems of quintessential quality. Traditional style and modern modular switches, sockets, dimmers, etc on brass, chrome, wood and polycarbonate plates. Designs include the familiar rope and bead edgings, sleek 'Linea' styles with concealed fixing, replicas of the original domed switch, System 45 with its continental influence. Bespoke configurations are a speciality. Lighting control systems for loads from 4 to 20 amp can be operated by digital or analogue control points mounted on plates matching their wiring accessories. Among the company's successes are National Trust and English Heritage sites plus prestige projects throughout the world. *See also: display entry above.*

▶ EXODUS ELECTRONIC LTD

Singleton Court, Wonastow Road, Monmouth NP25 5JA
Tel 01600 719444 Fax 01600 716744
E-mail info@exodus-electronic.com

LIGHTING CONTROLS: Exodus has developed a range of electronic lighting control products actuated by digital radio technology, the Smartswitch System. Control cabling is eliminated, with consequent savings in installation time, cost and inconvenience. Like conventional switches in appearance, the Exodus Smartswitch is available in traditional finishes, including white plastic, brass, stainless steel and wood. Receivers are designed to be permanently wired into the lighting circuit, and versions are available that control and dim all types of incandescent and fluorescent lights. Engineered to the highest quality the products are designed to meet the needs of engineers and architects working in professional environments.

EXTERIOR LIGHTING

▶ LIGHT & DESIGN ASSOCIATES

Unit 0615, Bell House, 49 Greenwich High Road, London SE10 8JL
Tel 020 8469 4000 Fax 020 8469 4005
E-mail design@lightanddesign.co.uk

INTERIOR AND EXTERIOR LIGHTING SPECIALISTS: Architectural lighting design consultants, established in 1990, specialising in the interior and exterior lighting of historic buildings including churches, museums and palaces. Lighting design awards have been gained for new lighting systems at St Luke's, Battersea, and Hinde Street Methodist Church, Marylebone. Recent commissions include concept proposals for exterior lighting at Buckingham Palace, the re-lighting of the interior to The Chapel Royal, St James's Palace, façade lighting to the Royal Exchange building, London and the re-lighting of the interior to St Bartholomew the Great, Smithfields.

BUILDING SERVICES ENGINEERS

▶ CHRIS READING & ASSOCIATES

6 Charfield Close, Winchester SO22 4PZ
Tel 01962 861496 Fax 01962 861496 E-mail consult@crassociates.freeserve.co.uk
MECHANICAL AND ELECTRICAL BUILDING SERVICES, ACOUSTICS: Specialists in the design of mechanical, electrical and ancillary building services engineering and acoustics within all types of listed and period buildings including those of national importance. A multi-disciplinary practice which aims to sensitively integrate building services into the structures in which they are placed.

▶ GIFFORD AND PARTNERS

Carlton House, Ringwood Road, Woodlands, Southampton SO40 7HT
Tel 023 8081 7500 Fax 023 8081 7600
MECHANICAL AND ELECTRICAL SERVICES FOR HISTORICAL BUILDINGS: Award winning specialist expertise in structural, mechanical and electrical services engineering and archaeology for historic buildings. This unique blend of multi-disciplinary engineering and archaeological services allows Gifford and Partners to provide a fully integrated service to their clients. Numerous commissions on buildings of national importance have been secured with clients in the Public and Private sectors including English Heritage, National Trust, Historic Royal Palaces Agency, the Royal Household, Ministry of Defence, church PCCs, along with a number of key developers and private owners. Specialisms include: post fire repair, fire safety services, environmental control, electronic security and surveillance, energy conservation and public health engineering.

▶ LIGHTWRIGHT ASSOCIATES

11 Mill Lane, Heatley, Lymm, Cheshire WA13 9SD
Tel 01925 755359 Fax 01925 756925
BUILDING SERVICES CONSULTING ENGINEERS: Established 1973, this small specialist practice is experienced in dealing with the sensitive and sympathetic design and specification of lighting, electrical, fire, security, heating and ventilation installations in listed and historic buildings in England and Wales. Clients include the National Trust, local authorities, owners of historic houses, cathedrals and churches, museums and art galleries. The practice also conducts mechanical and electrical quinquennial inspections, compiles reports on existing systems, undertakes feasibility studies and the preparation of drawings and budget costs for grant applications.

▶ OSCAR FABER

Marlborough House, Upper Marlborough Road, St Albans, Hertfordshire AL1 3UT
Tel 020 8784 5784 Fax 020 8784 5700
Website www.oscarfaber.com
Contact Doug Oughton
CONSULTING ENGINEERS: International consulting engineers providing a complete range of engineering skills including mechanical and electrical engineering; structural engineering; environmental monitoring; lighting design; acoustics; fire detection and protection and facilities management. Highly experienced staff work to quality assured standards and are supported by continuous research, development and training. Their conservation engineers provide a sensitive and holistic approach to the installation of services in historic buildings and provide a fully co-ordinated engineering service to clients. Projects include: Palace of Westminster, Windsor Castle, Chicksands Priory, Compton Verney, Oakhill Theological College, Royal Academy of Music, Courtauld and Gilbert Galleries at Somerset House, Bank of England, Royal Holloway College, Royal Exchange and Tower of London New Jewel House.

▶ ROBERT BLOXHAM-JONES ASSOCIATES

4 Lancaster Drive, Upper Rissington, Cheltenham, Glos GL54 2QZ
Tel 01451 822820 Fax 01451 822821
E-mail r.bloxham-jones@talk21.com
Contact Robert Bloxham-Jones
BUILDING SERVICES CONSULTING ENGINEERS: Attention to detail is the key to successful integration of engineering services into historic buildings. Coupled with an understanding of their uses and contents is essential in ensuring minimal intervention into the fabric. Clients include Leeds Castle, Kiplin Hall and Shakespeare Birthplace Trust.

LIGHTING CONSULTANTS

▶ LIGHTING DESIGN & CONSULTANCY

Newcombe House, 21 Market Place, Wolsingham, Co Durham DL13 3AB
Tel/Fax 01388 527809
E-mail ldc@lightingconsultants.co.uk
Website www.lightingconsultants.co.uk
Contact Michael Phillips I Eng MILE MSLL ACIBSE Lighting Diploma
INDEPENDENT LIGHTING CONSULTANTS: Founded in 1989, LDC has established extensive experience in interior and exterior lighting design for ecclesiastical and historic buildings, architecture and private residences. Commissions include over 150 cathedral and church lighting projects, many Grade I listed buildings, works of art and sculptures. LDC has a reputation for its unique approach to developing comprehensive lighting strategy reports for historic towns and conservation areas. Lighting technical expertise particularly in the control of light pollution and creative but sensitive design responses to planning and architectural values are key LDC design criteria. LDC is registered with the ILE and CIBSE as an independent lighting practice.

HEATING

▶ EMBERHEAT

The Heating Centre, 324 Battersea Park Road, London SW11 3BY
Tel 020 7223 5944 Fax 020 7223 7118
HEATING SPECIALISTS: Emberheat specialises in the design, supply and installation of heating systems to churches and other historic buildings. The unique and varied characteristics of church buildings are considered to be a challenge for the Emberheat environmental engineer. To achieve comfort, effectiveness and economy, whilst respecting the aesthetic character, requires the design of a system as individual as the building itself. From initial survey and consultation, through to installation by Emberheat's craftsmen, their many years of experience will be evident. Attention to detail, allied to that expertise, will result in an effective heating system which is visually and environmentally sympathetic to the building and its occupants.

CATHEDRAL COMMUNICATIONS LIMITED

▶ ALDERSHAW HANDMADE CLAY TILES LTD

Pokehold Wood, Kent Street, Sedlescombe, East Sussex TN33 0SD
Tel 01424 756777 Fax 01424 756888
E-mail sussexterracotta@wealden.freeserve.co.uk
Website www.sussexterracotta.co.uk www.aldershaw.co.uk

FIREPLACES: *See also: profile entry in Clay Tiles & Roof Features section, page 91.*

▶ H W POULTER & SON

279 Fulham Road, London SW10 9PZ
Tel 020 7352 7268 Fax 020 7351 0984

FIREPLACES: H W Poulter & Son specialise in the supply, installation and restoration of antique marble chimneypieces and works of art of quality of all periods. Work is executed through both its large London showrooms and workshops of highly skilled operatives. Additionally, the company supplies and restores antique artefacts associated with functional and decorative completion of the fireplace; for example: grates, fenders, fire irons, fire baskets and fire dogs. The company's expertise covers 50 years as established consultants to auction houses, insurance companies, private and business investment, architects, decorators and conservationists both in the United Kingdom and abroad.

▶ JONATHAN MURRAY FIREPLACES

358 Upper Richmond Road West, East Sheen, London SW14 7JT
Tel 020 8876 7934 Fax 020 8876 0869
E-mail jmf358@aol.com Website www.fireplaces-uk.com

ANTIQUE AND REPRODUCTION FIREPLACES: Jonathan Murray Fireplaces have been suppliers to the trade and public for over 15 years. They have earned a reputation as experts in period fireplaces, as well as in fine reproduction chimneypieces. Prestigious contracts include the Lanesborough Hotel at Hyde Park Corner, and they are well versed in requirements for listed buildings, working closely with English Heritage in the past. The East Sheen showrooms display a wide range of fireplaces and accessories, and for the many other pieces held off site, photo portfolios are available for browsing through, as is digital photography. A full consultancy and fitting service is offered.

▶ McMARMILLOYD LIMITED

Brail Farm, Wilton Road, Great Bedwyn, Marlborough, Wiltshire SN8 3LY
Tel 01672 870227 Fax 01672 870053
E-mail info@mcmarmilloyd.co.uk Website www.mcmarmilloyd.co.uk

SPECIALISTS IN MARBLE FIREPLACES: *See also: display entry in Marble & Granite section, page 96.*

▶ NOSTALGIA

Holland's Mill, Shaw Heath, Stockport, Cheshire SK3 8BH
Tel 0161 477 7706 Fax 0161 477 2267
Website www.nostalgiafireplaces.co.uk

RECLAIMED ANTIQUE FIREPLACES: Nostalgia was formed 25 years ago and specialises in the supply of reclaimed antique fireplaces in wood, marble, cast iron, slate and stone. With over 1,000 items, dating from 1750-1910 always available, it has one of the largest collections of these items in the country. There is always a selection of restored fireplaces on display, but customers are welcome to choose a piece which they wish to have restored to order. Nostalgia takes particular pride in the restoration of their marble chimneypieces. A photo service is available.

▶ STONEWORKS OF BATH LTD

Old Orchard, Walcot Street, Bath BA1 5AX
Tel 01225 311136 Fax 01225 481119 Website www.stoneworks.co.uk

MASTER STONE AND MARBLE MASONS, SPECIALISTS IN CHIMNEYPIECES AND FIRE SURROUNDS, DESIGN AND CONSULTANCY: Re-kindling the almost forgotten tradition of the design-led craft workshop, Stoneworks has been a leader in the renaissance of the use of natural stone and marble since 1990. The breadth of experience offered by Westminster Abbey trained master mason Ben Gale and J P Kuhnzack-Richards, formerly head of marble at Walcot Reclamation, is vast. Their design philosophy, which seeks to marry the wishes of the client to the needs of their building, has resulted in a comprehensive catalogue of beautiful and appropriate work, examples of which are always on display at their workshop in the centre of Bath.

▶ THISTLE & ROSE

Smailholm House, Nr Kelso, Roxburghshire TD5 7PQ
Tel/Fax 01573 460232

MANUFACTURERS OF FACSIMILE NEO-CLASSICAL CHIMNEYPIECES: The pine and gesso neo-classical chimneypiece was developed by Ramage and Ferguson of Leith and widely used by the Adam practise after Robert Adam's return from Italy in 1758. The scale of speculative building in Georgian Britain produced a high demand for well crafted fittings, and these had to be competitively priced. Thistle & Rose carries on this tradition, and also restores original chimneypieces, and manufactures items relevant to other periods. The company's approach is essentially conservation led, but as a general principle the aim is to sell at as low a cost as possible on a commissioned basis. Advice on related items such as grates and slips, is also available.

▶ WESTLAND LONDON

St Michael's Church, Leonard Street, London EC2A 4ER
Tel 020 7739 8094 Fax 020 7729 3620
E-mail westland@westland.co.uk
Website www.westland.co.uk

PERIOD AND PRESTIGIOUS CHIMNEYPIECES, ARCHITECTURAL ELEMENTS, PANELLING, FOUNTAINS, STATUARY, PAINTINGS AND FURNITURE: All these items are displayed in the vast interior of St Michael's Church in the Westland, London warehouse, and on the company's extensive website www.westland.co.uk. Westland liaise, co-operate and supply for projects of architects, decorators, developers and individuals worldwide. The company will also buy any suitable items. There are in-house workshops covering most product areas; please view the website and contact Westland for your present and future requirements. Approach by car via Great Eastern Street or by tube, Old Street, exit four. Open Monday-Saturday and by appointment. *See also: display entry above.*

PAUL SPENCER OF SPENCER AND RICHMAN
carving a marble chimneypiece at his London workshop

INTERIORS CONSULTANTS AND CONSERVATORS

COMPANY	CONSULTANT	CONSERVATOR	Page
Abbey Heritage Ltd		fa	104
Akers, TM		an	207
Alpha Mosaic & Terrazzo Ltd		mo	203
AMBO	id		24
Anthony Beech Furniture Conservation		an	206
Anthony Short and Partners	id		25
Archer Partnership	id		25
Arthur Brett & Sons Limited		an cb	206
Awakening Restorations		an	207
Clive Beardall		an cb fr gi	206
Belfield Timber Co		cb	205
Ben Norris & Co		an cb fr gi	207
Benjamin Tindall Architects	id	ic	26
Boshers (Cholsey) Ltd	pa		58
Bursurk UK Limited		gi ic wa	197
Campbell Smith & Co Ltd	wp	ic wl	197
Carden & Godfrey Architects	id		27
Carthy Conservation Ltd		gi ic mo	107
Carvers & Gilders		an gi	205
Church Conservation Limited		gi	84
Clifford J Tracy		an cb fr	206
Cliveden Conservation Workshop Ltd		mo wa	108
Cole & Son	wp		201
Compton & Schuster Ltd		an gi ic	208
Cyril John Decorators Ltd		gi ic	198
David Brown & Partners	id		28
Devereux Decorators Ltd	id		199
Donald Insall Associates Ltd	id		29
Elizabeth Holford Associates Ltd	pa	fa ic wa	197
David Everingham		fa	202
Feilden + Mawson	id		29
Fine Iron		gi	147
Fisher Decorations		gi ic	197
Flirok UK Limited		an	173
French Polishing Contracts		fr	206
G R Pearce		fr	206
Gibberd Conservation	id		29
Giltwood Restorations Ltd		gi	205
F G Guest Joinery Consultant		cb	140
H J Hatfield & Sons Ltd		an	206
Hamilton Weston Wallpapers	wp		201
Hanna Conservation		fa mo	114
Hare & Humphreys Ltd	pa	gi ic	198
Hearns (Specialised) Joinery Ltd		cb	140
Heritage Tile Conservation Ltd		ce mo	204
Hesp & Jones		gi ic	197
Hirst Conservation	pa	fa gi ic wa	202
Holden Conservation Ltd		ic	197
Holloway White Allom		ic	62
Howell & Bellion		gi ic	198
Huning Decorations		gi ic	199
Hutton+Rostron Environmental Investigations Limited		tx	176
International Fine Art Conservation Studios Ltd		fa ic wa	202
J A Harnett Antique Restoration		an	207
The Jackfield Conservation Studio		ce	204
James Brotherhood & Associates	id		30
Jameson Joinery		cb	73
Jimmy Gatt Ltd		an	207
John Hunt Associates, Interior Architecture	id		200
Jonathan Rhind Architects	id		31
Julian Harrap Architects	id		31
KAW Design	id		200
Fleur Kelly		ic	201
LASSCO		an	160
The Leather Conservation Centre		le	200
Luard Conservation		ic	198
Nola Marshall		ic	201
Michenuels of London		an fr	206
MBL		an	124
Peter Martindale		fa ic wa	203
Mosaic Marble & Granite		mo	96
Mott Graves Projects Limited		an cb ic	63
Mowlem Rattee & Kett		ic	62
N E J Stevenson		cb	205
National Federation of Terrazzo, Marble & Mosaic Specialists		mo	203
Nevin of Edinburgh	pa	ic gi	199
Lisa Oestreicher Architectural Paint Analysis	pa		200
Original Architectural Antiques Company Ltd		an	160
Original Features		ce	205
The Perry Lithgow Partnership		fa gi ic wa	202
Pew Corner Ltd		an cb	207
Plowden & Smith Ltd		ic fa le	202
Purcell Miller Tritton	id		34
Renaissance Weavers Ltd		tx	201
Retrouvius Architectural Salvage	id wp		160
Richard Griffiths Architects	id		34
Richard Ireland Period Restoration		wa	202
Rupert Harris Metalwork Conservation		fa gi	150
Sanderson	wp	wl	201
Sandiford and Mapes	wp	fa ic wl	201
Sylvie Schofield		gi ic	200
St Blaise Ltd		fa ic wa	199
The Stained Glass Specialist		mo	129
Stuart Page Architects	id		35
Sturge Conservation Studio		an	207
T P Bennett Limited	id		38
Tankerdale Ltd		an cb pi	142
Thistle & Rose		an	207
Tim Peek Woodcarving	id	an	205
Timothy Williams (Builders) Ltd		ic	66
Trevor Caley Associates Limited		mo	203
Walcot Reclamation Ltd		an	162
Wildwood Joinery		an cb	142
Wilkinson PLC		an	207
William Cook & Sons		an fr gi	207

KEY

an antique and furniture restoration
ca carpet suppliers and conservators
cb cabinet makers
ce ceramics
fa fine art conservation
fr French polishers
gc glass and crystalware
gi gilders
ic decorators
id interior designers
le leather conservation
mo mosaics
pa paint analysis
pi picture frames
tx textile conservators
wa wall painting conservators
wl wallpaper conservators
wp wallpapers

PLASTERWORK

COMPANY	MATERIALS	PLASTERERS	Page
Acanthus Plain & Decorative Plastering		pf pl	214
Alba Plastercraft		pf pl	214
Babylon Tile Works	bt		91
Bernard A Shepherd Ltd		pl	58
Between Time	re	pl	58
Boshers (Cholsey) Ltd		pf	58
Bursledon Brickworks Conservation Centre	bt re		214
C R Crane & Son Ltd		pl	60
Calder Traditional Building Services		pl	59
Carthy Conservation Ltd		pf pl	107
Cathedral Works Organisation (Chichester) Limited	re bt		107
Chalk Down Lime Ltd	bt re	pl	168
Cliveden Conservation Workshop Ltd		pl	214
Cornish Lime Company	re		169
Farthing & Gannon		pf pl sa	214
Fine Art Mouldings		pf pl	214
G & N Marshman	bt		214
G Cook & Sons Ltd		pf pl	214
George Jackson & Sons		pf pl	214
H J Chard & Sons	re		168
Hayles & Howe Ltd Ornamental Plasterers		pf pl sa	214
Hirst Conservation		pl	199
Hirst Conservation Materials Ltd	re		168
Historic Buildings Conservation Limited		pl	39
Holden Conservation Ltd		pf pl	215
I J P Building Conservation	bt	pf pl	215
J H Upton & Son		pf pl	215
Keyline	bt re	pf pl	115
Keymer Tiles Ltd	bt		91
The Lime Centre	re		169
London Fine Art Plaster		pf pl	215
London Plastercraft		pf pl	215
Luard Conservation		pl	198
Mike Wye & Associates	bt re		168
O'Reilly Period Cornice Restoration & Cleaning		pf pl	215
Plastercraft		pf pl	216
Richard Ireland Period Restoration		pf pl sa	216
S & J Whitehead Ltd		pf pl	66
St Blaise Ltd		pl	216
Standen Plastering		pf pl	216
T J Crump Oakwrights Limited		pl	74
Taylor Dalton, Heritage Building Contractors		pl	67
The Traditional Lime Co	bt re		170
Thomas & Wilson Limited		pf	216
Trumpers Limited		pf	216
Ty-Mawr Lime Ltd	bt re		170
Weald & Downland Open Air Museum	bt		221
Whippletree Hardwoods	bt		144
Yorkshire Decorative Plasterers Ltd	re	pf pl	216

KEY

bt battens, laths and tile pegs
pf fibrous plasterwork
pl lime plasterwork
re hair and fibre reinforcement
sa scagliola

PRODUCTS AND MATERIALS

WALLPAPERS, PAINTS, TEXTILE

		Page
Blackfriar Paints	pd	183
Carvers & Gilders	gi	205
Cole & Son	wp	201
Ecomerchant	pd	158
Farrow & Ball	pd wp	184
Hamilton Weston Wallpapers	wp	201
Hare & Humphreys Ltd	pd	198
Hirst Conservation Materials	pd	168
Keim Mineral Paints Ltd	pd	184
Nevin of Edinburgh	gi	199
Old House Store	pd	169
Renaissance Weavers Ltd	ts	201
Retrouvius Architectural Salvage	ts wp	160
Rose of Jericho	pd	170
Sandiford and Mapes	wp	201
Sanderson	pd ts wp	201
Ty-Mawr Lime Ltd	pd	170

KEY
gi gilding materials
pd paints and decorative finishes
wp wallpaper suppliers
ts textile suppliers

INTERIOR FITTINGS

		Page
Ace Demolition – The Reclamation Centre	fi ki	159
Ajeer Ltd	fi	159
Alpine Reclamation	li	159
Antique Buildings Limited	fi	162
Architectural Heritage	fi	189
Architectural Reclaim	fi ra	159
Artisan Oak Buildings Ltd	fi ki	142
Best & Lloyd Ltd	li	187
Boden & Ward Stonemasons Ltd	fi	106
Carvers & Gilders	fi	205
The Cast Iron Company	li	152
Chelsom Limited	li	187
Chris Reading & Associates	li	188
Illumin Glass Studio	li	129
H W Poulter & Son	fi ra	189
Haddonstone Limited	fi	117
Hearns (Specialised) Joinery Ltd	ki	140
The Hopton Wood Stone Firms Ltd	fi	109
I J P Building Conservation	fi	61
Jonathan Murray Fireplaces	fi ra	189
Keyline	fi	115
LASSCO	ba fi ki li	160
McMarmilloyd Limited	fi	96
Nostalgia	fi ba	189
Original Architectural Antiques Co Ltd	ba fi ki li	160
Original Features	fi	205
R Hamilton & Company Limited	li	187
Retrouvius Architectural Salvage	ba fi ki	160
Solopark Plc	fi	161
Spencer & Richman	fi	190
South West Reclamation	fi ki	162
StoneCraft	fi	113
Stoneworks of Bath Ltd	fi ra	189
Symonds Salvage	fi ki	162
Thistle & Rose	fi	190
Thorverton Stone Company Ltd	fi	116
Tim Peek Woodcarvers	fi	205
W J Haysom & Son	fi	114
Walcot Reclamation Ltd	ba fi ki li ra	162
Weldon Stone Enterprises Ltd	fi	114
Wells Cathedral Stonemasons Ltd	fi	113
Western Stone	fi	113
Westland	fi li	190
Wildwood Joinery	fi	142
Wilkinson Plc	li	207

KEY
ba baths, sinks and taps
fi fireplaces
li light fittings: antique and decorative
ki kitchen and bathroom fittings
ra radiators and stoves

MOSAICS, MARBLE AND GRANITE

		Page
Heritage Tile Conservation Ltd	mo	204
The Hopton Wood Stone Firms Ltd	mg	109
McMarmilloyd Limited	mg	96
Mosaic Marble and Granite Ltd	mg mo	96
The Stone Doctors	mg	112
Trevor Caley Associates Limited	mo	203
W J Haysom & Son	mg	114
Westland	mg	190

KEY
mg marble and granite
mo mosaic supplies

FLOOR TILES, TIMBER AND PARQUET FLOORING AND VENEERS

		Page
Aldershaw Handmade Clay Tiles Ltd	ft	204
Altham Hardwood Centre Ltd	tp	208
Belfield Timber Co	tp	208
Boshers (Cholsey) Ltd	tp	58
Capital Veneer Co Ltd	ve	144
Craven Dunnill Jackfield Limited	ft	204
Dennis Ruabon Limited	ft	204
Heritage Tile Conservation Ltd	ft	204
The Hopton Wood Stone Firms Ltd	ft	109
The Jackfield Conservation Studio	ft	204
Jameson Joinery	tp	208
McMarmilloyd Limited	ft	96
Morgan & Co (Strood) Ltd	tp	143
Mosaic Marble & Granite Ltd	ft	96
Original Architectural Antiques Co Ltd	ft tp	160
Original Features	ft	205
The Original Flooring Company	ft	205
Original Oak	ft tp	208
Retrouvius Architectural Salvage	ft	160
Robus Architectural Ceramics	ft	97
Ternex Ltd	tp	208
The Tile Gallery	ft	205
Tradewood Flooring	tp	208
Walden Joinery	tp	142
Weldon Flooring Limited	tp	209
West Meon Pottery	ft	98
Whippletree Hardwoods	tp	208

KEY
ft floor and wall tile suppliers
tp timber and parquet flooring
ve veneers

National Trust Photographic Library/James Mortimer

The Entrance Hall at Attingham *The columns are part of the original interior designed by Steuart and formed an open colonnade until 1805 when the wall between was added. The micrographs below show samples from the later wall (left, with two distinct schemes and a chunk of plaster) and from one of the original walls (right, with just four distinct paint layers and a small piece of the substrate). The top two layers of the original wall match those of the later wall.*

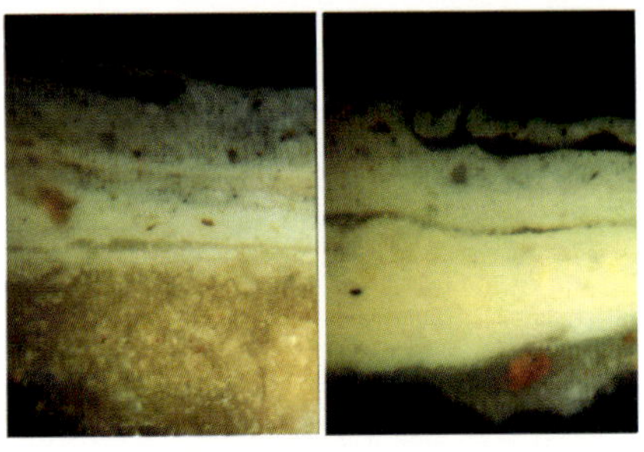

THE ARCHAEOLOGY OF DECORATION

LISA OESTREICHER

Most people will be familiar with the use of paint analysis to identify early schemes of decoration. Paint analysis is often undertaken so that an interior can be redecorated as it would have first appeared, or as it would have appeared at a particular point in its history. However, paint analysis can achieve much more than this. It is also used to discover how a room or a series of rooms evolved, how different architectural elements within them relate to one another chronologically, as well as how a room functioned or even how it was perceived by those who used it. In short, architectural paint analysis is a useful archaeological tool for exploring the past.

Paint analysis carried out at the three properties discussed here illustrates its potential.

HOW PAINT ANALYSIS WORKS

Historic houses are often the product of more than one period of construction as they expand to meet new requirements and are adapted to meet new fashions. Interior decorative finishes are even more vulnerable to changing tastes and it is rare for historic interiors to retain their original decorative finishes. Internally the result of this continual change in a room is a series of decorative schemes, one on top of another. Beneath the paint, architectural elements may be of different dates, including repairs and alterations, without any visible evidence for the change.

In the same way that an archaeologist can examine layers of earth on digs and identify the artefacts found within each datable strata, the paint analyst can chart the evolution of the ornamentation and structural changes using samples of wall decoration. But instead of layers of dirt, the analyst examines layers of paint for the information they contain.

This method of investigation is possible in part because it is known when some of the materials used in decoration were introduced or invented. For example, if emerald green, a pigment which was not discovered until 1814, is identified in a layer of paint, this would tell the analyst that the paint stratum which contains it can be no older than 1814.

It may also be possible to date some paint layers where contemporary accounts survive, such as a letter describing a room, an architect's proposal, or a decorators' bill. However, documentary sources are not always reliable: architect's proposals may not be carried out; and even descriptions of a room often differ from one person to another. Nevertheless, documentary research forms an important part of the paint analyst's work and often provides vital information. For example, documentary research may uncover a description of the room in 1825 after it was newly painted green. This might suggest that this green decoration is the same green identified under the microscope. If the description also includes the appearance of the doors, ceilings and remarks on the introduction of a new chimneypiece the analyst can really start to understand how all the paint layers of the various architectural features sampled relate to each other.

Paint analysis carried out in this way can be very useful in determining when architectural elements were introduced or altered within a room as an element added later is likely to have less paint layers (although the lack of early paint layers on one element may also mean a repair or that the piece was stripped prior to repainting).

The status of a room can also be deduced by identifying the decorative materials used: the more expensive the pigments and finishes, the greater the importance of the room.

GATHERING EVIDENCE

Many stages are involved in the paint analysis process. Typically it starts with a visit to the site in order to gain an appreciation of the room or series of rooms involved. At this point the opportunity can be taken to devise a brief for the analysis with the client. The more specific the brief, the better able the analyst is to tailor the investigations to the client's requirements, reducing the number of samples required and the cost of the project.

It is usually best to carry out the documentary research before taking samples in case there is any information that shows which areas are most likely to produce the information required and which are later additions. Where specific dates can be given for particular alterations, the first paint layer identified on the altered element might provide a useful bench-mark for dating other elements within the room.

The documentary research may also provide contemporary descriptions. If the identification of a particular scheme – perhaps a stencilled or gilded decoration – is sought, these may give the analyst an idea of where it might be most likely to survive.

Because it is important to know exactly where each sample has come from, samples are best taken by the analyst, usually with the aid of a scalpel or dental drill. Whoever takes the sample, it is most important that each one is numbered and stored separately, and the original location of each sample must be carefully recorded on a drawing or photograph immediately after it is taken. Each sample usually includes a small piece of the substrate so that there can be no doubt that the sample contains the first layer.

On returning to the studio, the samples are set in polyester resin and polished to expose a cross-section through the paint layers for examination under an incident light microscope (a microscope designed for use with reflected light). During this stage of investigation the 'stratigraphy pattern' and the materials used in the various samples can be charted and compared. The opportunity can also be taken to subject the samples to micro-chemical tests to aid in the further identification of materials, including both pigments and media. Polarising light microscopy can also be a useful tool because certain datable pigments are readily distinguished under this type of light. More sophisticated analytical equipment such as a scanning electron microscope, Fourier transform infrared spectroscopy and gas chromatography – mass spectroscopy (GCMS) can also be powerful aids if the materials found in the paint strata cannot be identified by the first three methods or if further verification is required. For example, whilst simple micro-chemical tests are capable of establishing whether a paint layer was oil based, they cannot determine the type of oil employed or the nature of many other materials which were sometimes included in the mixture.

Once an understanding of the stratigraphy pattern has been established including any datable materials, it is important to revisit the documentary research already undertaken. Only by marrying up the historical findings with the analytical findings can a full understanding of the development of the room be established.

THE ENTRANCE HALL AT ATTINGHAM PARK

Paint analysis carried out for the National Trust at Attingham Park, a fine Regency country house near Shrewsbury, illustrates how later alterations can be distinguished by comparing the stratigraphy of painted surfaces.

Attingham Park was designed by George Steuart in 1782 for Noel Hill (later the 1st Lord Berwick). The building was fronted by a large central portico which led directly into the Entrance Hall; from here the visitor could pass through a magnificent colonnade at the far side of the room into a top lit staircase. Twenty-three years later, in 1805, this colonnade was filled in by the 2nd Lord Berwick to create the Picture Gallery in the space formerly occupied by the staircase. The scheme, which was designed by John Nash, included the creation of a new circular staircase behind the Picture Gallery.

Evidence that the wall leading into the Picture Gallery was introduced as part of the John Nash alterations was given by illustrations and written records. This documentary evidence was then substantiated by analysis of the painted decoration. Three initial light cream decorations were recorded on all the walls of the Entrance Hall except for the wall leading into the Picture Gallery. The first decoration noted on the Picture Gallery wall – a grey marbled decoration – corresponded with the fourth scheme on the other walls.

The examination and comparison of stratigraphy layers in this instance was useful in a number of ways. First, it gave an idea of the original appearance of the room when first designed by Steuart as well as two subsequent decorations. Second, it allowed the Nash scheme, which related to the introduction of the new wall into the room, to be pinpointed. By doing so it was possible to establish that the Nash alterations in the first decade of the 19th century coincided with the first marbled decoration in the Entrance Hall.

Paint analysis was also useful in determining when the grisaille paintings were introduced to the Entrance Hall. The paintings were inserted into panels and niches created by Steuart in the 1780s. However, the paintings were smaller than the niches and an area had to be filled in and papered over to accommodate them, suggesting that the paintings were late arrivals. It is this papered area which was examined and found to retain only marbled decoration and none of the earlier schemes. This would suggest that the grisaille paintings were inserted as part of the Nash alterations in the room.

Finally, analysis was useful in determining the marbling techniques and materials employed by decorators at Attingham Park in the early part of the 19th century. In this instance the samples record an intricate accumulation of paint and varnish layers which resulted in a complex and beautiful Regency decorative finish.

THE NORFOLK HOUSE MUSIC ROOM AT THE VICTORIA & ALBERT MUSEUM

Norfolk House, St James's Square, London was designed between 1748 and 1756 by Matthew Brettingham the elder for the 9th Duke of Norfolk. On its completion, the Duke and Duchess held a grand opening: its Rococo interior caused a sensation in London, and many accounts survive describing its decoration.

The Norfolk House Music Room at the Victoria & Albert Museum
Nobody suspected that the fine joinery of this fine interior, which is now at the Victoria & Albert Museum was anything other than 18th century original until paint analysis carried out for the Museum showed that the joinery could not be older than 1816. Only the carving is likely to be the original.

©V&A Picture Library

Montacute House *The garden elevation (above) and (right) the Screens Passage from the Great Hall from a watercolour by C J Richardson, 1834 (National Trust)*

Shortly before the demolition of the building in 1938, the interior structure of the first floor Music Room was given to the Victoria & Albert Museum. Three of its walls were re-erected in the same year and have until recently been on display in the Victoria & Albert Museum, London. The room was dismantled in 1998 and stored in preparation for its conservation and redecoration as part of the new British Galleries at the Museum, and the opportunity was taken to carry out a study of its decoration.

This project is a good example of how the identification of a datable pigment can radically alter our understanding of a room's development. Analysis of the Music Room revealed evidence of up to seven off-white decorations on the ceiling and joinery together with over seven gilding applications. At the commencement of the project it was assumed that the joinery was part of the original design. However, analysis of the paint layers on the joinery gave everyone a surprise. The pigment cobalt blue was identified in the first scheme to survive on the timber and in the second phase of ceiling decoration. Cobalt blue was only discovered in 1802 and in England it was being used in the execution of fine art by 1816. Obviously, this date does not correspond with the interior designed by Brettingham in the 1750s when, according to contemporary accounts, the joinery was painted white. Although over 100 samples were taken from the joinery and analysed, no trace of their earlier paint layers was identified on any of them. As it would not have made sense to have had all the 18th century joinery systematically stripped of its decoration, it is more likely that the majority of white painted joinery was replaced prior to the second decoration. Wholesale replacement of large areas would have been a lengthy undertaking and was presumably required as a result of some major problem. Whatever the cause of the problem, it seems likely that only the joinery had to be replaced, as the more expensive finely carved mouldings appear to retain evidence of the primary scheme, such as priming and undercoats, implying that they were indeed original.

During transfer of the room to the Museum it was noted that grooves had been chased into the walls at Norfolk House which bore no relationship to the joinery fixings used. In the past it has been suggested that this mismatch indicated that a dado has been used initially, with a frame above for wall hangings. However, in the light of paint research it may be that the altered position of fixings was simply the result of joinery replacement.

MONTACUTE HOUSE

In 1999 the National Trust initiated a programme of historical research into the architectural development of Montacute House, an Elizabethan manor house in Somerset. Paint analysis was seen as an important component of this research. The National Trust had taken over the management of the property in 1931 and it was unclear how much of the original decorative schemes had survived under later decorative applications, or the dates when certain architectural elements were introduced into the house.

Montacute was constructed by the Phelips family in the 1690s. The importance of Montacute House lies in its completeness as one of the least altered examples of Elizabethan architecture in the country. Nevertheless, alterations were made, the first major programme being in 1786 when, in a bid to create a corridor to interconnect the rooms internally, the entrance was repositioned on the west facade and ornamental stonework brought from nearby Clifton Maybank House, a mid 16th century house, was re-erected and placed into the space formed by the wings at either end. Further alterations followed in the 19th century particularly in the 1820s and in the period 1845-52. The building then fell into a long period of decline which lasted until it was eventually acquired by the Society for the Protection of Ancient Buildings in 1931 who passed it on to the National Trust.

Perhaps not surprisingly, paint analysis showed that a considerable degree of paint stripping had been undertaken at Montacute: only eight schemes of decoration remained in many of the rooms. Fortunately, sufficient evidence survived to allow Montacute's 19th and 20th century decoration schemes to be pieced together from paint samples and documentary evidence.

At Montacute, analysis was also a valuable tool in determining how rooms functioned as well as how they developed. For example, it showed that the walls of the Screens Passage and the Great Hall were treated in a similar fashion throughout their surviving decorative history. In turn, this suggests that these two rooms were viewed as one space and may have been occupied as such.

Paint analysis was also helpful in forming an understanding of how rooms were seen and treated by their occupants. In the Old Kitchen a total of 15 decoration schemes were identified, all executed in inexpensive water-based distempers. Although this was a greater number than generally found in other rooms at Montacute, it was still insufficient for a building constructed in 1601. Furthermore, the identification of the pigment French ultramarine, which was invented in 1828, in the first decorative scheme indicates that the paintwork is no older than this date. Bearing in mind the ease with which distempers can be removed, the loss of early schemes in the kitchen is not surprising, and the accumulation of layers may simply be due to regular repainting, perhaps for sanitary reasons.

Thus the investigation demonstrated that the kitchen area was painted more frequently and in less expensive coatings than the grander spaces in the rest of the house, illustrating how paint analysis can be used to gain an understanding of the status of a room and the more functional manner in which some rooms were treated.

The three paint analysis projects outlined above show just how useful paint layers arewas an archive charting the history of the development of an interior and the way rooms were seen and used. However, the importance of this archive is easily overlooked and painted surfaces are regularly stripped in refurbishment schemes, destroying the information they contain. So please think carefully before reaching for the paint stripper in any historic building!

LISA OESTREICHER is both an architectural historian and an analyst of architectural paint, having studied at the Architectural Association and the Victoria & Albert Museum. She now runs a specialist paint analysis consultancy for historic buildings. For details see page 200.

▶ CAMPBELL SMITH & CO LTD

99 Fleet Road, Fleet, Hampshire GU51 3PJ
Tel 01252 618000 Fax 01252 618001
E-mail cousins@cousins-cem.co.uk

DECORATIONS: Decorations to the highest standard since 1873. Multi colour glaze effects, marbling, graining, stenciling, hand painted ornament, gilding and the hanging of wall covering are carried out by fully trained experienced craftsmen. Restoration and conservation projects have been carried out in many fine buildings at home and abroad in co-operation with English Heritage, the National Trust, PSA, the Royal Household and many respected architects. Small and large projects, plain or elaborate, all receive the same attention to detail.

▶ ELIZABETH HOLFORD ASSOCIATES LTD

The Tabaco Factory, Raleigh Road, Southville, Bristol BS3 1TF
Tel 0117 902 0269 Fax 0117 902 0243 Mobile 07977 997207
E-mail elizabeth.holford@cableint.co.uk

CONSERVATION AND RESTORATION OF PAINTINGS AND PAINTED SURFACES: The company has many years of experience specialising in the conservation and restoration of wall paintings, easel paintings and other painted surfaces. Work includes surveys, fully documented reports and analytical services; ultra-violet and infra-red reflectography, x-radiography, microscopy for cross-sections, pigment dispersions, and access to further extensive research facilities. Work has been undertaken for a range of heritage organisations, churches, museums, art galleries and private individuals both in the UK and abroad.

▶ FISHER DECORATIONS

157 Marston Road, Stafford ST16 3BD
Tel 01785 251300 Fax 01785 251500 E-mail fisherdecs@aol.com

INTERIOR CONSULTANTS AND CONSERVATORS: Fisher Decorations, established in 1932, has an award winning team that continually carries out well-researched projects to a high standard. The company has vast specialist knowledge of hanging the most difficult types of papers and fabrics. It also identifies and reproduces historic stencils, designs suitable alternatives, and undertakes loose-leaf gilding. Fisher Decorations carries out consultancy work with a philosophical approach, always bearing in mind the customer's requirements. Technical reports incorporating the latest digital photography can be produced to show detail and colours, and advice on paints used can also be given. The restoration and conservation of churches, large public buildings and private houses are the main areas of work.

▶ HESP & JONES

The Cedars, Beningbrough, York YO30 1BY
Tel 01904 470256 Fax 01904 470937

SPECIALIST DECORATORS AND CONSERVATORS: Charles Hesp heads a small team of skilled craftsmen that carry out decoration and restoration work in stately homes, churches and large private houses throughout England and Europe. Over the last few years their contracts have included Harewood House, Warwick Castle, Castle Howard, the British Embassy in Paris and Prague, St Paul's Cathedral and numerous National Trust stately homes. They are skilled in all aspects of decoration with an emphasis on graining and gilding. They also advise and write reports for historic colour schemes.

▶ HOLDEN CONSERVATION LTD

6 Warple Mews, Warple Way, London W3 0RF
Tel 020 8740 1203 Fax 020 8749 8356
▶ Dalshangan House, Dalry, Castle Douglas, Scotland DG7 3SZ
Tel/Fax 01644 460233

MARBLE, STONE, TERRACOTTA AND PLASTER: *See also: display entry in Statuary section, page 115.*

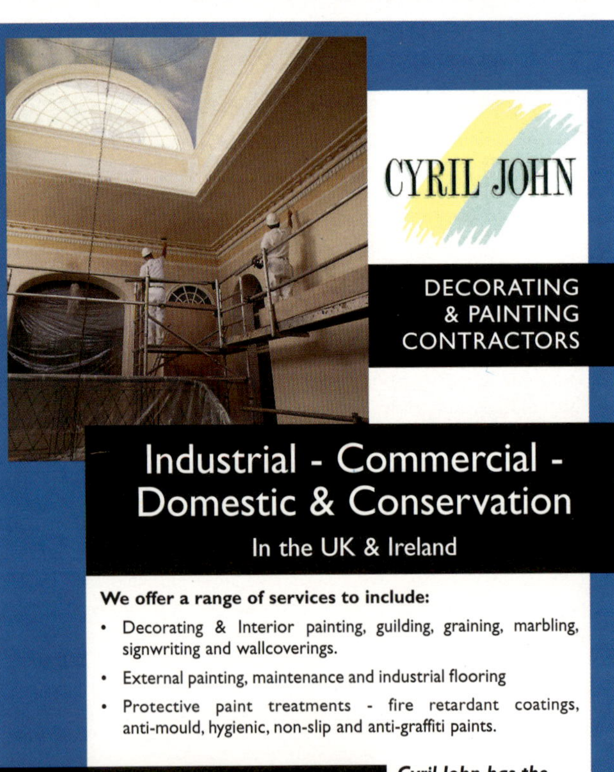

▶ HOWELL & BELLION

66A High Street, Saffron Walden, Essex CB10 1EE
Tel 01799 522 402 Fax 01799 525 696

CHURCH INTERIOR DECORATION, CONSERVATION AND RESTORATION: Howell & Bellion have many years experience of the decoration of churches and other fine buildings. Projects undertaken throughout the country have resulted in a prestigious client list, from local churches to buildings of national importance. Works include the cleaning and conservation of existing decorative schemes, restoration of lost or damaged decoration and the execution of new work such as gilding, stencilling, heraldry, hand painted ornament and the application of traditional materials. To provide a comprehensive service to clients projects often include associated small works such as repairs to carving, joinery, metalwork, stonework and the refurbishment of church metalware. Angels for riddel posts also supplied. Colour leaflets illustrating recent work are available upon request.

▶ INTERNATIONAL FINE ART CONSERVATION STUDIOS LTD

43-45 Park Street, Bristol BS1 5NL
Tel 0117 929 3480 Fax 0117 922 5511

CONSULTANTS, RESTORERS AND CONSERVATORS OF PAINTINGS AND MURALS: *See also: display entry in Fine Art Conservation section, page 202.*

▶ LUARD CONSERVATION

23 Harlesden Gardens, London NW10 4EY
Tel 020 8961 7544 Fax 020 8961 7545 Mobile 07973 741117
E-mail david.luard@virgin.net

CONSERVATION AND RESTORATION OF WOOD, CARVINGS, LIME PLASTER, PANELLING, DECORATIVE AND STRUCTURAL WOODWORK, AND ASSOCIATED FINISHES: Luard Conservation provides a comprehensive service from initial enquiry to completion of contract; consultancy services are available as well as hands-on conservation. Recent contracts include the conservation of Grinling Gibbons carvings at Windsor Castle, Burghley House and Lyme Park, consultant conservator to the Grinling Gibbons exhibition at the V & A; surveying 18th century joinery at Christ Church Spitalfields (Hawksmoor) and St Paul's Deptford (Archer); restoration of the fire damaged reredos (stone) at St Margaret's church Downham, Essex; restoration of the plaster ceilings at Longleat House.

▶ PLOWDEN & SMITH

190 St Ann's Hill, London SW18 2RT
Tel 020 8874 4005 Fax 020 8874 7248
E-mail info@plowden-smith.com
Website www.plowden-smith.com

CONSERVATION AND RESTORATION: *See also: display entry in Fine Art Conservation section, page 202.*

▶ ST BLAISE LTD – FINE CONSERVATION

Westhill Barn, Evershot, Dorchester, Dorset DT2 0LD
Tel 01935 83662 Fax 01935 83017
E-mail stblaise@compuserve.com
Contact Lynne Humphries HND Cons, PG Dip Arch Cons, MA (RCA) Cons

CONSERVATION OF ARCHITECTURAL WORKS OF ART: Fine Conservation is supported by St Blaise's well respected in-house trades, contractual competence and resources to offer a uniquely professional service. New ground broken in analysis of materials, plaster and stone conservation, and the writing of BoQs and conservation specifications for tender documents. Scope also covers gesso, gilding, composition, papier mache, carton Pierre, metals, painted surfaces, polychromy and their substrates. Also, cost-of-replacement valuations. Full insurance and professional indemnity. Projects at Uppark House, Windsor Castle, Buckingham palace and many less grand houses and churches. *See also: display entry in this section, page 199.*

INTERIORS CONSULTANTS & CONSERVATORS

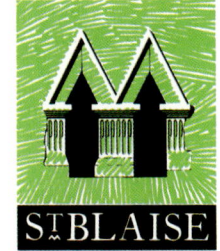

INTERIORS

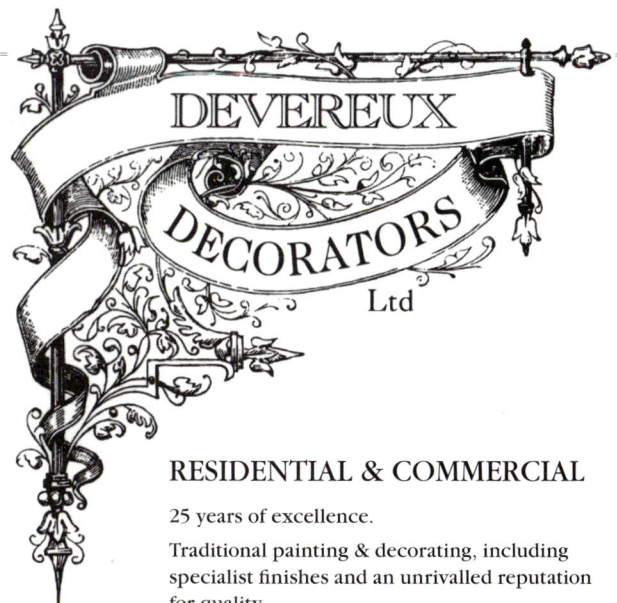

INTERIORS CONSULTANTS & CONSERVATORS

▶ SYLVIE SCHOFIELD

83 St Leonard's Road, Exeter, Devon EX2 4LS
Tel 01392 437428 Mobile 07710 750553

EXPERIENCED SPECIALIST PAINTER IN HISTORIC HOUSES AND ECCLESIASTICAL BUILDINGS: All types of decorative finishes undertaken including limewash, distemper, stencilling, gilding, colour matching, lincrust and paperhanging. Regularly contracted by the National Trust (Devon & Cornwall) and diocesan commissions. *See also: display entry above.*

HISTORIC PAINT ANALYSIS

▶ LISA OESTREICHER

Jubilee House, High Street, Tisbury, Wiltshire SP3 6HA
Tel 01747 871717 Fax 01747 871718

ARCHITECTURAL PAINT ANALYSIS: Lisa Oestreicher provides a full range of analytical skills and techniques for the study of paint and interior finishes within historic buildings. These include the identification of pigments and media as well as archival research. Full reports are prepared to provide a detailed insight into the historical development of interior and exterior decorative schemes, for documentation purposes, conservation or accurate restoration. Assistance can also be given in the design and implementation of programmes of conservation and decoration. Recent clients include the National Trust, The Victoria & Albert Museum, architects, conservators and owners. *See also: The Archaeology of Decoration on page 194.*

INTERIOR DESIGNERS

▶ KAW DESIGN

38 Meadow Road, London SW8 1QB Tel/Fax 020 7735 6088
E-mail kate@kawdesign.co.uk Website www.kawdesign.co.uk
Contact Kate Ainslie-Williams, IIDA, MSc Historic Conservation, associate member IHBC

INTERIOR ARCHITECTURE AND DESIGN: KAW DESIGN offers specialist advice and interior design in historic and traditional interiors, with experience stretching back 20 years over a wide range of building types. Work undertaken includes private residential property, hotel refurbishment, company headquarters and institutional buildings. Kate Ainslie-Williams has a particular interest in ensuring the continued viable use of old buildings and the sympathetic adaptation of their interiors to modern-day requirements. She combines design flair with extensive knowledge of interiors including the more practical aspects of a project and works closely with clients to achieve the optimum results.

LEATHER CONSERVATION

▶ THE LEATHER CONSERVATION CENTRE

University College Campus, Boughton Green Road, Moulton Park, Northampton NN2 7AN
Tel 01604 719766 Fax 01604 719649 Registered Charity No. 276485

CONSERVATORS OF OBJECTS MADE WHOLLY OR PARTLY OF LEATHER: Established in 1978, A non-profit-making organisation with charitable status offering world-wide service in leather conservation. The Centre undertakes practical conservation, training, research and analysis, consultancy work, and produces a wide-ranging series of publications on all aspects of leather conservation. Its principal activity of fee-earning practical conservation work covers a wide range of historical items, including decorated screens, wall-hangings, car and carriage upholstery, harness and saddlery, costume, and archaeological and ethnographic items. The LCC is included on the UKIC Conservation Register.

▶ PLOWDEN & SMITH

190 St Ann's Hill, London SW18 2RT Tel 020 8874 4005 Fax 020 8874 7248
E-mail info@plowden-smith.com Website www.plowden-smith.com

CONSERVATION AND RESTORATION: *See also: display entry in Fine Art Conservation section, page 202.*

FINE ART

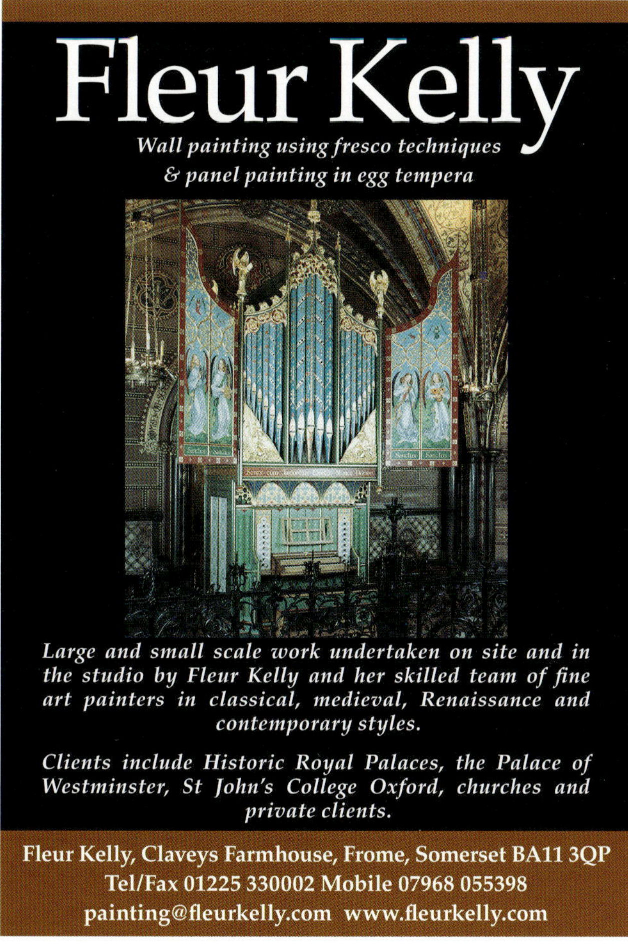

Fleur Kelly
Wall painting using fresco techniques & panel painting in egg tempera
Large and small scale work undertaken on site and in the studio by Fleur Kelly and her skilled team of fine art painters in classical, medieval, Renaissance and contemporary styles.
Clients include Historic Royal Palaces, the Palace of Westminster, St John's College Oxford, churches and private clients.
Fleur Kelly, Claveys Farmhouse, Frome, Somerset BA11 3QP
Tel/Fax 01225 330002 Mobile 07968 055398
painting@fleurkelly.com www.fleurkelly.com

▶ FLEUR KELLY

Claveys Farmhouse, Frome, Somerset BA11 3QP
Tel/fax 01225 330002 Mobile 07968 055398
E-mail painting@fleurkelly.comWebsite www.fleurkelly.com

FINE ART PAINTERS: Fleur Kelly completed her training in Italy under Leonetto Tintori. Specialist in wall painting in fresco and panel painting in egg tempera. Both site and studio commissions undertaken, small and large scale. Fleur Kelly and her team of fine-art painters work in a variety of idioms: classical, medieval, Renaissance and contemporary. Previous commissions undertaken for: Historic Royal Palaces, Palace of Westminster, St John's College, Oxford, Royal Caribbean International, churches and private clients. *See also: display entry above.*

PAINT EFFECTS

▶ NOLA MARSHALL

Little Pen-y-glog, Ton Road, Llangibby, Usk, Monmouthshire NP15 1PH
Tel 01633 450395

PAINT EFFECTS: Specialist decorator carrying out traditional and contemporary schemes of the highest quality using historical paint techniques and materials, hand-made from the finest ingredients. Site- and studio-based commissions undertaken for public and private sector clients, architects and designers. Examples range from small private houses to Windsor Castle, including a gesso dining room, hand-coloured in tempera; a gilded and lacquered private library; a hand-coloured lime-painted chapel; and 300 heraldic shields; as well as graining, marbling and gilding. Nola Marshall works within a realistic budget and time schedule. *See also: faux snake skin by Nola Marshall on the cover of this edition of The Building Conservation Directory.*

WALLPAPERS

▶ COLE & SON (WALLPAPERS) LTD

Showroom: Chelsea Harbour Design Centre, Lots Road, Chelsea, London SW10 0XE
Tel 020 7376 4628 Fax 020 7376 4631

▶ HAMILTON WESTON WALLPAPERS LTD

18 St Mary's Grove, Richmond, Surrey TW9 1UY
Tel 020 8940 4850 Fax 020 8332 0296
E-mail info@hamiltonweston.com
Website www.hamiltonweston.com

HISTORIC WALLPAPERS: Hamilton Weston specialise in reproducing wallpapers for the restoration and refurbishment of period interiors. Hand and machine printed designs date from c1690. Documents for specific projects may be selected from the archive or reproduced from clients' designs. Handprints can be coloured to order in small quantities to suit. Clients include the National Trust, leading architectural and design practices, museums, film and television companies. Robert Weston is an architectural historian and interior designer with over 25 years experience. A design consultancy service is available. Corporate members of IDDA and The Wallpaper History Society.

▶ SANDERSON

House, Oxford Road, Denham, Middx UB9 4DX
Tel 01895 830127 Fax 01895 830031
E-mail contractsales@a-sanderson.co.uk
Website www.sanderson-uk.com
Contact Andrew Keer, Contracts Manager

HISTORIC WALLPAPER MANUFACTURE AND DESIGN: Sanderson holds over 250 sets of wallpaper hand-blocks dating from 1860 to the 1970s. As well as its own extensive Sanderson Archives containing over 12,000 wallpaper documents, the company holds key designs from Morris & Co, William Woollams, Charles Knowles, Heffer Scott, and Jeffery & Co. Sanderson has worked on projects with the National Trust, leading interior designers, hotels, West End theatres, and private and royal residences worldwide. Designs can be printed to specific requirements including clients' original documents and colourations, utilising Sanderson's archive resources and research. Coordinating fabrics, paints, and machine-printed wallpapers are also available.

▶ SANDIFORD AND MAPES

Tower Farm, Grimsthorpe, nr Bourne, Lincolnshire PE10 0NF
Tel/Fax 01778 591239 Mobile 0468 732100

HISTORIC WALLPAPER CONSERVATION: Sandiford and Mapes are conservators, accredited by the IPC, specialising in wallpapers and large paper artworks. Clients include the National Trust, English Heritage, major stately homes, public bodies and private individuals. Clients abroad include heritage institutions and consultancy work for international colleagues. Trained at the Victoria & Albert Museum, Sandiford and Mapes are the only conservators with a Master of Arts qualification in wallpaper conservation. This in conjunction with an apprenticeship and ten years experience in the decorating trade provides them with an in-depth knowledge of essential wall lining systems and application techniques for historic wall coverings. They are at the forefront of developments in this field. The high standard of their work has won them a Conservation Award from the Museums and Galleries Commission.

TEXTILES

▶ RENAISSANCE WEAVERS LTD

Staple Court, Hockworthy, Wellington, Somerset TA21 0NH
Tel/Fax 01398 361543

TEXTILE DESIGNERS AND WEAVERS: Renaissance Weavers Ltd was established in 1982 and has remained an independent family business. They specialise in reproducing textiles for properties of historic importance worldwide and their work has included the wool/silk damask curtains for Uppark, and projects at Newstead Abbey, Temple Newsam, Manchester Town Hall, Caerphilly Castle, and Wightwick Manor. They provide a design service and all textiles are woven on the premises. Renaissance Weavers also weave a full range of textiles including silk brocades, worsted damasks, tapestries and flat woven carpets.

FINE ART CONSERVATION

International Fine Art Conservation Studios Ltd

Restorers and Conservators of Paintings & Murals

Conservation in practice at our studios in Bristol

For 30 years IFACS has been involved in important conservation projects throughout the United Kingdom and overseas.

Recently completed works include the conservation of a rarely seen painting by G F Watts entitled 'Alfred inciting the Saxons to prevent the landing of the Danes', also conservation and decoration of St John's Roman Catholic Cathedral in Portsmouth, conservation of 17th century painted panels in Bolsover Castle, Derbyshire together with numerous overseas projects.

IFACS is able to offer clients a team of fully trained conservators and can accommodate easel paintings and murals of practically any size. Our services also include consultancy, decoration and conservation of contemporary and fine historic interiors, paint investigation and analysis.

Accredited Member of UKIC Fellow of ABPR

43–45 PARK STREET, BRISTOL BS1 5NL Tel. (0117) 929 3480 Fax. (0117) 922 5511

CONSERVATION & RESTORATION

With over 35 years experience in conservation and restoration, Plowden & Smith is established as one of the world's leading conservators and restorers of fine art objects, furniture and paintings.

Each department is staffed by highly trained specialists. Departments include; furniture, paintings, ceramics, marble and stone, metalwork and decorative arts including ivories, waxes, gilding and ancient lacquer. The range of objects and paintings is diverse with early pieces from antiquity through to contemporary sculpture. Most objects are worked on at the company's two London premises but work is also carried out both on site and abroad.

In 1998 ***Plowden & Smith Exhibitions*** *was launched to build on the substantial experience in object mounting and exhibition displays.*

Plowden & Smith Projects *one of the leaders for on site conservation and restoration projects with a proven track record for completion on time and to budget. Projects are undertaken in all our specialist areas including painted ceilings, architectural and decorative metalwork, sculpture, gilding, stone and wood. Full project management facilities are available.*

PLOWDEN & SMITH LIMITED 190 ST ANN'S HILL LONDON SW18 2RT
Telephone: 020 8874 4005 Facsimile: 020 8874 7248
E-mail: info@plowden-smith.com www.plowden-smith.com

▶ DAVID EVERINGHAM

39A Harlow Oval, Harrogate, North Yorkshire HG2 0DR
Tel 01423 530340 E-mail david_everingham@hotmail.com

CONSERVATION AND RESTORATION OF EASEL PAINTINGS: David Everingham BSc, MA (Conservation of Fine Art) specialises in the conservation and restoration of easel paintings for galleries and private clients, and is covered by full professional indemnity insurance. Services offered include the treatment of paintings and decorative schemes in oil, acrylic or tempera on canvas, wood, metal and plaster supports; technical and scientific examination of paintings and surface cleaning of paintings and other decorative surfaces. Recent projects include the restoration and reconstruction of decorative wall scheme, Siemens' Manor, Tunbridge Wells, Kent; the cleaning and restoration of an 18th century panelled room at Clifton Hall, Nottingham, and cleaning and restoration of a 19th century frieze, Uppingham School, Rutland.

▶ HIRST CONSERVATION

Laughton, Sleaford, Lincolnshire NG34 0HE
Tel 01529 497449 Fax 01529 497518
E-mail hirst@hirst-conservation.com Website www.hirst-conservation.com

SPECIALIST BUILDING AND ART CONSERVATORS: Consultancy and conservation work to painted and applied decoration on plaster, stone, canvas, wood and metal substrates. Restoration and recreation of historic decorative schemes. Also specialist building works including joinery, sculpture, marble, stonework, stone cleaning, stucco, pargetting, wall and floor plasters. Surveys, specifications and analysis services available. Hirst's policy is to produce work of the highest quality, the approach to which is fully justified with regard to contemporary conservation ethics, both during the period of work on site and in documentation which follows the completion of projects. Their team of conservators represents many different skills and disciplines and their combined knowledge and experience is used to develop comprehensive conservation practices. *See also: display entry on the inside front cover.*

WALL PAINTING CONSERVATION

▶ HIRST CONSERVATION

Laughton, Sleaford, Lincolnshire NG34 0HE
Tel 01529 497449 Fax 01529 497518

CONSULTANTS AND CONSERVATORS: *See also: display entries on the inside front cover and opposite, and profile entry in Building Contractors section, page 62.*

▶ THE PERRY LITHGOW PARTNERSHIP

1 Langston Lane, Station Road, Kingham, Oxon OX7 6YA
Tel 01608 658067 Fax 01608 659133

CONSERVATORS OF WALL PAINTINGS AND OTHER POLYCHROME DECORATION: Established 1983, the partnership operates throughout the UK and Eire specialising in wall paintings, panel paintings and paintings on canvas and stone. Clients include; English Heritage, the National Trust, Council for the Care of Churches, architects and local governments. Extensive experience with 12th to 20th century schemes including recently: the 16th century High Great Chamber Frieze, Hardwick Hall; the 13th century Nave Ceiling, Peterborough Cathedral and; 13th to 17th century wall paintings, St Albans Cathedral. Conservation services include: project consultancy, condition surveys, monitoring, technical analysis, photographic and digitised graphic recording. Sympathetic methods and materials.

▶ PLOWDEN & SMITH

190 St Ann's Hill, London SW18 2RT
Tel 020 8874 4005 Fax 020 8874 7248
E-mail info@plowden-smith.com
Website www.plowden-smith.com

CONSERVATION AND RESTORATION: *See also: display entry in Fine Art Conservation section on this page.*

▶ RICHARD IRELAND • PERIOD RESTORATION

22 Avenue Road, Isleworth, Middlesex TW7 4JN
Tel 020 8568 5978 Fax 020 8568 5978

HISTORIC PLASTERWORK AND POLYCHROMATIC DECOR CONSERVATION: *See also: profile entry in Plasterwork section, page 216.*

WALL PAINTING CONSERVATION

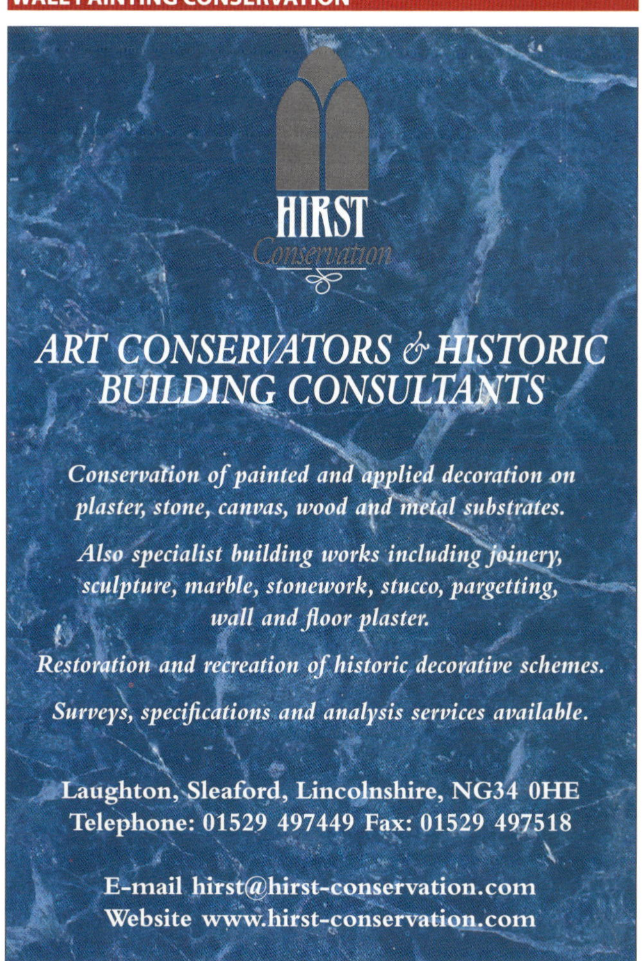

MOSAICS & TERRAZZO

▶ ALPHA MOSAIC & TERRAZZO LTD

Unit 2, Munro Drive, Cline Road, London N11 2LZ
Tel 020 8368 2230 Fax 020 8368 2301

SPECIALIST CONTRACTORS FOR IN SITU TERRAZZO, MARBLE MOSAIC AND GLASS MOSAIC FINISHES: Alpha Mosaic & Terrazzo Ltd offers a wide range of skills, expertise and experience in the repair, restoration and conservation of terrazzo, marble mosaic and glass mosaic finishes. They have carried out work in commercial, public and private buildings, many of which are listed. Alpha has been involved in many notable projects including the V&A Museum, Old War Office, Foreign Office, Harris Museum Preston, Croydon Town Hall, St Marylebone Church, Bank of England and British Museum.

▶ PLOWDEN & SMITH

190 St Ann's Hill, London SW18 2RT Tel 020 8874 4005 Fax 020 8874 7248
E-mail info@plowden-smith.com
Website www.plowden-smith.com

CONSERVATION AND RESTORATION: *See also: display entry in Fine Art Conservation section, page 202.*

▶ TREVOR CALEY ASSOCIATES LIMITED

Woodgreen, Fordingbridge, Hampshire SP6 2AU
Tel 01725 512320 Fax 01725 512420

MOSAIC CONSERVATION AND RESTORATION, DESIGN AND EXECUTION OF NEW WORKS: Trevor Caley Associates Limited has expertise in mosaic conservation and restoration, design, fabrication, contract management and historical research. In addition to meticulous mosaic conservation and documentation, the company also custom-designs and executes new mosaics in a wide range of traditional and contemporary styles for both major public buildings and private homes. Winner of the RICS Building Conservation Award 1999 for mosaic work to The Albert Memorial, London, for English Heritage, other recent projects include the design of new mosaics for Westminster Cathedral, London, and St Patrick's Cathedral, Ballarat, Australia.

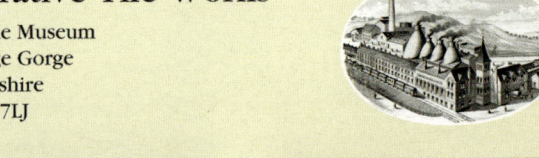

▶ ALDERSHAW HANDMADE CLAY TILES LTD

Pokehold Wood, Kent Street, Sedlescombe, East Sussex TN33 0SD
Tel 01424 756777 Fax 01424 756888
E-mail sussexterracotta@wealden.freeserve.co.uk
Website www.sussexterracotta.co.uk www.aldershaw.co.uk

HANDMADE CLAY TILES: *See also: profile entry in Clay Tiles & Roof Features section, page 91.*

▶ CLASSICAL FLAGSTONES

Lyncombe Vale Farm, Lyncombe Vale, Bath Avon BA2 4LT
Tel 01225 316759 Fax 01225 482076
Website www.classical-flagstones.com

REPRODUCTION FLAGSTONES: Classical Flagstones has gained a reputation for producing excellent copies of traditional flagstone floors at affordable prices. All products are individually made and virtually indistinguishable from the real thing. Deliveries throughout the UK and abroad.

▶ DENNIS RUABON LIMITED

Hafod Tileries, Ruabon, Wrexham LL14 6ET
Tel 01978 843484 Fax 01978 843276
E-mail project@dennisruabon.co.uk
Website www.dennisruabon.co.uk

CLAY QUARRY TILES AND PAVERS: Dennis Ruabon Limited is a leading British manufacturer of clay quarry tiles and pavers. Established for over 120 years the quarry and factory are located in Ruabon, near Wrexham in North Wales. Equally suited to either domestic or commercial use, clay quarry tiles are the traditional choice for kitchens, conservatories, passages and high traffic areas. As they are also suitable for external use they are ideal for pathways and patios. Special pieces include cove skirtings, borders and decorative panels. Dennis Ruabon Limited remains a natural choice for both new build and renovation where traditional character is desired.

▶ HERITAGE TILE CONSERVATION LTD

Unit 1, Stretton Road Industrial Estate, Much Wenlock, Shropshire TF13 6AS
Tel/Fax 01952 728157
E-mail heritagetile@msn.com Website www.heritagetile.co.uk

CONSULTANTS AND SPECIALISTS IN THE CONSERVATION AND RESTORATION OF ARCHITECTURAL CERAMICS AND GLASS: The company offers a comprehensive service for the restoration of pictorial tiled panels, geometric, encaustic and mosaic pavements, Opus Sectile, wall mosaic, tiling and faience facades. Restoration work is undertaken throughout the UK, or at their workshop, where they have pioneered the development of specialised techniques for the careful removal of architectural ceramics, wherever this proves necessary. Ceramic and glass material can be manufactured to match the original or provided from the company's stock of reclaimed materials. Clients include: Royal Museum of Scotland, Victoria & Albert Museum, Lichfield Cathedral, Foreign Office, London Transport Museum, Birmingham Museum & Art Gallery, Guy's and St Thomas' Hospital, Chatham Historic Dockyard Trust and the Highland Council.

▶ THE JACKFIELD CONSERVATION STUDIO

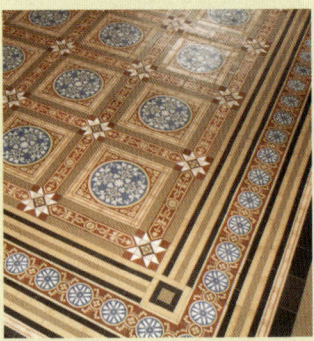

Jackfield Tile Museum, Ironbridge, Telford, Shropshire TF8 7AW
Tel/Fax 01952 883720

CONSERVATION AND RESTORATION OF ARCHITECTURAL CERAMICS: A wide range of skills, expertise and experience in the conservation and restoration of architectural tile schemes from the medieval period through to the art deco period of the 1930s. The company can offer a complete tile service at Jackfield Tile Museum, including bespoke manufacture of both glazed wall tiles and encaustic floor tiles. They also undertake detailed reports and surveys concerning the condition and treatment of historic schemes or collections. Included on the Conservation Register, member of IHBC and PACR accredited. Established 1990, projects since 1996 include, Boot's D6 Building, House of Commons, E H Medieval Tile Project, Royal Courts of Justice, St Albans Cathedral and Osgoode Hall, Toronto.

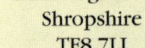

FLOOR & WALL TILES

▶ ORIGINAL FEATURES

155 Tottenham Lane, Crouch End, London N8 9BT
Tel 020 8348 5155 Fax 020 8341 4744
E-mail sales@originalfeatures.co.uk
Website www.originalfeatures.co.uk

VICTORIAN AND EDWARDIAN BUILDING RESTORATION: Original Features is a specialist restoration company increasingly working on geometric and encaustic tiled floors. Even floors in an advanced state of deterioration can be repaired with underlying defects in the sub-structure being treated without recourse to wholesale removal. New tiles can be produced by the same methods used in the original. In addition to floor tiles the company restores and repairs fireplaces (cast iron, stone and marble), fibrous plaster and stained glass. A large stock of original fireplaces is available and on display at the company showroom at the above address.

▶ THE ORIGINAL FLOORING COMPANY

230A Grange Road, Kings Heath, Birmingham B14 7RS
Tel 0121 605 8898 Fax 0121 605 8828 Mobile 07831 300628

DESIGN, INSTALLATION AND RESTORATION OF FLOOR AND WALL TILES: Brothers Michael and Mario Puopolo formed The Original Flooring Company after serving an apprenticeship with Italian craftsmen, learning traditional skills in tiling. Since the company's inception in 1987 the brothers' specialist skills have taken them all over the United Kingdom. They offer a complete service which includes the design, installation and restoration of Victorian and Edwardian floor tiles. Large stocks of original tiles are available. Recent work includes restoration to floors of The Birmingham College of Art and St James Church, Hartlebury, both Grade I listed.

▶ ORIGINAL OAK

Ashlands, Burwash, East Sussex TN19 7HS Tel 01435 882228

FLOOR TO WALL TERRACOTTA TILES: Quality handmade 6″/12″ terracotta floor tiles. *See also: profile entry in Timber Flooring section, page 208.*

▶ THE TILE GALLERY

1 Royal Parade, 247 Dawes Road, Fulham, London SW6 7RE
Tel 020 7385 8818 Fax 020 7381 1589

FLOOR AND WALL TILES, CERAMIC AND NATURAL STONE

CABINET MAKERS

▶ BELFIELD TIMBER CO

Pen-Y-Bryn, Glascoed Road, St Asaph, Denbighshire, North Wales LL17 0LH
Tel 01745 585929 Fax 01745 583984
E-mail belfieldtimber@cwcom.net

QUALITY HARDWOODS: *See also: profile entry in Timber Frame Builders section, page 72.*

▶ N E J STEVENSON

Church Lawford Business Centre, Limestone Hall Lane, Church Lawford, Coventry CV23 9HD
Tel 024 765 44662 Fax 024 765 45345

DESIGNERS AND MAKERS OF DISTINCTIVE COMMISSIONED FURNITURE: This company has established itself over many years as one of the finest furniture making companies in this country, and their skills are recognised as far afield as Korea, Russia and the USA. Their work encompasses free standing furniture through to fitted interiors, and has included major projects for the National Trust and English Heritage. They have undertaken many commissions for the Royal Household including the reproduction of the rosewood and gilt Pugin sideboard for Windsor Castle. Neil Stevenson and his workforce are committed to the very highest standards and delivering those standards on time.

WOOD CARVERS

▶ CARVERS & GILDERS

9 Charterhouse Works, Eltringham Street, London SW18 1TD
Tel 020 8870 7047 Fax 020 8874 0470 E-mail bcd@carversandgilders.com

RESTORATION AND CONSERVATION: Restoration and conservation of fine decorative wood carving and giltwood. Designers and makers of carved and giltwood furniture, mirror frames and other decorative pieces in both period and contemporary style. Commissions undertaken. Consultancy service offered. Members of The Master Carvers Association, Guild of Master Craftsmen. Clients include: The Royal Collection, The Historic Royal Palaces, English Heritage, the National Trust, Harewood House Trust, The Wallace Collection, and a wide range of British and international designers and private collectors.

▶ GILTWOOD RESTORATIONS LTD

Covers Studio, Westerham Road, Westerham, Kent TN16 2EY
Tel/Fax 01959 564936 E-mail rococo1750@freenet.co.uk

WOODCARVING AND GILDING RESTORATION: Specialists in high quality restoration and replacement of architectural wood carvings. The company's work can involve many aspects from replacement of coats of arms to carved skirtings, dado rails, cornices, balustrades, overdoors and mantelpieces. Major works have been in the royal palaces, House of Lords, House of Commons, Brighton Pavilion, National Trust houses and city livery companies. They are also known in England and America for the restoration of fine carved giltwood furniture. Established in 1968, they are a small competent business capable of gathering a large skilled workforce for large commissions.

▶ TIM PEEK WOODCARVING

The Woodcarving Studio, Highfield Avenue, Booker, High Wycombe, Buckinghamshire HP12 4ET
Tel/Fax 01494 439629

SPECIALIST WOOD CARVING ARCHITECTURAL AND FURNITURE: Following a family tradition, spanning many generations, I undertake all types of wood carving exacted to the very highest standards. I also have access to a selection of highly skilled craftsmen to call upon to design, make and finish pieces. My studio is located close to Junction 4 of the M40, not far from Central London. Please contact me on the above telephone number to discuss your requirements.

FRENCH POLISHERS

▶ G R PEARCE

5, Ancton Drive, Middleton on Sea, West Sussex PO22 6NA
Tel 01243 583272 Fax 01243 584926 Mobile 0956 237293

SPECIALIST FRENCH POLISHING CONTRACTORS: G R Pearce is a company engaged in high quality contracts in London, throughout the United Kingdom and overseas. Traditional hand French polishing and contemporary wood finishing is undertaken by experienced, reliable craftsmen who also specialise in colour matching new timbers to existing joinery. 30 years practical experience. Technical advice and colour/finish samples are provided by the company. Please contact Mr Geoffrey Pearce.

▶ MICHENUELS OF LONDON

Unit 7, Titan Business Estate, Ffinch Street, London SE8 5QA
Tel 020 8694 9206 Fax 020 8694 9201

FRENCH POLISHERS AND FURNITURE RESTORERS: The partnership team has firmly established itself as among the leading restorers in the UK. Widely known to the industry for their unusual and amazing techniques in colour matching and particularly recognised for their depth of knowledge and expertise on the conservation of listed buildings. Previous clients include Harrods, the MOD at Whitehall and the Royal Naval College, Greenwich.

ANTIQUE & FURNITURE RESTORATION

▶ ANTHONY BEECH FURNITURE CONSERVATION AND RESTORATION

The Stable Courtyard, Burghley House, Stamford, Lincolnshire PE9 3JY
Tel 01780 481199

SPECIALIST CONSERVATION OF HISTORIC HOUSE FURNITURE COLLECTIONS: Anthony Beech provides consultation and conservation services to protect important collections from deterioration. All aspects of furniture conservation are undertaken, fully documented at all stages. Accredited member of The British Antique Furniture Restorers Assocation and member of the United Kingdom Institute For Conservation.

ANTIQUE & FURNITURE RESTORATION

▶ ARTHUR BRETT & SONS LIMITED

Hellesdon Park Road, Drayton High Road, Norwich, Norfolk NR6 5DR
Tel 01603 480700 Fax 01603 788984 Website www.arthur-brett.com

CABINET MAKERS, ANTIQUE FURNITURE RESTORERS AND MANUFACTURERS OF FINE JOINERY: A fifth generation family business, established in 1870, Arthur Brett manufactures the finest quality traditional furniture of English 18th and early 19th century styles. Much of their work involves special commissions, and in recent times they have expanded their capabilities into classic joinery work for leading architects. In addition they undertake certain restoration and re-polishing work for both antique furniture and interior fitments, and can offer replica manufacturing facilities where an original item must be copied in every detail. *See also: display entry in Fine Joinery section, page 140.*

▶ CLIVE BEARDALL

Workshop and Showroom, 104b High Street, Maldon, Essex CM9 5ET
Tel 01621 857890 Fax 01621 850753 Website www.clivebeardall.co.uk

ANTIQUE FURNITURE RESTORATION AND FRENCH POLISHING: Established since 1982, Clive Beardall's attention to detail and authenticity of restoration and materials have earned him and his highly skilled team of craftsmen many prestigious commissions and a first class reputation. The team is experienced in all aspects of the antique furniture trade. Services include period furniture restoration, traditional hand French polishing, re-upholstery, marquetry, carving, gilding, leather desk lining, rush and cane seating, decorative finishes, valuations and made to order furniture. Clive Beardall is a member of The British Antique Furniture Restorers Association and his workshops are included on the register of conservators maintained by UKIC.

▶ CLIFFORD J TRACY

Unit 3, Shaftesbury Industrial Centre, Icknield Way, Letchworth, Herts SG6 1HE
Tel 01462 684855 Fax 01462 684833

CABINET MAKERS AND ANTIQUE FURNITURE RESTORERS: Clifford J Tracy specialises in the restoration of fine antique furniture including marquetry, inlaid ivory brass and tortoiseshell, wall panelling, wood carving, wax and French polishing, clock case restoration, desk top leather lining, keys and locks. A complete re-upholstery service is also provided. Worm infested furniture treated by non toxic methods and guaranteed clear. The firm also specialises in designing and making items of furniture to customers' specifications. Member of the British Antique Furniture Restorers Association, the standard of their work is of the highest order. Established in 1961 their clients number many from around the world.

▶ COMPTON & SCHUSTER LTD

Studio A133, Riverside Business Centre, Haldane Place, London SW18 4UQ
Tel 020 8874 0762 Fax 020 8870 8060 Mobile 0831 757 504
E-mail mail@comptonandschuster.com Website www.comptonandschuster.com
Contact Dominic Schuster or Lucinda Compton (BAFRA member)

RESTORATION AND CONSERVATION OF DECORATIVE FURNITURE, INTERIOR AND EXTERIOR ARCHITECTURAL GILDING: *See also: display entry in Gilders section, page 208.*

▶ H J HATFIELD AND SONS LTD

42 St Michael's Street, London W2 1QP Tel 020 7723 8265 Fax 020 7706 4562

CONSERVATION AND RESTORATION OF ANTIQUE FURNITURE AND INTERIOR ARCHITECTURAL FEATURES: Founded in 1834, Hatfield and Sons has been responsible for the conservation and restoration of many collections of furniture and metalwork. In 1860 the company was commissioned to re-frame the miniatures in the Royal Collection. Hatfield and Sons will undertake conservation in all areas of the fine arts, specialising in furniture, especially boulle and marquetry, ormolu lighting, furniture mounts and interior architectural features. Advice is offered on conservation and restoration to museums, private and trade clients.

▶ PLOWDEN & SMITH

190 St Ann's Hill, London SW18 2RT Tel 020 8874 4005 Fax 020 8874 7248
E-mail info@plowden-smith.com Website www.plowden-smith.com

CONSERVATION AND RESTORATION: *See also: display entry in Fine Art Conservation section, page 202.*

▶ THERMO LIGNUM UK LTD

19 Grand Union Centre, West Row, London W10 5AS
Tel 020 8964 3964 Fax 020 8964 2969 Mobile 07946 830013
E-mail thermolignum@btinternet.com Website www.thermolignum.com

CHEMICAL FREE ERADICATION OF INSECT PESTS: *See also: profile entry in Insect Eradication section, page 178.*

ANTIQUE AND FURNITURE RESTORERS

❶ AKERS, T M
The Forge, 39 Chancery Lane, Beckenham, Kent BR3 2NR
Tel 020 8650 9179
Established 1978. All aspects of period furniture conservation carried out to exacting standards. Members of the British Antique Furniture Restorers Association.

❷ ANTHONY BEECH FURNITURE CONSERVATION AND RESTORATION
The Stable Courtyard, Burghley House, Stamford, Lincolnshire PE9 3JY
Tel 01780 481199
Specialist conservation of historic house furniture collections: See also: profile entry on page 206.

❸ ARTHUR BRETT & SONS LIMITED
Hellesdon Park Road, Drayton High Road, Norwich, Norfolk NR6 5DR
Tel 01603 480700 Fax 01603 788984
Website www.arthur-brett.com
Cabinetmakers, antique furniture restorers and manufacturers of fine joinery. See also: profile entry in Antique & Furniture Restoration section, page 206.

❹ AWAKENING RESTORATIONS
Abbey Lane, Torry, Aberdeen AB11 9QR
Tel 01224 898778 Fax 01224 870810
Associate of Earlier World Interests Emporium. Architectural antiques and furniture. Doors, fire surrounds, balustrades, radiators. Specialist wood restorers. Paint stripping.

❺ BEN NORRIS & CO
Knowl Hill Farm, Knowl Hill, Kingsclere, Newbury, Berkshire RG20 4NY
Tel 01635 297950 Fax 01635 299851
Established since 1980, Ben Norris & Co has a wide range of experience in all aspects of period furniture restoration including architectural fittings, panelling and staircases.

❻ CARVERS & GILDERS
9 Charterhouse Works, Eltringham Street, London SW18 1TD
Tel 020 8870 7047 Fax 020 8874 0470 E-mail bcd@carversandgilders.com
Restoration and conservation. Specialists in fine water gilding. See also: profile entry on page 205.

❼ CLIVE BEARDALL
Workshop and Showroom, 104b High Street, Maldon, Essex CM9 5ET
Tel 01621 857890 Fax 01621 850753
Website www.clivebeardall.co.uk
Antique furniture restoration and French polishing. See also: profile entry on page 206.

❽ CLIFFORD J TRACY
Unit 3, Shaftesbury Industrial Centre, Icknield Way, Letchworth, Herts SG6 1HE
Tel 01462 684855 Fax 01462 684833
Cabinetmakers and antique furniture restorers: See also: profile entry on page 206.

❾ COMPTON & SCHUSTER LTD
Studio A133, Riverside Business Centre, Haldane Place, London SW18 4UQ
Tel 0208 874 0762 Fax 0208 870 8060 Mobile 0831 757 504
E-mail mail@comptonandschuster.com
Website www.comptonandschuster.com
Contact Dominic Schuster or Lucinda Compton (BAFRA member)
Restoration and conservation of decorative furniture, interior and exterior architectural gilding: See also: display entry on page 208.

❿ FRENCH POLISHING CONTRACTS
12 Trundle Street, London SE1 1QT
Tel 020 7407 6954 Fax 020 7407 6954
Website www.frenchpolishinglondon.co.uk
French polishers specialising in finishing and restorative cleaning of furniture, doors and fine interior joinery. See also: display entry on page 206.

⓫ G R PEARCE
5, Ancton Drive, Middleton on Sea, West Sussex PO22 6NA
Tel 01243 583272 Fax 01243 584926
Mobile 0956 237293
Specialist French polishing contractors: See also: profile entry on page 206.

⓬ J A HARNETT ANTIQUE RESTORATION
13-15 Newbury Mews, Malden Road, London NW5 3HP
Tel 020 7482 4675
Antique furniture restoration, cabinet making, French polishing, upholstery, gilding, carving and metalwork. Private, trade and museum work undertaken. Highest standards guaranteed.

⓭ H J HATFIELD AND SONS LTD
42 St Michael's Street, London W2 1QP
Tel 020 7723 8265 Fax 020 7706 4562
Conservation and restoration of antique furniture and interior architectural features: See also: profile entry on page 206.

⓮ JIMMY GATT LTD
11-13 Balmoral Terrace, Aberdeen AB10 6HH
Tel 01224 582288 Fax 01224 582299
Specialists in French polishing, cabinet making, furniture restoration and upholstery with over 20 years experience from a dedicated and sympathetic team.

⓯ MICHENUELS OF LONDON
Unit 7, Titan Business Estate, Ffinch Street, London SE8 5QA
Tel 020 8694 9206 Fax 020 8694 9201
French polishers and furniture restorers. See also: profile entry on page 206.

⓰ PEW CORNER LTD
Artington Manor Farm, Old Portsmouth Road, Guildford, Surrey GU3 1LP
Tel 01483 533337 Fax 01483 535554
E-mail pewcorner@pewcorner.co.uk
Website www.pewcorner.co.uk
Buyers and suppliers of old church furnishings, fixtures and fittings, from pews to pulpits, and lecterns to light fittings. Also makers of traditional furniture. See also: profile entry on page 160.

⓱ PLOWDEN & SMITH
190 St Ann's Hill, London SW18 2RT
Tel 020 8874 4005 Fax 020 8874 7248
E-mail info@plowden-smith.com
Website www.plowden-smith.com
See also: display entry in Fine Art Conservation section, page 202.

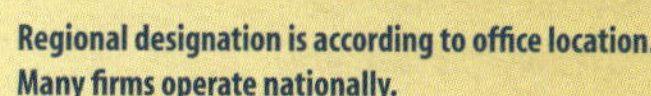

Regional designation is according to office location. Many firms operate nationally.

⓲ STURGE CONSERVATION STUDIO
Theo Sturge BA FIIC AMUKIC ACR
6 Woodland Avenue, Abington, Northampton NN3 2BY
Tel 01604 717929 Fax 0870 284 5267
E-mail sturge@primex.co.uk
Website www.sturgeconservation.co.uk
Specialist restoration of gilt leather and historic leather. Also antiquities, archaeological material, mosaics, replicas, condition reports. An accredited conservator restorer.

⓳ THISTLE & ROSE
Smailholm House, Nr Kelso, Roxburghshire TD5 7PQ
Tel/Fax 01573 460232
Restorers of original and manufacturers of facsimile pine and gesso neo-classical chimneypieces. See also: profile entry in Fireplaces section, page 190.

⓴ WILKINSON PLC
5 Catford Hill, London SE6 4NU
Tel 020 8314 1080 Fax 020 8690 1524
Specialists in the manufacture, repair and restoration of chandeliers and crystal tableware. Antique and fine reproduction chandeliers available from stock.

㉑ WILLIAM COOK & SONS
High Trees House, Savernake Forest, Marlborough, Wiltshire SN8 4NE
Tel 01672 513017 Fax 01672 514455
E-mail william.cook@virgin.net
William Cook & Sons is a family run antique furniture restoration business established over 40 years and with a worldwide reputation for excellent quality of work.

GILDERS

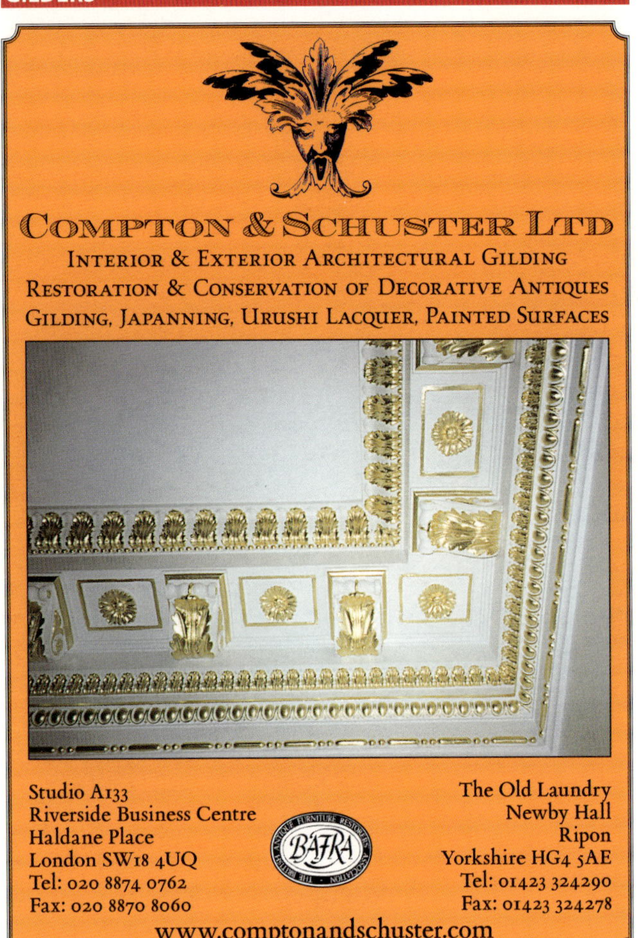

COMPTON & SCHUSTER LTD
INTERIOR & EXTERIOR ARCHITECTURAL GILDING
RESTORATION & CONSERVATION OF DECORATIVE ANTIQUES
GILDING, JAPANNING, URUSHI LACQUER, PAINTED SURFACES
Studio A133
Riverside Business Centre
Haldane Place
London SW18 4UQ
Tel: 020 8874 0762
Fax: 020 8870 8060
BAFRA
The Old Laundry
Newby Hall
Ripon
Yorkshire HG4 5AE
Tel: 01423 324290
Fax: 01423 324278
www.comptonandschuster.com

TIMBER FLOORING

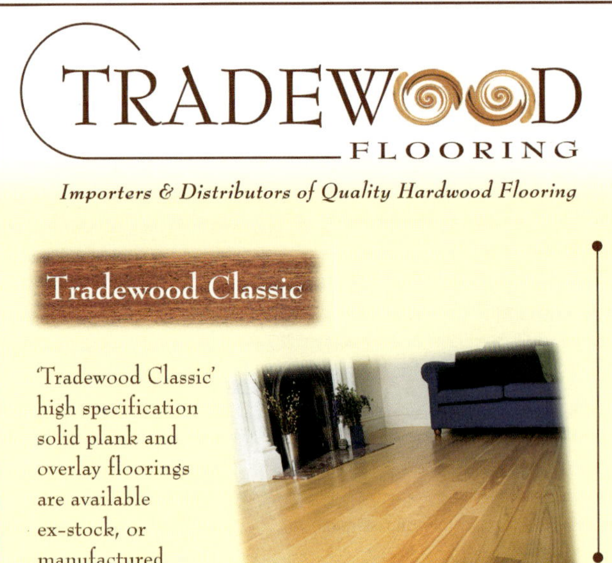

TRADEWOOD
FLOORING
Importers & Distributors of Quality Hardwood Flooring
Tradewood Classic
'Tradewood Classic' high specification solid plank and overlay floorings are available ex-stock, or manufactured bespoke to your own requirements:
• All major species available
• Face widths 89mm to 205mm
• Finished thickness 15mm to 50mm
• Lengths to 3.6 meters
• T&G or square edge
Tel: 01772 811811
Fax: 01772 812601
e-mail: sales@tradewoodflooring.com

▶ CARVERS & GILDERS

9 Charterhouse Works, Eltringham Street, London SW18 1TD
Tel 020 8870 7047 Fax 020 8874 0470
E-mail bcd@carversandgilders.com
RESTORATION AND CONSERVATION: Specialists in fine water gilding. *See also: profile entry in Wood Carvers section, page 205.*

TIMBER FLOORING

▶ ALTHAM HARDWOOD CENTRE LTD

Altham Corn Mill, Burnley Road, Altham, Accrington, Lancs BB5 5UP
Tel 01282 771618 Fax 01282 777932
Website www.oak-beams.co.uk
KILN DRIED FLOORBOARDS: *See also: profile entry in Timber Frame Builders section, page 72.*

▶ BELFIELD TIMBER CO

Pen-Y-Bryn, Glascoed Road, St Asaph,
Denbighshire, North Wales LL17 0LH
Tel 01745 585929 Fax 01745 583984
E-mail belfieldtimber@cwcom.net
HARDWOOD FLOORING: *See also: profile entry in Timber Frame Builders section, page 72.*

▶ JAMESON JOINERY

Hook Farm, West Chiltington Lane, Billingshurst, West Sussex RH14 9DP
Tel 01403 782868 Fax 01403 786766
SPECIALISTS IN TIMBER FLOORING, FURNITURE, STAIRCASES AND KITCHENS: *See also: display entry in Timber Frame Builders section, page 73.*

▶ MORGAN & CO (STROOD) LTD

Knight Road, Strood, Rochester, Kent ME2 2BA
Tel 01634 290909 Fax 01634 290800
E-mail info@morgantimber.co.uk
RESTORATION TIMBER: Morgan Timber, established in 1923, has one of the largest stocks of oak logs in the country. *See also: profile entry in Timber Suppliers section, page 143.*

▶ ORIGINAL OAK

Ashlands, Burwash, East Sussex TN19 7HS Tel 01435 882228
BUILDING AND RESTORATION TIMBER SUPPLIERS: Original Oak supplies old oak, pine, elm and beech floorboards. These are often superb wide (12") floorboards like those found in 17th century houses, and can be viewed *in situ* at their previous project sites. They also supply oak beams for restoration and construction, which meet all stress grading criteria if necessary. Doors, both traditional plank and ledged and braced, and Victorian four panelled are also available, as are original solid oak staircases, made to order. Quality handmade terracotta floor tiles/parquet flooring are also supplied. Traditional oak timber-framed buildings are constructed using traditional methods. Please ring Original Oak on 01435 882228 for prompt, personal attention.

▶ TERNEX LTD

The Sawmill, 27 Ayot Green, Welwyn, Herts AL6 9BA
Tel 01707 324606 Fax 01707 334371
TRADITIONAL T&G FLOORING IN ENGLISH AND IMPORTED HARDWOODS AND SOFTWOODS: *See also: profile entry in Timber Suppliers section, page 143.*

▶ WELDON FLOORING LIMITED

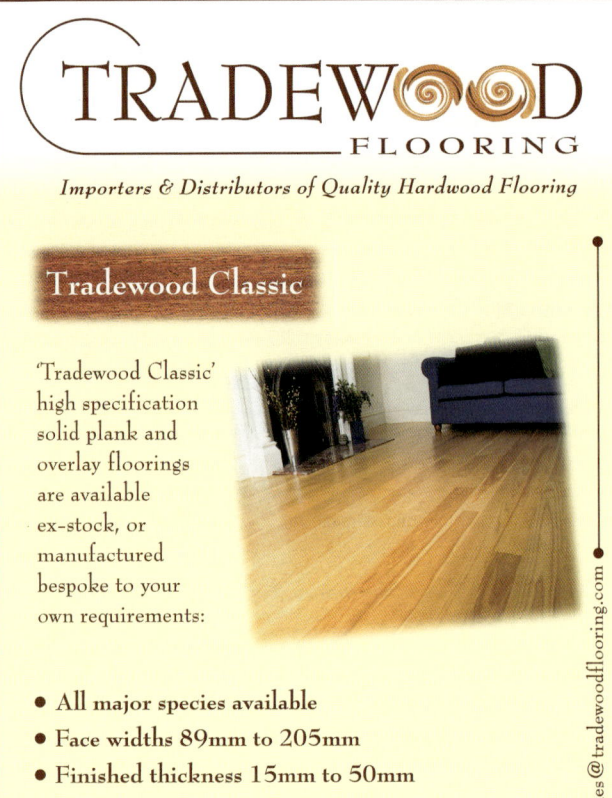

Glebe Farm Workshops, Caunton, Norwell, Nottingham NG23 6LB
Tel 01636 636962 Fax 01636 636961
PROFESSIONAL HARDWOOD FLOORING: The firm specialises in the design, specification, supply and fitting of elaborate marquetry and parquetry floors. They are a team of highly skilled craftsmen with many years of experience in both the restoration and creation of hardwood floors. They use only the finest materials and pay particular attention to detail, their emphasis is on superior quality and they are totally dedicated to their art. *See also: display entry opposite.*

▶ WHIPPLETREE HARDWOODS

Milestone Farm, Barley Road, Flint Cross, nr Royston, Herts SG8 7QD
Tel 01763 208966
TIMBER MERCHANTS: *See also: display entry in Timber Suppliers section, page 144.*

Setting out the Rococo ceiling at Uppark

DECORATIVE LIME PLASTER

TREVOR PROUDFOOT

The subject of decorative plaster is really a tale of two plasters – two plasters that are often confused as one, both having the same appearance but each having very different qualities.

The first, the one plaster that is usually associated with early decorative work is lime plaster. Made from lime putty, lime plaster has wonderful versatility, but its reward is gained at a price, for lime is deceptively difficult substance to use and its behaviour is often unpredictable.

The other plaster comes from an easier to use, more popular material, a fine white powder capable of a quick predictable set. This is gypsum plaster. It is the most common material used today for plain and decorative plasterwork, but prior to cheap mass-produced gypsum plaster in the late 19th century, both gypsum and lime were used for decorative plasterwork, at times combined side by side in one decorative scheme where the two methods and materials complement each other. Indeed it is highly unusual to find early decorative plasterwork to be the product of strictly one plaster, and lime plaster was often used with an additive of gypsum to aid the set.

Lime plaster remained in widespread use for traditional vernacular buildings beyond the advent of fibrous plaster and cheap gypsum plaster, mainly because of ease of availability in the countryside.

LIME

To make lime plaster, a limestone of almost pure calcium carbonate has to be chosen. This is fired in a limekiln at a temperature of about 1,000°C. The burnt stone taken out of the limekiln is quick lime (calcium oxide), a very caustic material that is difficult to keep, so it is almost immediately turned into lime putty (calcium hydroxide) by adding water, a process known as 'slaking' which generates a great deal of heat and steam.

Putty lime will harden slowly when exposed to air as the lime reacts with carbon dioxide to form calcium carbonate once again – a process known as 'carbonation'. Fresh lime putty is therefore protected from hardening by being stored in waterproof containers in a damp state, permanently covered by a thin film of water.

Lime can be used by a mason to bed stones or modelled by a sculptor once the necessary aggregates have been added. (In plasters, aggregates such as sand are added in the proportions of up to around three-to-one for all but the finishing coat, principally to reduce shrinkage.) A modeller using lime plaster, or 'stucco' as it is often known, has time to change his mind some time after he has used it, for lime plaster will set over a five to ten day period. During this period it must be protected from drying out too quickly or it will crack. Once set, stucco will last for centuries.

GYPSUM

Gypsum plaster behaves very unlike lime plaster. It is made simply by heating gypsum rock or alabaster – both of which are mineral forms of hydrated calcium sulphate – and grinding the result to a fine flour-like powder. At a relatively low temperature some of the water which makes up the crystalline mineral structure is driven off, forming calcium sulphate hemihydrate, which is then ground to a fine powder.

Gypsum plaster will set rapidly – within 15 minutes once it has been 'knocked up' with water – forming interlocking crystals of gypsum. This is not a material for modelling with, more a material for casting with, as it sets so quickly. So we have two completely different materials, for different purposes. A slow setting lime plaster and a fast setting gypsum plaster.

One of the earliest and most renowned sources of relatively pure gypsum rock was Montmartre, Paris, from which the material takes perhaps its most common name, plaster of Paris.

Plasterers, particularly since the late 18th century, have generally used gypsum plaster both to imitate earlier lime plasterwork and to create their own contemporary plasterwork of varying quality.

EARLIEST ORIGINS

Perhaps the earliest known examples of

decorative plasterwork are from the Old Kingdom in Egypt. Painted plaster masks adorned the linen wrapped head of a mummy, and stone walls would have their irregular surface smoothed with plaster before being carved or shaped and painted. This plasterwork was formed with fast setting gypsum plaster.

Roman stucco work, though mainly painted, shows widespread use of lime plaster, for example; as a wall covering for landscape painting, as can be seen in Hadrian's villa in Tivoli in the 1st century AD; or as a theatrical backdrop of mythological figures and theatrical figures in the upper class *Hang* houses in Ephesus, *c* 5th century AD.

Instructions by the Roman architect Vitruvius on the means of ensuring that stucco relief decoration remains sound and firmly attached to the wall are as relevant today as they were in the 1st century BC. His advice on the need for cane and metal support for relief work to prevent distortion is, of course, common sense, as are the rules he describes for obtaining a flat wall surface using three coats of plaster: a coarse base coat of rough sand and lime reinforced with hair to prepare the wall surface, followed by a levelling coat of medium graded sand, lime and hair to level the wall, and finally, a finer finish coat, much thinner than the rest, of fine lime, sand and possibly goat hair.

Vitruvius' advice on how to make lime plaster adhere to a damp wall has a particular resonance today. To combat wet conditions, he recommended a pozzolanic additive of brick shards and brick powders for the first of the three layers of lime plaster. The combination of brick and lime, well mixed, provides a hydraulic set for the plaster ('hydraulic' literally means the ability to set under water), enabling the mortar to set whilst still wet, without carbonation.

In addition to brick dust, a multitude of other additives were used to accelerate the set of lime, but perhaps the one ingredient that carries the most historical significance must be marble flour. This aggregate was the key ingredient of the finest mid 18th century plaster work, *Stucco duro*, which was largely confined to Italy and southern Europe. Marble flour allegedly aids both the plasticity and the set of stucco. Although it was never widely used as an additive by English plasterers, the style of the stuccodurists was much admired and imitated. The twists and turns of a fine Rococo ceiling, with all its convoluted curves and intertwining shapes, could not have been easily made without the setting properties of marble dust or, as was later discovered by the English imitators of the stuccodurist, a lacing of gypsum plaster.

TECHNIQUES

Modelling for decorative work is made up in many layers in much the same manner as for flatwork. To minimise shrinkage, graded sands of various particle sizes are added to lime putty for each layer: coarse sand is used for the hidden core and a very small proportion of fine sand is added for the top layer. On the Continent, hair of differing strength and thickness is also added for reinforcement; coarse cattle hair for the base layer and fine goat hair for the finish coat.

Plaster additives used for decorative work are legion. The setting time of lime plaster can be speeded up with crystalline additives of alum and potassium sulphate, or retarded with animal glues and urine, and its strength can be increased with the mineral additives, magnesium and fluorosilicate; but there were many others, and those found in historic plasters can be difficult to identify from analysis, particularly if they are organic in origin.

As stucco is pliable while it cures and hardens, it generally requires some kind of support or reinforcement. This may simply be the wall itself or an armature set within it, particularly where the modelling is in high relief.

Lengths of ornament, or 'runs' are made by pushing a metal form cut to the profile of the moulding required through wet lime plaster. This profile is carried on a simple wooden frame called a 'horse' and it is guided by battens set out in the ceiling or walls. Alternatively, moulding can be run in much the same way but on the bench, for fixing to the ceiling or wall later. Sections of runs are then cut for corners, mitres and awkward returns, and fixed in position with nails or screws and fresh plaster used as an adhesive.

Repeated ornament is cast in the workshop using moulds, traditionally of hard material such as lead or boxwood, lead moulds being cast from a hand-modelled plaster original, boxwood being carved in the reverse. The moulds, which are usually of one piece, are coated with a releasing agent such as olive oil. The stiff but pliable lime plaster is then forced into it and left until firm enough to be removed. If no gypsum has been added, this may take around five days. On partially setting, the ornament is pulled out for a final attention with the modelling tool.

Confirmation that these working methods are the same as those used in the 18th century was given by the discovery in 1983 of a selection of tools, moulds and trial casts left under the floorboards at Audley End in the 18th century by Joseph Rose, the travelling Yorkshire plasterer.

EDINBURGH CASTLE

Choice of form and material are closely linked in decorative plasterwork. Jacobean plaster has a coarser, less intricate appearance than later work, partly because it also involves coarser materials. In 1997 and 1998 two ceilings in the Royal Apartments at Edinburgh Castle were reinstated by Historic Scotland.

The decorative plasterwork at the Royal Apartments at Edinburgh Castle was made using a 'horse' (above) to create the basic pattern of long ribs of lime plaster onto which ornamental casts were applied (below).

Work on the overmantle in the King's Dining Room at Edinburgh Castle, modelled in lime plaster

Time constraints of the project prevented suitable experimentation for strict use of this plaster mix, and so the mix was adapted with the more conventional materials chalk and gypsum. However, four lengths of cornice, several casts and the modelling of the overmantle in the King's Dining Room were executed in a mix similar to that used at Thirlestane, containing lime, coarse local limestone dust, old lime, and smithy waste of clinker and iron filings. The mix was so thick that the decorative modelling had to be almost pressed out of the lime or even carved. The result was a highly successful copy in the style of the original, confirming that the original Jacobean plasterwork really was formed from plaster with seemingly impossible quantities of lime – impossible, that is, until you discover how much of the lime present in samples is derived from limestone and recycled old lime used as aggregate. This plaster could be described as particularly rugged, its component aggregates being somewhat too large for fine modelling, and its rather thick consistency and quick setting time ruling out elaborate designs.

In 1617 a suite of five rooms was hastily made ready by imported London plasterers for the inauguration of James VI of Scotland. Cromwell made short work of the ceilings later, but examples of the plasterers' work remains at contemporary houses elsewhere, including the Scottish castles of Muchalls, Glamis, Thirlestane and Graigievar. This evidence together with the account of the 'Master of Works' which details the tradesmen, plasterers and materials of the decorating programme gave Historic Scotland more than a glimpse of the missing plaster scheme. Historic Scotland's aim was to reconstruct the missing plaster scheme, in both technique and material, to match those used by the original plasterers.

Uncovering their methods and materials proved more difficult and time consuming than had been envisaged. Ordinary chemical analysis of surviving contemporary plaster from Thirlestane failed to extract and differentiate the amorphous mixture of lime and lime-based aggregate and additives. However, petrographic analysis under a polarised microscope was more successful. This method of analysis is founded on the principle that each known mineral has different optical properties which enable them to be identified under cross polarised light. Using this technique, Professor Graham Morgan of Leicester University was able to determine that the plaster contained up to 50 per cent aggregate and other 'rubbish', including kiln ash, old plaster and limestone sand. Kiln waste, like the brick shards recommended by Vitruvius, promotes a reactive set in lime plaster, allowing fast curing of the hundred or so moulded pieces that made up the ceilings, and for the large and weighty mouldings of the ribs and cornices. The old lime plaster was introduced as an aggregate, supplementing the limestone sand. As both these materials and the proportion of the lime that was active when added all contain calcium carbonate, the three materials were indistinguishable by chemical analysis.

The neo-classical 'Adam style' saloon ceiling at Uppark (top). Its geometric and ordered patterns are made possible by the use of gypsum plaster, and are in complete contrast to the earliest style of plasterwork to be found in the Red Drawing Room (bottom), which is hand modelled in lime plaster.

THE DECORATIVE CEILINGS AT UPPARK

The reinstatement of the fine decorative stucco work at Uppark, West Sussex by the National Trust after the disastrous fire in 1989 provided essential information on the craft of traditional decorative plastering. Here was a house of great importance that had a decorative plaster scheme spanning a cross-over in styles in the late 18th century.

On the west side three ceilings survived in the earliest style used here, the flamboyant Rococo style, for which the plasterers used a little fine local 'Harting' sand (less than five per cent) together with a small amount of gypsum to make the plaster mix flow and remain fluid for several hours. Thick egg and dart runs are intersected by modillions and dentils along the cornices with sunburst and Apollo masks, together with grape, sunflower and goat motifs for the Little Drawing Room, and two large masks with cornucopia basket hats for the Red Drawing Room Ceiling. The Staircase Hall has what must surely be one of the largest acanthus style roses which is set down from the ceiling centre in a cone shape some half metre in depth.

On the east side two ceilings had survived that had been undertaken twenty years later, 1770, by Sir Matthew Featherstonehaugh to designs by Paine. These ceilings, which included the Saloon, were in the neo-classical style that had become fashionable by then; a 'tighter', more repetitive form of decoration, later to be known as the Adam style. The thinner, linear design of harebell swags, numerous paterae and arabesques owed much to the craft of setting out with chalk lines and trammel (netting), as well as intricate modelling. The central elipse was a triumph of flexible casting: lime and sand together with considerable amounts of pearl glue enabled the cast of egg and dart to be curved to fit the changing shape of the coffers.

And finally, there was the dining room. This room had been altered and decorated in fine painted wainscot and plaster statuary by Repton in 1812-13. Here the plasterwork is almost purely cast. Gypsum casts painted bronze form overdoors of hind and horse with busts of Napoleon, Fox, Bedford and Bathine by George Garrard.

Regency busts by George Garrard cast in gypsum plaster

Gypsum casts of modillions for the Staircase Hall (top) and of paterae for the saloon ceiling (bottom) at Uppark

REPAIR POLICY

The National Trust bases its approach to building conservation on the principle that the cause of the decay must be identified properly first before treatment is embarked on. It also adheres to the two dictums to 'repair like with like' and to 'preserve as much of the original as possible'.

Recently repair techniques have been devised in particularly sensitive locations such as the Jacobean decorative plaster scheme in Chastleton and the ceiling in the medieval chapel at Petworth. In these cases, wherever surviving plaster had been severely weakened by material or structural decay but remained in position, the repair programme was designed to provide hidden structural support. The consolidation work was designed to secure plasterwork to sound building fabric using mechanical ties, without compromising the flexibility of traditional building material and building design. For example ceilings are tied with flexible anchors set in the back of the plasterwork, connected to metal bars fixed to the ceiling joists above, and loose plaster walls are fixed to repaired stud work behind with penny sized washers and tie rods. The bottom line, as in all conservation work, is minimum intervention.

TREVOR PROUDFOOT is Managing Director of Cliveden Conservation and advises the National Trust on stone and plaster conservation. For details of Cliveden Conservation see page 108.

PLASTERWORK

▶ ACANTHUS PLAIN & DECORATIVE PLASTERING

5 Hansford Square, Combe Down, Bath BA2 5LQ
Tel/Fax 01225 837223

SPECIALIST PLASTERING, CONSERVATION AND TRADITIONAL LIME PLASTERING: As seasoned craftsmen in their field, Acanthus has earned a reputation for providing a personal service and quality in every detail of their work. They work with private and public sector clients, direct or as sub-contractor, and were instrumental in helping to win the 1995 Silver Salver for plasterwork carried out at Prior Park, Bath. They are conversant with all types of plasterwork including external renders, *in situ* mouldings and freehand modelling, through to fibrous work and more modern forms such as GRG. For technical advice and more information please telephone 01225 837223.

▶ ALBA PLASTERCRAFT

12 Russell Road, Lee-on-the-Solent, Hampshire PO13 9HP
Tel 023 9255 3027 Fax 023 9279 9290 Contact Alan Bailey MCPG

FIBROUS PLASTERWORK, ALSO SAND/CEMENT MOULDINGS: Providing accurate, high quality craftsmanship, working on projects of all sizes across the UK. 28 years experience in restoration and specialist refurbishment work for commercial, public, private and listed buildings. Manufacturing and fixing mouldings, working from existing pieces, salvaged fragments, models, drawings or photographs. Being involved with the main plastering contractors at Lloyds Register of Shipping, London, also at Windsor Castle and Uppark after the extensive fires. Offering a wide range of mouldings, some most unusual, they create Classical, period and themed styles in new-build properties as well. Projects completed for: councils, the National Trust, Wimpey (UK) Ltd, Costain, Granada Hotels, Kier Southern, film industry, many smaller builders and private owners. They are members of the Craft Plasterers Guild and League of Professional Craftsmen.

▶ BURSLEDON BRICKWORKS CONSERVATION CENTRE

Bursledon Brickworks, Coal Park Lane, Swanwick, Southampton SO31 7GW
Tel/Fax 01489 576248 E-mail bursledon@ndirect.co.uk

SUPPLIERS OF MATERIALS FOR TRADITIONAL LIME PLASTERWORK: *See also: profile entry in Mortars & Renders section, page 167.*

▶ CLIVEDEN CONSERVATION WORKSHOP LTD

The Tennis Courts, Cliveden Estate, Taplow, Maidenhead, Berkshire SL6 0JA
Tel 01628 604721 Fax 01628 660379

SCULPTURE, STONE AND WALL PAINTINGS CONSERVATION: *See also: profile entry in Stone section, page 108.*

▶ FARTHING & GANNON

Mounts Lane, Newnham, Northants NN11 3ES
Tel/Fax 01327 310146

SCAGLIOLA AND TRADITIONAL PLASTERWORK RESTORATION AND CONSERVATION: Experienced specialist in the use of lime and gypsum. *In situ* moulding, casting and mouldmaking, freehand modelling, flooring, plaster consolidation and grouting. New works and restoration of scagliola. Full pigment and material analysis available. Public and domestic clients.

▶ FINE ART MOULDINGS

Unit 6, Roebuck Road Trading Estate, 15-17 Roebuck Road, Hainult, Ilford, Essex IG6 3TU
Tel 020 8502 7602 Fax 020 8502 7603

SPECIALISTS IN FIBROUS PLASTERING: Established in 1982, Fine Art Mouldings has a vast amount of experience in the renovation and replication of all plaster features. From drawings and modelling, all work is carried out at their workshops where they produce fine plasterwork to be truly proud of. The company is very sympathetic to the wider uses of limes and putties and have a great deal of experience using lime products. Recent contracts include The Palace of Westminster and St James Palace, both carried out in 1999. Please contact Andrew Barry to discuss your project or to request a catalogue showing the Fine Art Mouldings standard mouldings range.

▶ G & N MARSHMAN

1 Nell Ball, Plaistow, Billingshurst, West Sussex RH14 0QB
Tel 01798 342427

MANUFACTURERS OF RIVEN OAK, CHESTNUT PLASTERERS' LATHS AND TILE AND STONE SLATE BATTENS: George & Nicholas Marshman have been supplying riven oak, chestnut plasterers' laths and tile and stone slate battens for over 20 years. They take pride in the quality of the materials they produce and treat every project as a challenge. Examples of projects are: Petworth House, Ightham Mote, Kings College, the Maritime Museum, Windsor Castle and the Globe Theatre. George & Nicholas Marshman supply all Great Britain and Ireland. Although they produce 320,000 feet annually, at times demand is greater than stock so please place your orders well in advance of the required delivery date.

▶ G COOK & SONS LTD

37 Montague Road, Cambridge CB4 1BU
Tel 01223 359511 Fax 01223 323966

SPECIALISTS IN FIBROUS PLASTERWORK AND TRADITIONAL LIME PLASTER: Founded in 1887, this family business has a reputation for high quality workmanship based on sound traditional principles gained over many years working in the Cambridge Colleges and surrounding country houses. Recent projects include the remodelling of Gonville and Caius College, Cambridge, Uppark and Ightham Mote for the National Trust, the new gallery for the Gilbert Collection at Somerset House and the Church of the Holy Sepulchre in Jerusalem. Continuity of experience has enabled the company to pass on the traditional skills of the craft and today it offers a highly trained work force capable of undertaking the most demanding work in plain or decorative plasterwork and external rendering.

▶ GEORGE JACKSON & SONS

Unit 19, Mitcham Industrial Estate, Streatham Road, Mitcham CR4 2AP
Tel 020 8685 5000 Fax 020 8640 1986

ORNATE PLASTERWORK AND COMPOSITION MOULDINGS: For over 200 years George Jackson & Sons has occupied a unique position 'by appointment' to provide much of the fine decoration, ornate plasterwork and composition mouldings in many notable buildings including Royal Palaces. The company is a division of Clark & Fenn and part of Skanska Construction. Services offered: interior conservation and restoration of historic buildings; composition enrichment, polished plaster, specialist glass reinforced gypsum products and fibrous plaster mouldings available from stock. For a catalogue call 020 8685 5020.

▶ HAYLES & HOWE LTD ORNAMENTAL PLASTERERS

25 Picton Street, Montpelier, Bristol BS6 5PZ
Tel 01179 246673 Fax 01179 243928
E-mail info@hayles-and-howe.co.uk
Website www.hayles-and-howe.co.uk

▶ Hayles-and-Howe Inc
3500 Parkdale Avenue, Suite C1, Baltimore 21211-1408, Maryland USA
Tel (00 1) 410 462 0986 Fax (00 1) 410 462 0989
E-mail h&h@erol.com

CRAFTSMEN IN CONSERVATION: Hayles & Howe Ltd combine traditional skills with up to the minute technology to provide service which is second to none. The company specialises in scagliola, consolidation of early plasterwork, fire prevention and insurance advice, fine finishes and modern plasterwork from their standard range to new design. For technical advice please fax 01179 243928. For an immediate response in all aspects of conservation please call 01179 246673. Traditional ceilings to order. Please call for their new brochure.

▶ HOLDEN CONSERVATION LTD

6 Warple Mews, Warple Way, London W3 0RF
Tel 020 8740 1203 Fax 020 8749 8356, and
▶ Dalshangan House, Dalry, Castle Douglas, Scotland DG7 3SZ
Tel/Fax 01644 460233
PLASTERWORK AND SCULPTURAL MATERIALS: *See also: display entry in Statuary section, page 115.*

▶ I J P BUILDING CONSERVATION LTD

Hampstead Farm, Binfield Heath, Nr Henley-on-Thames, Oxfordshire RG9 4LG
Tel 0118 969 6949 Fax 0118 969 7771
E-mail info@ijp.co.uk Website www.ijp.co.uk
LIME AND EARTH PLASTERING INCLUDING ORNATE WORKS: *See also: display entry in Building Contractors section, page 61.*

▶ J H UPTON & SON

189 London Road, Temple Ewell, Dover, Kent CT16 3DG
Tel 01304 823540/825456
DECORATIVE PLASTERWORK: J H Upton & Son specialises in all aspects of plain and ornamental plasterwork using traditional techniques and materials. The company can run mouldings *in situ* or in fibrous plaster, and more specialist aspects of the craft such as the *in situ* hand modelling of lime stucco and the repair and conservation of historic decorative plasterwork. J H Upton & Son can also restore or replace fibrous plasterwork to match its original state. All work involving historic plasterwork, plain or decorative is thoroughly researched regarding materials and methods originally employed. The company holds an extensive range of mouldings and can design authentic ceilings, cornices etc for any period or style.

▶ LONDON PLASTERCRAFT LTD

314 Wandsworth Bridge Road, Fulham, London SW6 2UF
Tel 020 7736 5146 Fax 020 7736 7190
▶ 1 Poplar Court Parade, St Margaret's Road, Twickenham TW1 2DT
Tel 020 8744 2965
Website www.londonplastercraft.com
SPECIALIST PLASTERING: London Plastercraft Ltd is a specialist plastering company carrying out all types of interior and exterior moulded work including restoration and cleaning of cornices, ceilings and ornate plasterwork. Reproduction of any type of period moulding in either traditional materials such as lime mortar and lime putty or new modern equivalents. The company employs a team of highly skilled time- served craftsmen who carry out work of the highest quality throughout the UK and Europe. In addition to production from a large range of stock moulds, London Plastercraft Ltd offers a tailor made service, often working from photographs or architect drawings. Visit the newly refurbished Central London showroom to view the large range of decorative items such as cornices, ceiling roses, fire surrounds, columns and pilasters. Recent projects include Lambeth Palace, London; Royal College of Music, London; Kredit bank, Brussels; Bank of California and Museum of Mankind. *See also: display entry above.*

▶ O'REILLY PERIOD CORNICE RESTORATION & CLEANING

141 Pennine Drive, London NW2 1NG
Tel/Fax 020 8458 5917
E-mail oreilly@beeb.net
PERIOD CORNICE RESTORATION AND CLEANING: Mouldings cleaned using steam, repaired, replicated, fitted and expertly restored to original condition by craftsmen. Accumulated experience of father and son in fibrous and solid lath plasterwork using both traditional and compatible modern methods. Plain face cornice run *in situ*, enriched mouldings remodelled, replicated and matched to existing, multiple layers of paint removed *in situ*. The firm has many satisfied clients.

Yorkshire Decorative Plasterers Ltd
PLASTER RESTORATION SPECIALISTS
IN LIME, PUTTY, PLASTER & MODELLING
MANUFACTURERS & INSTALLERS OF
NEW DECORATIVE PLASTER
Over 150 years experience
FREEPHONE 0800 035 1044
FAX: (0114) 285 2811
E-mail: ydp@ydp-it.co.uk www.ydp-it.co.uk
OFFICE & SHOWROOMS:
UNIT 25/27, NUTWOOD TRADING ESTATE,
LIMESTONE COTTAGE LANE, WADSLEY BRIDGE,
SHEFFIELD S6 1NJ
PHONE NOW FOR A FREE BROCHURE

▶ PLASTERCRAFT

63b Henleaze Road (entrance rear of Gateway), Henleaze, Bristol BS9 4JT
Tel/Fax 0117 962 8108

CONSERVATION AND ORNAMENTAL PLASTERING SPECIALISTS: This family business was established in 1959 and all staff are time-served craftsmen and have a reputation for high quality work. There are no facets of the Plasterers' craft beyond the expertise of Plastercraft, including limework, external and internal in-situ running, mouldings, rendering and fibrous plasterwork. For advice and consultancy nation-wide on new work, restoration and conservation please telephone Alan, Stan or Bill Flood, 0117 962 8108

▶ RICHARD IRELAND • PERIOD RESTORATION

22 Avenue Road, Isleworth, Middlesex TW7 4JN
Tel 020 8568 5978 Fax 020 8568 5978

HISTORIC PLASTERWORK AND POLYCHROMATIC DECOR CONSERVATION: Consultant and practitioner specialising in the conservation, repair and restoration of decorative plaster, architectural paintwork and wall paintings. Applied ornament – lime, gypsum, scagliola, composition and papier-mache. Decoration – recreation of historic paintwork, graining, marbling and gilding. Ceiling and wall paintings – conservation, cleaning, repair and restoration.
A full pigment and materials analysis is available together with detailed conservation reports. Clients include English Heritage, the National Trust, Historic Royal Palaces Agency, ecclesiastical and private clients. Contracts throughout the UK and overseas from small scale projects to large commissions varying from the Church of the Holy Sepulchre, Jerusalem, to domestic interiors.

▶ ST BLAISE LTD

Westhill Barn, Evershot, Dorchester, Dorset DT2 0LD
Tel 01935 83662 Fax 01935 83017
E-mail stblaise@compuserve.com

SOLID, PLAIN, ORNAMENTAL AND FREEHAND, CONSERVATION OF HISTORIC PLASTERWORK: *See also: entries in Building Contractors section, pages 65 and 66.*

▶ STANDEN PLASTERING LIMITED

6b College Place, Brighton BN2 1HN
Tel 01273 680316 Fax 01273 676610

SPECIALIST AND TRADITIONAL PLASTERERS: Standen Plastering is a family run business having highly skilled craftsmen with over 40 years experience in traditional interior and exterior finishing. They have experience in *in situ* run cornice work and casting work. Recent projects have involved lath and plaster, wattle and daub work, restoration and conservation work as well as external stucco and internal enriched cornice work to various buildings and areas.

▶ THOMAS AND WILSON LIMITED

903 Fulham Road, London SW6 5HU
Tel 020 7384 0111 Fax 020 7384 0222
E-mail sales@thomasandwilson.com Website www.thomasandwilson.com

SPECIALISTS IN FIBROUS PLASTERWORK: Established in 1919, Thomas and Wilson specialise in refurbishment work in period buildings. They have been involved in a number of notable projects in buildings including Buckingham Palace, Windsor Castle, The British Museum, The Royal Opera House and The National Gallery. Manufacturing takes place in their workshops where mouldings are carefully reproduced from existing pieces or modelled from drawings and photographs. They have an extensive range of period mouldings and also specialise in the manufacture of external mouldings in GRP and GRC.

▶ TRUMPERS LIMITED

1 Greswolde Road, Sparkhill, Birmingham B11 4DJ
Tel 0121 777 6201 Fax 0121 702 2784 Contact Brian or Chris Trumper

CRAFTSMEN IN PLASTERING: The name Trumper has been associated with both quality and craftsmanship in the plastering trade for over 150 years. The company covers all aspects of both new and restoration plasterwork such as internal/external in-situ run cornices and traditional sand/lime mortars. A high level of technical competence is provided along with attention to detail and quality craftsmanship on all contracts no matter what age or size the building. Winners of the Crabb Trophy in 1996 and 1997 and the Humber Silver Salver (solid) in 2000.

A WATTLE AND DAUB DEMONSTRATION AT THE CHILTERN OPEN AIR MUSEUM
Photograph © Steve Norris

CONSERVATION COURSE LISTING 2001/2002

NVQ AND DIPLOMA COURSES

Institution	Course	Duration
BOURNEMOUTH UNIVERSITY Tel 01202 595444	HND Production and Conservation of Architectural Stonework	2 years FT
BUILDING CRAFTS COLLEGE, LONDON Tel 020 7636 0480	C&G Dip/NVQ 2&3 Stonework or Woodwork (conservation modules available at NVQ 3)	1 year FT
CITY OF BATH COLLEGE Tel 01225 312191 ext343	NVQ 2&3 Stone Masonry. Options: Banker Mason, Memorial Mason, and Architectural Stone Carver C&G Dip Fine Woodwork – includes architectural joinery	1–2 years FT, 3 years block/ flexible day release 2 years FT
COUNTRYSIDE AGENCY, THE Tel 01242 521381	Training in Rural Trades – courses in various trades: Forgework, Furniture Making and Restoration, Thatching etc. Can lead to NVQ	
EDINBURGH'S TELFORD COLLEGE Tel 0131 332 2491	HNC Architectural Conservation NC Building Conservation	1 year FT 1 year FT
GLASGOW COLLEGE OF BUILDING AND PRINTING Tel 0141 332 9969	HNC/D Furniture Restoration NC Heritage HNC/D Architectural Conservation	1/2 years FT 1 year FT 1/2 years FT
HEREFORDSHIRE COLLEGE OF TECHNOLOGY Tel 01432 352235	BTEC ND Blacksmithing and Restoration Crafts (Metal)	2 years FT
LAMBETH COLLEGE, LONDON Tel 020 7501 5000	C&G Diploma Mastercrafts	3 years PT (1 day a week)
LAUDER COLLEGE Tel 01383 845010	HNC Antique Furniture Restoration HND Antique Furniture Restoration Business	1 year FT, 2 years PT 2 years FT, 4 years PT
LEEDS COLLEGE OF ART AND DESIGN Tel 0113 202 8000	BTEC (ND Design) Furniture Making and Restoration Advanced C&G/NVQ 3 Furniture Restoration	2 years FT 6 hours a week PT
PEMBROKESHIRE COLLEGE Tel 0800 716236	HNC/D Architectural and Building Conservation	FT and PT options
RYCOTEWOOD COLLEGE Tel 01844 212501	HND Design: Furniture Restoration and Conservation	2 years FT
UNIVERSITY OF GLAMORGAN Tel 01443 482289	HND Architectural and Building Conservation	2 years FT
WEYMOUTH COLLEGE Tel 01305 208946	HNC/D Architectural Stonework, Stone Masonry, Carving and Letter Cutting	1/2 years FT
YORK COLLEGE Tel 01904 770400	NVQ 3 Stone Masonry Conservation and Restoration C&G Antique Furniture Restoration	8 weeks, block release 1–2 years PT

UNDERGRADUATE COURSES

Institution	Course	Duration
ANGLIA POLYTECHNIC UNIVERSITY Tel 01245 493131 ext 3461	BSc (Hons) Building Conservation and Heritage	3 years FT, 5 years PT
BOURNEMOUTH UNIVERSITY Tel 01202 595444	BSc (Hons) Heritage Conservation	3 years FT
BUCKINGHAMSHIRE CHILTERNS UNIVERSITY COLLEGE Tel 01494 522141	BA (Hons) Furniture Restoration and Craftsmanship	3 years FT
CAMBERWELL COLLEGE OF ARTS Tel 020 7514 6302	BA Conservation	3 years FT
CITY AND GUILDS OF LONDON ART SCHOOL Tel 020 7735 2306/5210	BA (Hons) Conservation Studies	3 years FT
DE MONTFORT UNIVERSITY (LINCOLN) Tel 01522 512912	BA (Hons) Conservation & Restoration (NB historic artefacts, not buildings)	3 years FT
MANCHESTER COLLEGE OF ARTS AND TECHNOLOGY Tel 0161 953 4290	BA (Hons) Furniture Restoration/ Conservation	3 years FT
UNIVERSITY OF BRIGHTON Tel 01273 642390	BSc (Hons) Urban Conservation and Environmental Management	3 years FT, 4 years sandwich, 4–8 years PT
UNIVERSITY OF DERBY Tel 01332 591766	BSc (Hons) Architectural Conservation	3 years FT
UNIVERSITY OF GLAMORGAN Tel 01443 482289	BSc (Hons) Architectural and Building Conservation	3 years FT, 4 years sandwich – PT also available
UNIVERSITY OF HUDDERSFIELD Tel 01484 422288	BSc (Hons) Building Conservation	3 years FT
UNIVERSITY OF LUTON Tel 01582 489242	BSc (Hons) Building Conservation	3 years FT, 5 years PT
UNIVERSITY OF NORTHUMBRIA AT NEWCASTLE Tel 0191 227 4722	BSc (Hons) Architectural and Urban Conservation	3 years FT
UNIVERSITY OF WOLVERHAMPTON Tel 01902 321000	BSc (Hons) Built Environment Conservation and Management	3 years FT

POSTGRADUATE COURSES

Institution	Course	Duration
ANGLIA POLYTECHNIC UNIVERSITY Tel 01245 493131 ext 3410	MSc Conservation of Buildings	2 years and 3 months PT
ARCHITECTURAL ASSOCIATION, LONDON Tel 020 7887 4011	Graduate Diploma Conservation of Historic Buildings Graduate Diploma Historic Landscape and Garden Conservation	2 years, day release 2 years, day release
BOURNEMOUTH UNIVERSITY	MSc/PgDip Building Conservation MSc Architectural Materials Conservation MSc/PGDip Timber Building Conservation	2 years PT – extended weekend study 1 year FT, 2 years PT 2 years PT – extended weekend study
BURLESDON BRICKWORKS CONSERVATION CENTRE, SOUTHAMPTON Tel 01489 576248	Various courses to postgraduate level in traditional materials, techniques and conservation of the built environment	
CAMBERWELL COLLEGE OF ARTS Tel 020 7514 6302	MA/PgDip Conservation	1 year FT, 2 years PT
CITY AND GUILDS OF LONDON ART SCHOOL Tel 020 7735 2306/5210	Graduateship Diploma Architectural Stone Carving Graduateship Diploma Ornamental Wood Carving and Gilding	2 years FT, 3 years PT 2 years FT, 3 years PT
COLLEGE OF ESTATE MANAGEMENT, READING Tel 0118 986 1101	RICS PGDip Building Conservation	2 years PT (distance learning)
COURTAULD INSTITUTE OF ART, DEPT OF CONSERVATION AND TECHNOLOGY Tel 020 7872 0220	PgDip Conservation of Paintings PgDip Conservation of Wall Paintings	3 years FT 3 years FT
DE MONTFORT UNIVERSITY (LEICESTER AND LINCOLN) Tel 0116 253 2781	MSc Conservation Science MA Architectural Building Conservation	1 year FT, 2 years PT 1 year FT, 2 years PT
EDINBURGH COLLEGE OF ART, HERIOT WATT UNIVERSITY Tel 0131 221 6071	MSc/PGDip Architectural Conservation	1 year FT, 3 years PT
HAMILTON KERR INSTITUTE, (UNIVERSITY OF CAMBRIDGE) Tel 01223 832040	Cert/Dip in the Conservation of Easel Paintings *option for 1 year internship leading to award of the diploma*	3 years FT, *with*
INSTITUTE OF MARITIME AND HERITAGE STUDIES AT THE UNIVERSITY OF PORTSMOUTH Tel 023 9284 2421	MSc/PgDip/PgCert Historic Buildings Conservation	1 year FT, 2 years PT
IRONBRIDGE INSTITUTE (UNIVERSITY OF BIRMINGHAM) Tel 01952 432751	MA/PGDip Heritage Management MA/PGDip Industrial Heritage	1–4 years 1–4 years
OXFORD BROOKES UNIVERSITY AND UNIVERSITY OF OXFORD Tel 01865 483458	MSc/PGDip Historic Conservation (awarded by Oxford Brookes University)	1 year FT, 2 years PT /Diploma 9 months FT, 12 months PT
OXFORD BROOKES UNIVERSITY Tel 01865 483458	Certificate in Historic Conservation	9 months PT
ROBERT GORDON UNIVERSITY, THE Tel 01224 263716	MSc/PGDip/PGCert in Built Heritage Conservation	PGCert 15 weeks, PGDip 30 weeks, MSc additional 15 weeks, PT options available
SHEFFIELD HALLAM UNIVERSITY Tel 0114 225 4267	MA/PGDip/PGCert in Heritage Management	1 year FT, 2 years PT
UNIVERSITY OF BATH Tel 01225 826908	MSc/PGDip Conservation of Historic Buildings	1 year FT, 2 years PT
UNIVERSITY OF BRISTOL Tel 0117 954 6073	MA/PGDip Architectural Conservation	1 year FT, 2 years PT
UNIVERSITY COLLEGE, DUBLIN Tel 00 3531 269 3244	MUBC Master of Urban and Building Conservation	1 year FT
UNIVERSITY OF DUNDEE Tel 01382 345236	MSc/PGDip European Urban Conservation	Variable (research based course)
UNIVERSITY OF NEWCASTLE Tel 0191 222 6810	PG Cert in Conservation and Planning PG Cert in Conservation Principles and Techniques MA/PGDip Urban Conservation MA/PGDip Town Planning (Urban Conservation)	1 year FT 1 year FT 1 year FT, 2 years PT 1 year FT, 2 years PT
UNIVERSITY OF NORTHUMBRIA AT NEWCASTLE Tel 0191 227 3250	MA Conservation of Fine Art	2 years FT
UNIVERSITY OF PLYMOUTH Tel 01752 233630	MA/PGDip Architectural Conservation	1 year FT, 2 years PT
UNIVERSITY OF THE WEST OF ENGLAND Tel 0117 344 3210	MA/PgDip/PgCert Conservation of Buildings and their Environments	1 year FT, 2 years PT
UNIVERSITY OF YORK Tel 01904 433963	MA in Conservation Studies (Historic Buildings) MA in Conservation Studies (Historic Landscapes and Gardens)	1 year FT 1 year FT

SHORT COURSES

Institution	Course	Duration
BRITISH WATERWAYS – HERITAGE SKILLS CENTRE 01788 566030	Various intensive, practical building conservation courses. C&G accredited and CITB registered	
BURLESDON BRICKWORKS CONSERVATION CENTRE, SOUTHAMPTON Tel 01489 576248	Various short courses in building conservation, traditional materials and crafts	
CENTRE FOR CONSERVATION AND URBAN STUDIES AT THE UNIVERSITY OF DUNDEE Tel 01382 345236	Dundee Conservation Lectures – arranged by The Architectural Heritage Society of Scotland	Evenings
CENTRE FOR EARTHEN ARCHITECTURE, UNIVERSITY OF PLYMOUTH Tel 01752 233630	The Care and Conservation of Cob Buildings	
CHARLESTOWN WORKSHOPS AT THE SCOTTISH LIME CENTRE TRUST Tel 01383 872722	Charlestown Workshops–SVQ modules in: preparing and using basic lime mixes; preparing hydraulic lime mortars; conservation masonry	1–5 days
ENGLISH HERITAGE Tel 020 7973 3821	Some courses in association with West Dean College – *see below*	
ESSEX COUNTY COUNCIL Tel 01245 437666	Various short courses – *for details see Events on page 221*	1–3 days
INTERNATIONAL CENTRE FOR THE STUDY OF THE PRESERVATION AND RESTORATION OF CULTURAL PROPERTY, ROME (ICCROM) Tel 00 39 06 585 531	Various postgraduate short courses and training programmes in specific aspects of conservation principles and practice	
LEAD SHEET ASSOCIATION, KENT Tel 01892 513351	Short courses in leadwork	
LIME CENTRE, WINCHESTER Tel 01962 713636	Introduction to the Use of Lime in Historic Buildings (CITB approved) Advanced Use of Lime (CITB approved)	1 day 1 day
NATIONAL MUSEUMS & GALLERIES ON MERSEYSIDE Tel 0151 478 4904	An Introduction to Laser Cleaning in Conservation	3 days
ORTON TRUST, THE Tel 01536 761303	Practical courses in all aspects of stone masonry	3 days (Fri, Sat & Sun)
RYCOTEWOOD COLLEGE Tel 01844 212501	Professional Development Award: Design – Furniture Restoration	4 months FT, 1 year PT
SOCIETY FOR THE PROTECTION OF ANCIENT BUILDINGS (SPAB), LONDON Tel 020 7377 1644	The Repair of Old Buildings – a short course of lectures and site visits for contractors and building professionals – *for details see Events on page 222* SPAB Technical Days: subjects vary	1–2 days (Spring and Autumn) 1 day
UNIVERSITY OF BRISTOL Tel 0117 954 6073	Maintenance of Period Houses	8 afternoons or evenings
UNIVERSITY OF YORK Tel 01904 433963	Various short courses on the conservation of historic buildings and places, and historic landscapes and gardens	3 days
WEALD AND DOWNLAND OPEN AIR MUSEUM Tel 01243 811363	Various short courses in building conservation – *for details see Events on page 221*	1–7 days
WEST DEAN COLLEGE, NEAR CHICHESTER Tel 01243 818294	A range of Building Conservation Masterclasses, in collaboration with English Heritage and Weald and Downland Open Air Museum – *for details see Events on page 221*	1–4 days
WOODCHESTER MANSION TRUST, GLOUCESTER Tel 01453 750455	Various, regular courses on stone masonry, lime plasters, stone repair, dry stone walling etc	1 day
YORK COLLEGE 01904 770400	Antique Furniture Restoration and General Woodwork	2 hours a week over 2 terms

MISCELLANEOUS COURSES

Institution	Course	Duration
BRUCE LUCKHURST AT THE LITTLE SURRENDEN WORKSHOPS Tel 01233 820589	Conservation and Restoration of Antique Furniture	1 year
PRINCE OF WALES INSTITUTE OF ARCHITECTURE Tel 020 7916 7380	Architecture and the Building Arts – foundation course	1 year
SOCIETY FOR THE PROTECTION OF ANCIENT BUILDINGS (SPAB), LONDON Tel 020 7377 1644	SPAB Scholarship – programme for young qualified architects, surveyors and others William Morris Craft Fellowship – craft training for qualified craftsmen or women	9 months FT 3 blocks, 2 months each FT
THE DEPT OF CONTINUING EDUCATION, UNIVERSITY OF NOTTINGHAM Tel 0115 846 6466	Certificate and Advanced Certificate in Heritage Conservation	2 years PT/ 3 years PT
THE THOMAS CHIPPENDALE SCHOOL OF FURNITURE Tel 01620 810680	Furniture Design, Making and Restoration	30 weeks

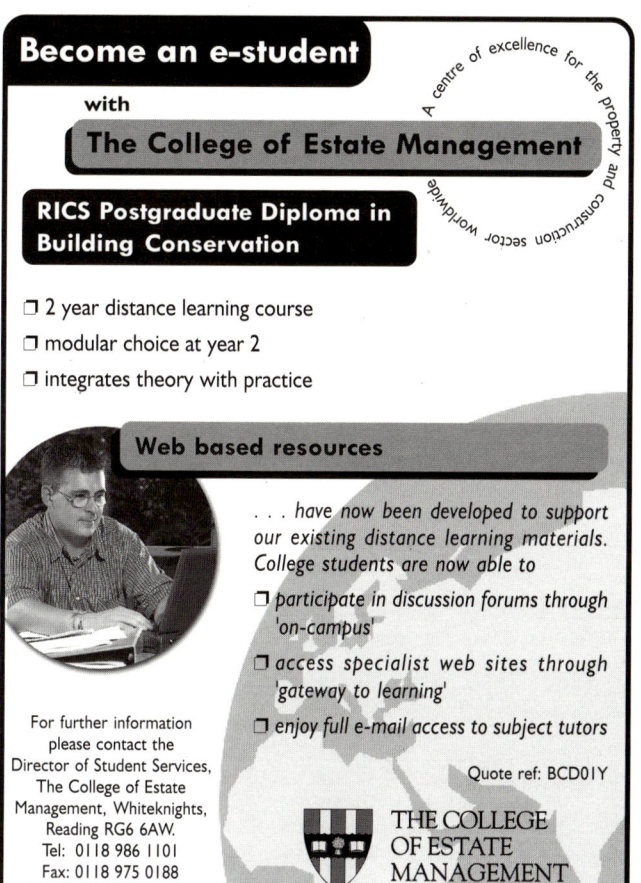

ARCHITECTURAL ASSOCIATION SCHOOL OF ARCHITECTURE

34-36 Bedford Square, London WC1B 3ES
Tel 020 7636 0974 Fax 020 7414 0782

BUILDING CONSERVATION COURSE: The Architectural Association School of Architecture offers a two-year, day-release postgraduate course intended for qualified architects and other professionals who are involved in the care and conservation of historic buildings. It was set up in 1975 by the RIBA and COTAC and, now in its 25th year, continues to offer all the advantages of a central London venue. Students meet one day per week for lectures, seminars and discussions, and also visit craft workshops, studios and buildings in varying states of decay or conservation. It is an intensive course.

BOURNEMOUTH UNIVERSITY

School of Conservation Sciences, Fern Barrow, Poole, Dorset BH12 5BB
Tel 01202 595444 Fax 01202 595255

COURSES IN BUILDING CONSERVATION:
Bournemouth University offers the following courses:
MSc Building Conservation (2-3 years part time, extended weekend study; CPD also available) for professionals working in historic building conservation who wish to improve their practical skills and theoretical background in all aspects of building conservation. There is an emphasis on the scientific analysis of building materials.
MSc Timber Building Conservation (2 years part time, extended weekend study) delivered in partnership with the Weald & Downland Open Air Museum, in West Sussex, this Master's degree enables building conservators to gain both practical and theoretical knowledge of timber building conservation.
BSc Heritage Conservation (3 years, full time).

THE BUILDING CRAFTS & CONSERVATION TRUST

Kings Gate, Dover Castle, Kent CT16 1HU
Tel 01227 451795 Fax 01233 861265

BUILDING CONSERVATION TRAINING: The Trust's work is aimed at providing the building professional and craftsman with the skills necessary to effect the accurate repair of traditional buildings. The Trust provides building conservation training in both workshop and site environments. The Trust manages conservation projects where training is an essential element. Project evaluation and structuring is undertaken by the Trust to develop training programmes.

BUILDING CRAFTS COLLEGE

153 Great Titchfield Street, London W1W 5BD(until August 2001)
Kennard Road, London L15 1AH (from August 2001)
Tel 0207 636 0480 Website www.thecarpenterscompany.co.uk

CITY AND GUILDS DIPLOMA COURSES IN WOODWORK OR STONEMASONRY: *See also: display entry in this section, page 219.*

BURSLEDON BRICKWORKS CONSERVATION CENTRE

Bursledon Brickworks, Coal Park Lane, Swanwick, Southampton SO31 7GW
Tel/Fax 01489 576248
E-mail bursledon@ndirect.co.uk

SHORT COURSES IN BUILDING CONSERVATION, TRADITIONAL MATERIALS AND TRADE TECHNIQUES: *See also: profile entry in Mortars & Renders section, page 167.*

CITY & GUILDS OF LONDON ART SCHOOL

124 Kennington Park Road, London SE11 4DJ
Tel 020 7735 2306/020 7735 5210 Fax 020 7582 5361
E-mail info@cityandguildsartschool.ac.uk
Website www.cityandguildsartschool.ac.uk

CONSERVATION COURSES:
BA (Hons) in Conservation Studies – The crafts, technical skills and theoretical knowledge needed for the conservation and re-establishment of works of art and artefacts made of wood, stone and related materials with polychromed and gilded surfaces.
Graduate/Diploma in Architectural Stone Carving – For specialists in replacement stone carving and the training of designer craftsmen.
Graduate/Diploma in Ornamental Wood Carving and Gilding – For specialists in replacement or reproduction carving and gilding and the training of designer craftsmen.
P/T Courses in Lettering and Letter Carving – The understanding of letter forms and the tools, implements and materials involved.

THE COLLEGE OF ESTATE MANAGEMENT

Whiteknights, Reading RG6 6AW
Tel 0118 986 1101 Fax 0118 975 5344
Website www.cem.ac.uk

RICS POSTGRADUATE DIPLOMA IN BUILDING CONSERVATION: *See also: display entry, page 219.*

EDINBURGH COLLEGE OF ART

Heriot Watt University, Dept of Architecture, Lauriston Place, Edinburgh EH3 9DF
Tel 0131 221 6072/6168 Fax 0131 221 6006/6157

DIPLOMA AND MASTERS DEGREE IN ARCHITECTURAL CONSERVATION: Established in 1969, The Scottish Centre for Conservation Studies at Edinburgh College of Art, Heriot Watt University Edinburgh runs a one year full time diploma course and two year part time diploma course in Architectural Conservation which provides a thorough grounding in conservation principles and methods (admittance by first degree in Architecture or related subject). Students may continue to MSc degree by dissertation if they achieve the necessary standard in diploma work. Theory and practice are inseparable throughout the course. The teaching, by lectures and seminars, is tested by projects, written and drawn, involving all the problems of practical conservation. Main subjects: theory; sources; technology; design intervention, and; area conservation. No government funding available. The College aims to promote creativity in the arts and the environment.

HERITAGE SKILLS CENTRE

British Waterways, Hatton, Warwick CV35 7SL
Tel 01788 566030 Fax 01788 541076
E-mail david.sleight@britishwaterways.co.uk

PRACTICAL BUILDING CONSERVATION COURSES:
British Waterways offers short intensive one and two day courses, concentrating on 'hands on' practical training for both professionals and craftsmen. Courses are City and Guilds accredited and CITB registered. Courses include Basic Masonry Repairs, Lime Mortars, Historic Metalwork, Brick Repairs, Painting Historic Structures and Rural Carpentry.

THE ORTON TRUST

20 Copelands Road, Desborough, Kettering, Northants NN14 2QF
Tel 01536 761303
Website www.ortontrust.org.uk

STONE MASONRY TRAINING COURSES

SOCIETY FOR THE PROTECTION OF ANCIENT BUILDINGS

37 Spital Square, London E1 6DY
Tel 020 7377 1644 Fax 020 7247 5296

CONSERVATION COURSES:
The Repair of Old Buildings – a six day course of lectures and visits held each Spring and Autumn, for architects, surveyors, engineers, planners, builders and craftsmen.
SPAB Owners' Courses – weekend courses of lectures for owners of old houses, showing how to care for and sensitively repair old buildings, one in London, others regional.
The SPAB Scholarships – a nine-month specialist training for young qualified architects, building surveyors and structural engineers.
The William Morris Craft Fellowships – a six-month specialist training in three blocks for qualified building craftsmen or women of any trade. Further details are available from SPAB.

UNIVERSITY OF BATH

Department of Architecture and Civil Engineering, Bath BA2 7AY
Tel 01225 323016 Fax 01225 826691
E-mail E.S.J.Greeley@bath.ac.uk

MASTERS DEGREE IN THE CONSERVATION OF HISTORIC BUILDINGS (MSc): The course provides technical training within an academic framework including the teaching of classical architecture and the philosophy of conservation. Teaching units include: structural conservation; materials, construction and skills; history and theory; and the law relating to conservation and urban management. Taking place within the world-heritage city of Bath, the course may be taken over one year full time or two years part-time. Architects, engineers, surveyors and suitably qualified candidates from other fields with first degree or equivalent are eligible.

COURSES & TRAINING

▶ WEALD AND DOWNLAND OPEN AIR MUSEUM

Singleton, Chichester, West Sussex PO18 0EU
Tel 01243 811363 Fax 01243 811475
E-mail wealddown@mistral.co.uk
Website www.wealddown.co.uk

CONSERVATION TRAINING, SERVICES AND SUPPLIES: The Museum has an established reputation as a provider of specialist training/education in historic building conservation and the use of traditional building materials and processes, led by Research Director Richard Harris. The 46 historic buildings reconstructed on its site give the museum an unrivalled teaching resource in this specialised field. Courses for surveyors, architects, conservation officers and craftspeople are suitable for CPD. MSc in Timber Building Conservation with Bournemouth University. English Heritage Building Conservation Masterclasses in partnership with nearby West Dean College. Research library designed for use by professionals. Supplies of materials for use in timber-framing, traditional building methods and the care of ancient buildings eg cleft battens and laths for roofing and plasterwork, spars, longstraw and wheat reed for thatching, woven wattle panels and fencing.

▶ WEST DEAN COLLEGE

West Dean College, Chichester, West Sussex PO18 0QZ
Tel 01243 811301 Fax 01243 811343
E-mail westdean@pavillion.co.uk
Website www.westdean.org.uk

BUILDING CONSERVATION MASTERCLASSES: West Dean College in collaboration with English Heritage and the Weald and Downland Open Air Museum offers short intensive training courses in building conservation, a development of English Heritage's previous Masterclass programme. Courses offer 'hands-on-training and are designed for professionals and craftsmen to enhance their understanding of suitable materials, methods and techniques for sympathetic repairs. Most courses are residential, over four days. Professional development courses are also offered for conservators, as are Post Graduate Diploma courses in conservation of historic artefacts: ceramics, clocks, furniture and fine metalwork.

▶ WEYMOUTH COLLEGE

BEE Department, Newstead Road, Weymouth, Dorset DT4 0DX
Tel 01305 208946 Fax 01305 208990
E-mail bee@weycoll.ac.uk

HND AND NVQ COURSES IN ARCHITECTURAL STONEWORK, MASONRY, CARVING AND LETTERCUTTING: Weymouth college has over 70 years experience training individuals with an interest in the traditional and current uses of natural stone. The responsibility to preserve our architectural heritage and the use of stone in contemporary projects requires qualified artisans able to operate effectively in an exacting environment. The college's aim is to enable the students to meet the developing criteria whilst fulfilling a personal need for highly skilled and creative work.

▶ WOODCHESTER MANSION TRUST

1 The Old Town Hall, High Street, Stroud, Glos GL5 1AP
Tel 01453 750455

SHORT COURSES IN BUILDING CONSERVATION PARTICULARLY THE USE OF STONE AND LIME

SCHOLARSHIPS

▶ THE PRINCE'S FOUNDATION

19-22 Charlotte Road, London EC2A 3SG
Tel 020 7613 8500
E-mail jeade@princes-foundation.org
Website www.princes-foundation.org

THE PRINCE OF WALES'S CRAFT SCHOLARSHIP SCHEME:
See also: display entry on page 226.

EVENTS

JUNE

5–6 **Pestex-Protex 2001.** Open to all those with an interest in the wood preserving, damp-proofing and pest control industries. The Pavilion, NEC, Birmingham. **Contact** Christine Selden, BWPDA, 1 Gleneagles House, Vernon Gate, South Street, Derby DE1 1UP Tel 01332 225100

5–7 **Traditional Painting and Decorating Techniques.** A three-day course at the Weald & Downland Open Air Museum led by Wilm and Joy Huning. **Contact** – see below

5–8 **Cleaning Masonry Buildings.** A Building Conservation Masterclass led by John Ashurst with Graham Abrey. West Dean College/Weald & Downland Open Air Museum. **Contact** West Dean College – see below

7 **English Historic Towns Forum: Town Visit to Gloucester.** Urban Design Strategy Seminar. Open to all. **Contact** EHTF – see below

7 **The Fourth National Conservation Conference: 'From Finials to Footings'.** A one-day cross-professional conference, on practical solutions for projects large and small. Keynote speaker; Sir Neil Cossons, Chairman of English Heritage. 1 Great George Street, London SW1. **Contact** National Conservation Conference Tel 01342 410242

7 **Landscape Institute Conference: International Regeneration.** A joint Conference with the British Urban Regeneration Association (BURA). Hilton Metropole, London. **Contact** The Landscape Institute Tel 020 7350 5200

8–9 **Earth Buildings: Sustaining the Tradition.** Kelmarsh Hall. **Contact** SPAB – see below

10 **LHPGT London Garden Squares Day.** Various garden squares open to the public around London. **Contact** LHPGT, Duck Island Cottage, c/o The Store Yard, London SW1A 2BJ Tel 020 7839 3969

15–22 **Landmark Trust Open Days,** The Ancient House, Clare, Suffolk. Entry is free and all are welcome. **Contact** – see below

17–22 **The Care and Conservation of Wallcoverings.** West Dean College. **Contact** – see below

21 **Repair of Timber-framed Buildings.** A Weald & Downland Open Air Museum Day School led by Richard Harris and Roger Champion. **Contact** – see below

26–29 **Conservation and Repair of Timber.** A Building Conservation Masterclass led by Richard Harris with Peter McCurdy at West Dean College/Weald & Downland Open Air Museum. **Contact** West Dean College – see below

EVENTS

JULY

4–Oct 14 **Archaeology in Guernsey Exhibition,** Guernsey Museum and Art Gallery. **Contact** Peter Sarl, Director of Museums Tel 0148 172 6518

5 **English Historic Towns Forum: Elected Members' Seminar,** Lincoln. **Contact** EHTF– see below

6 **English Historic Towns Forum: Town Visit to Lincoln.** 'Reconnecting the Inner City' – innovative transport strategies. **Contact** EHTF– see below

12 **IHBC Ipswich Walkabout.** The Willis Corron and Isaac Lord buildings with Bob Kindred. IHBC East Anglia Branch Event, Ipswich. **Contact** Robert Scrimgeour, Maldon District Council Tel 01621 875725

20–22 **Annual Symposium of The Church Monuments Society.** Including visits to Tong, Wroxeter and Candover. Harper Adams College, Newport, Shropshire. **Contact** Mark Adams, 15 Astley Court, Astley, Shrewsbury, Shropshire

23–27 **Our Protected Past: Managing the historic environment in Europe's National Parks and protected areas.** A major international conference organised by English Heritage, UK National Parks Authorities, Countryside Council for Wales, RCAHM Wales and others. Exeter. **Contact** Our Protected Past, Centre for Education, Development and Co-operation, School of Education, University of Exeter, Heavitree Road, Exeter, Devon EX1 2LU

AUGUST

17–24 **Association for Industrial Archaeology Annual Conference,** Fitzwilliam College, Cambridge. **Contact** Miss I Wilson, AIA Liaison Officer, AIA Office, School of Archaeological Studies, University of Leicester, Leicester LE1 7RH Tel 0116 252 5337

21 **IHBC Northern Ireland Meeting and Site Visit,** Market Hill Court House, Co Armagh. **Contact** Charlie Morrison Tel 028 7127 1366

SEPTEMBER

Scottish Archaeology Month. Guided walks, excavation open days, lectures, hands-on workshops, exhibitions throughout September, open to all, free of charge. Venues across Scotland. **Contact** Fiona Davidson Tel 0131 247 4119

1–2 **Frame 2001.** UK Carpenters Fellowship Convention at the Weald & Downland Open Air Museum. Open to anyone interested in Timber Framing. **Contact** Weald & Downland Open Air Museum – see below

2–9 **Heritage Week, Ireland.** A full week of more than 400 events and activities around the country. **Contact** Teresa Clarke Tel +353 (0)1 647 2455 or Catherine O'Connor Tel +353 (0)1 647 2466

7–10 **Landmark Trust Open Days,** Wilmington Priory, East Sussex. Open 10am-5pm, admission is free and all are welcome. **Contact** – see below

9 **Landmark Trust Open Days:** Beckford's Tower, Bath; New Inn, Peasenhall, Suffolk; The Music Room, Lancaster; Crownhill Fort, Plymouth. Open 10am-5pm, admission is free and all are welcome. **Contact** – see below

11–14 **Conservation and Repair of Stone Masonry.** A Building Conservation Masterclass at West Dean College/Weald & Downland Open Air Museum. **Contact** West Dean College – see below

14–16 **The Building Limes Forum Annual Gathering,** Charlestown Workshops, Fife. **Contact** James Simpson, Simpson & Brown Architects Tel 0131 555 4678

16–20 **The Care of Miniature Paintings on Ivory and Vellum.** In collaboration with the Victoria and Albert Museum. West Dean College. **Contact** – see below

19–21 **English Historic Towns Forum: Annual Conference and AGM.** 'Tourism in Historic Towns – Supporting Wise Growth'. Cambridge. **Contact** EHTF– see below

21–23 **Association of Diocesan and Cathedral Archaeologists Annual Conference.** Open to professional archaeologists concerned with churches. York. **Contact** Simon Ward Tel 01244 402026

tbc **IHBC Scotland Seminar: Buildings at Risk and Redundant Listed Buildings – The Partnership Approach,** Inverness. **Contact** John Clare, Vennel Cottage, Goose Green, Gullane, East Lothian EH31 2BA Tel 01620 842086

21–23 **Council for Independent Archaeology Congress,** Nottingham University. **Contact** Mike Rumbold Tel 01327 340855

OCTOBER

1–3 **2001: A Pest Odyssey – No collection is safe from pests!** Conference, organised by English Heritage, the Science Museum and the National Preservation Office at the British Library, London. **Contact** Belinda Sanderson, Information Officer, National Preservation Office, The British Library, 96 Euston Road, London NW1 2DB Tel 020 7412 7724

1–6 **SPAB Autumn Repair Course,** London. **Contact** – see below

4 **IHBC Day School on Ecclesiastical Conservation.** IHBC East Anglia Branch Event. Ely. **Contact** Robert Scrimgeour, Maldon District Council Tel 01621 875725

6 **Institute of Horticulture AGM,** Royal Botanic Gardens, Kew. **Contact** Institute of Horticulture Tel 020 7245 6943

9–12 **The Association of Building Engineers: Annual Conference.** Open to all. Copthorne Hotel, Cardiff. **Contact** Gillian McKenzie, ABE, Lutyens House, Billing Brook Road, Weston Favell, Northampton NN3 8NW Tel 01604 404121

9–12 **Understanding and Using Architectural Paint.** A Building Conservation Masterclass at West Dean College/Weald & Downland Open Air Museum. **Contact** West Dean College – see below

16–18 **AI 2001: Architectural Ironmongery Exhibition.** Open to all. Excel Exhibition Centre. **Contact** Peter Spill, The Guild of Architectural Ironmongers, 8 Stepney Green, London E1 3JU Tel 020 7790 3431

30–Nov 2 **Specifying for Conservation Work.** A Building Conservation Masterclass at West Dean College/Weald & Downland Open Air Museum. **Contact** West Dean College – see below

31–Nov 5 **National Trust for Historic Preservation: Conference on Historic Preservation,** Los Angeles, USA.

NOVEMBER

7–9 **3rd International Seminar on Structural Analysis of Historical Constructions.** Organised by The Department of Civil Engineering of the University of Minho, Portugal and the Polytechnical University of Catalonia, Barcelona, Spain at Guimarães, Portugal. **Contact** Prof Paulo Lourenço, Organising Committee, Dept of Civil Engineering, School of Engineering, University of Minho, Azurém, P-4800-058 Guimarães, Portugal Tel +351 253 510200

13 **IHBC Northern Ireland Meeting and Site Visit,** Verbal Arts Centre, Londonderry. **Contact** Charlie Morrison Tel 028 7127 1366

20 **Historic Houses Association AGM.** Historic Buildings, Parks and Gardens Exhibition, and Smiths Gore Lecture. Entry by application. Queen Elizabeth II Conference Centre, London SW1. **Contact** David Lewis Tel 01462 896688

20–22 **Mortars for Repair and Conservation.** A Building Conservation Masterclass at West Dean College/Weald & Downland Open Air Museum. **Contact** West Dean College – see below

21–23 **ICCROM 22nd General Assembly,** Rome, Italy. **Contact** ICCROM, 13 Via de San Michele, 1-00153, Rome Tel 00 39 06 585531

22 **English Historic Towns Forum: Norwich Town Visit.** 'Delivering Urban Renaissance'. Open to all. **Contact** EHTF– see below

tbc **IHBC Scotland Seminar: Best Value and Buildings Conservation,** Edinburgh. **Contact** John Clare, Vennel Cottage, Goose Green, Gullane, East Lothian EH31 2BA Tel 01620 842086

25–28 **The Care and Conservation of Historic Floors.** A Building Conservation Masterclass led by Jane Fawcett and Trevor Proudfoot at West Dean College/ Weald & Downland Open Air Museum. **Contact** West Dean College – see below

DECEMBER

4–6 **Conservation Engineering.** A Building Conservation Masterclass at West Dean College/Weald & Downland Open Air Museum. **Contact** West Dean College – see below

CONTACTS

ENGLISH HISTORIC TOWNS FORUM PO Box 22, Bristol BS16 1RZ Tel 0117 975 0459 Fax 0117 975 0460 E-mail ehtf@uwe.ac.uk

ESSEX COUNTY COUNCIL Pauline Hudspith, Heritage Conservation, Essex County Council, County Hall, Chelmsford, Essex CM1 1QH Tel 01245 437672 Fax 01245 258353 E-mail pauline.hudspith@essexcc.gov.uk

THE LANDMARK TRUST Shottesbrooke, Maidenhead, Berkshire SL6 3SW Tel 01628 825920 Fax 01628 825417

WEALD & DOWNLAND OPEN AIR MUSEUM Diana Rowsell, Weald and Downland Open Air Museum, Singleton, Chichester, West Sussex PO18 0EU Tel 01243 811363 E-mail wealddown@mistral.co.uk

WEST DEAN COLLEGE Isabel Thurston, West Dean College, West Dean, Chichester PO18 0QZ Tel 01243 818294 Fax 01243 811343 E-mail isabel_thurston@westdean.org.uk

THE SOCIETY FOR THE PROTECTION OF ANCIENT BUILDINGS (SPAB) 37 Spital Square, London E1 6DY Tel 020 7377 1644

ABBREVIATIONS

EHFT	English Historic Towns Forum
ICCROM	The International Centre for the Study of the Preservation and Restoration of Cultural Property
IHBC	Institute of Historic Building Conservation
LHPGT	The London Historic Parks and Gardens Trust
SPAB	The Society for the Protection of Ancient Buildings

See www.buildingconservation.com for the most up to date information

Our website is updated monthly which means that we can add new events when we hear about them. In addition to information about events the site has over 1,300 pages of articles, companies, courses and books all related to building conservation.

To get a free listing for an event in The Building Conservation Directory or on our website, contact Cathedral Communications with the details: E-mail hannah.moffat@cathcomm.demon.co.uk

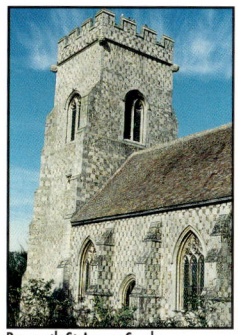

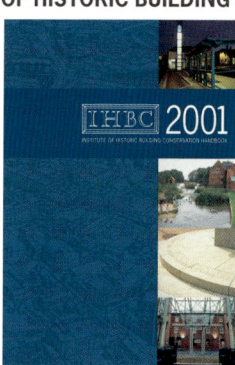

Founded 1877

THE SOCIETY FOR THE PROTECTION OF ANCIENT BUILDINGS

The SPAB works with **Professionals, Builders, Craftsmen, Planners, Students** and **Home Owners** to help them care for old buildings.

This means that the SPAB:

Runs educational courses ranging from one day to nine months,
Publishes low cost practical pamphlets,
Gives free technical advice via its telephone line,
Comments on hundreds of applications affecting listed buildings every year,
Trains the next generation of conservation specialists through the Scholarship and Fellowship,
Compiles a quarterly list of historic buildings in need of repair and for sale,
Lobbies the government on issues affecting old buildings,
Adheres to a straightforward set of principles concerning the protection of old buildings,
But **keeps** conservation thought moving by providing a forum for debate and discussion,
Needs more members!

Further information is available from our offices:

SPAB, 37 Spital Square, London E1 6DY
Phone: 020 7377 1644 Fax: 020 7247 5296
info@spab.org.uk www.spab.org.uk
Charity N°: 231 307

the Victorian

The Victorian Society publishes ***the victorian*** its new three times a year magazine replacing the twice yearly newsletter and annual journal.

The Society is an influential pressure group for the protection and conservation of Victorian and Edwardian architecture plus the decorative arts of the period 1837–1914. Supported by Central Government it is one of the four national amenity societies with a legal role in the listed buildings consent system.

the victorian is read by 3,250 members who will be architects, other conservation professionals as well as homeowners.

For details of membership contact:
The Victorian Society,
1 Priory Gardens, Bedford Park, London W4 1TT
Telephone: (020) 8994 1019 Fax: (020) 8995 4895

For information on *the victorian* contact:
Hall-McCartney Limited,
PO Box 21, Baldock, Hertfordshire SG7 5SH
Telephone: (01462) 896688 Fax: (01462) 896677

HISTORIC CHURCHES

The Conservation & Repair of Ecclesiastical Buildings

BCD Special Report

Now in its ninth annual edition, *Historic Churches* provides hard hitting and practical technical information and case studies on the conservation and maintenance of old church buildings.

£5.95 including delivery from Cathedral Communications Limited

Tel 01747 871717
Fax 01747 871718

▶ THE COLLEGE OF ESTATE MANAGEMENT

Whiteknights, Reading RG6 6AW
Tel 0118 986 1101 Fax 0118 975 5344
Website www.cem.ac.uk

BOOKS: Building Conservation Philosophy. VIDEO TAPES: British Domestic Architecture AD800-1800, British Domestic Architecture AD1800-1990, Countryside in the Making, Towns in the Making, Urban Conservation in Practice. CPD STUDY PACKS: CDM Regulations, Dilapidations, Facilities Management, Planning Appeals, Structural Failures in Domestic Buildings. COURSES: The College provides the two year RICS Postgraduate Diploma in Building Conservation by distance learning. *See also: display entry in Courses & Training section, page 219.*

WWW.BUILDINGCONSERVATION.COM

The Building Conservation Directory's website *www.buildingconservation.com*, in association with Blackwell's on-line bookshop, offers a unique selection of recommended book titles for historic building conservation, refurbishment and maintenance. If you want to learn more about the history and protection of our architectural heritage then click in to *www.buildingconservation.com*. Avoid those time consuming trips to specialist bookshops by selecting and ordering your books direct via the Internet using Blackwell's secure credit card purchasing system. *See also: inside back cover.*

PUBLICATIONS

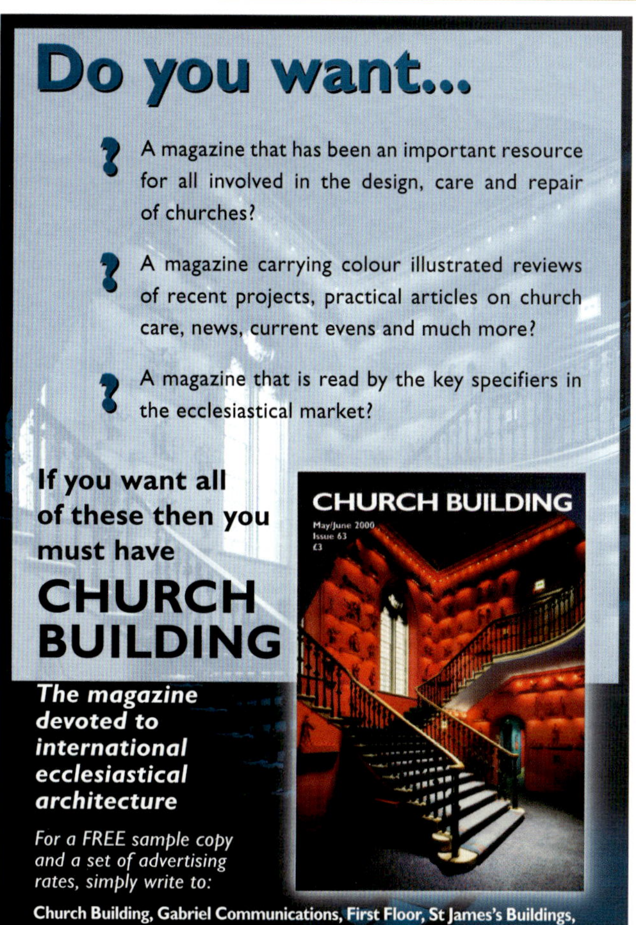

The definitive biographical reference tool for the architecture of an era

Directory of British Architects 1834-1914

An updated and expanded edition

Compiled for the British Architectural Library, Royal Institute of British Architects by Antonia Brodie, Alison Felstead, Jonathan Franklin, Leslie Pinfield and Jane Oldfield

More than 11,000 British architects working between 1834 and 1914 are included in the updated edition of the *Directory*. Its scope extends beyond members of the RIBA (Royal Institute of British Architects) to cover all those who practised architecture in Great Britain during the period, as well as many who practised in the former colonies. It increases the 1993 edition by some 4,000 new entries, it updates and expands entries from the earlier edition with significant new information, and it extends the coverage by fourteen years.

June 2001 • 2 vols., 2,248 pages • cloth • 276 x 219 mm
0 8264 4963 8 set of 2 vols. • £400.00 the set

CONTINUUM
CONTINUUM
CONTINUUM

Order from your bookseller or, in case of difficulty, telephone: Continuum, Orca Book Services, +44 (0)1202 665432

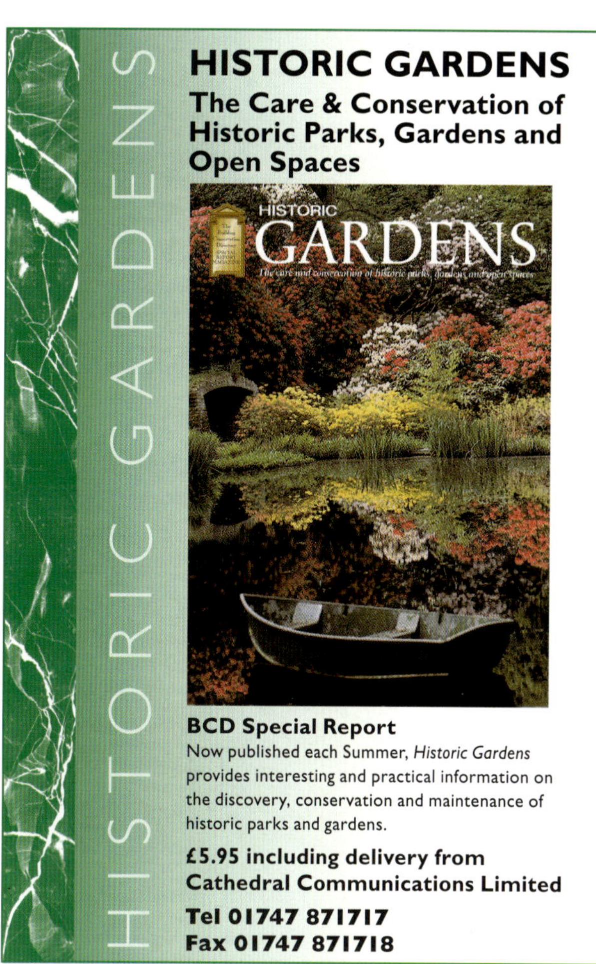

HISTORIC GARDENS
HISTORIC GARDENS
The Care & Conservation of Historic Parks, Gardens and Open Spaces
HISTORIC GARDENS
BCD Special Report
Now published each Summer, *Historic Gardens* provides interesting and practical information on the discovery, conservation and maintenance of historic parks and gardens.
£5.95 including delivery from Cathedral Communications Limited
Tel 01747 871717
Fax 01747 871718

USEFUL INFORMATION

USEFUL CONTACTS Further information about the organisations in this section, including links to website and e-mail addresses, can be found at www.buildingconservation.com

ADVISORY BOARD FOR REDUNDANT CHURCHES, THE
Cowley House, 9 Little College Street, London SW1P 3XS
Tel 020 7898 1870

ALLCHURCHES TRUST LTD
Beaufort House, Brunswick Road, Gloucester GL1 1JZ
Tel 01452 528533 Fax 01452 308860

ALMSHOUSE ASSOCIATION
Billingbear Lodge, Carters Hill, Wokingham, Berkshire RG40 5RU
Tel 01344 452922 Fax 01344 862062

AMBERLEY MUSEUM, THE
Amberley, Arundel, West Sussex BN18 9LT
Tel 01798 831370 Fax 01798 831831

ANCIENT MONUMENTS BOARD FOR WALES
National Assembly for Wales, Crown Building, Cathays Park, Cardiff CF10 3NQ
Tel 029 2082 6376

ANCIENT MONUMENTS SOCIETY
St Ann's Vestry Hall, 2 Church Entry, London EC4V 5HB
Tel 020 7236 3934 Fax 020 7329 3677

ARCH FOUNDATION (ART RESTORATION FOR CULTURAL HERITAGE)
Hellbrunn 17, A-5020 Salzburg, Austria
Tel +43 662 833340 Fax +43 662 822867

ARCHITECTS ACCREDITED IN BUILDING CONSERVATION REGISTER
33 Macclesfield Road, Wilmslow, Cheshire SK9 2AF
Tel 01625 523784 Fax 01625 548328

ARCHITECTURAL HERITAGE FUND, THE (AHF)
Clareville House, 26/27 Oxendon Street, London SW1Y 4EL
Tel 020 7925 0199 Fax 020 7925 0199

ARCHITECTURAL HERITAGE SOCIETY
Beech House, Cotswold Avenue, Lisvane, Cardiff CF14 0TA
Tel 029 2076 3060 Fax 029 2076 3070

ARCHITECTURAL HERITAGE SOCIETY OF SCOTLAND
The Glasite Meeting House, 33 Barony Street, Edinburgh EH3 6NX
Tel 0131 557 0019 Fax 0131 557 0049

ARCHITECTURAL SALVAGE INDEX, THE
c/o Hutton + Rostron, Netley House, Gomshall, Surrey GU5 9QA
Tel 01483 203221 Fax 01483 202911

ART LOSS REGISTER, THE
12 Grosvenor Place, London SW1X 7HH
Tel 020 7235 3393 Fax 020 7235 1652

ASSOCIATION FOR INDUSTRIAL ARCHAEOLOGY
AIA Office, School of Archaeological Studies, University of Leicester, Leicester LE1 7RH
Tel 0116 252 5337 Fax 0116 252 5005

ASSOCIATION FOR STUDIES IN THE CONSERVATION OF HISTORIC BUILDINGS
Institute of Archaeology, 31-34 Gordon Square, London WC1H 0PY
Tel 020 7973 3326 Fax 020 7973 3090

ASSOCIATION OF BUILDING ENGINEERS, THE
Lutyens House, Billing Brook Road, Weston Favell, Northampton NN3 8NW
Tel 01604 404121 Fax 01604 784220

ASSOCIATION OF DIOCESAN AND CATHEDRAL ARCHAEOLOGISTS
Chester Archaeology, 27 Grosvenor Street, Chester CH1 2DD
Tel 01244 402026

ASSOCIATION OF PRESERVATION TRUSTS
Clareville House, 26-27 Oxendon Street, London SW1Y 4EL
Tel 020 7930 1629 Fax 020 7930 0295

ASSOCIATION OF SMALL HISTORIC TOWNS AND VILLAGES
7-9 Gerrard Street, Warwick CV34 4HD
Tel 01926 400717 Fax 01926 400717

ASSOCIATION OF TOWN CENTRE MANAGEMENT
1 Queen Annes Gate, Westminster, London SW1H 9BT
Tel 020 7222 0120 Fax 020 7222 4440

AVONCROFT MUSEUM OF HISTORIC BUILDINGS
Stoke Heath, Bromsgrove, Worcestershire B60 4JR
Tel 01527 831363 Fax 01527 876934

BEAMISH, THE NORTH OF ENGLAND OPEN AIR MUSEUM
Beamish, Co Durham DH9 0RG
Tel 01207 231811 Fax 01207 290933

BRICK DEVELOPMENT ASSOCIATION
Woodside House, Winkfield, Windsor, Berkshire SL4 2DX
Tel 01344 885651 Fax 01344 890129

BRITISH ANTIQUE FURNITURE RESTORERS' ASSOCIATION
The Old Rectory, Warmwell, Dorchester, Dorset DT2 8HQ
Tel 01305 854822 Fax 01305 854822

BRITISH ARCHITECTURAL LIBRARY
RIBA, 66 Portland Place, London W1N 4AD
Tel 020 7580 5533 Fax 020 7631 1802

BRITISH ARTIST BLACKSMITHS ASSOCIATION
Yew Tree Cottage, Bredenbury, Bromyard, Herefordshire HR7 4TJ
Tel 01885 482572

BRITISH FOUNDRY ASSOCIATION
6th Floor, The McLaren Building, 35 Dale End, Birmingham B4 7LN
Tel 0121 200 2100 Fax 0121 200 1306

BRITISH GEOLOGICAL SURVEY
Keyworth, Nottingham NG12 5GG
Tel 0115 936 3171 Fax 0115 936 3593

BRITISH INSTITUTE OF ARCHITECTURAL TECHNOLOGISTS
397 City Road, London EC1V 1NH
Tel 020 7278 2206 Fax 020 7837 3194

BRITISH INSTITUTE OF NON-DESTRUCTIVE TESTING, THE
1 Spencer Parade, Northampton NN1 5AA
Tel 01604 630124 Fax 01604 231489

BRITISH METAL CASTING ASSOCIATION, THE
Boardesley Hall, The Holloway, Alvechurch, Birmingham, West Midlands B48 7QB
Tel 01527 585222 Fax 01527 590990

BRITISH MUSEUM, DEPT MEDIEVAL AND MODERN EUROPE
Department of Medieval and Modern Europe, Great Russell Street, London WC1B 3DG
Tel 020 7323 8741 Fax 020 7323 8496

BRITISH SLATE ASSOCIATION
Construction House, 56-64 Leonard Street, London EC2A 4JX
Tel 020 7608 5094 Fax 020 7608 5081

BRITISH SOCIETY OF MASTER GLASS PAINTERS – CONSERVATION COMMITTEE
5 Tivoli Place, Ilkley, West Yorkshire LS29 8SU
Tel 01943 602521 Fax 01943 602521

BRITISH STANDARDS INSTITUTION
389 Chiswick High Road, London W4 4AL
Tel 020 8996 9001 Fax 020 8996 7001

BRITISH SUNDIAL SOCIETY
4 New Wokingham Road, Crowthorne, Berkshire RG45 7NR
Tel 01344 772303 Fax 01344 772303

BRITISH URBAN REGENERATION ASSOCIATION
33 Great Sutton Street, London EC1V 0DX
Tel 020 7253 5054 Fax 020 7490 8735

BRITISH WATERWAYS
The Locks, Hillmorton, Rugby, Warwickshire CV21 4PP
Tel 01788 566030 Fax 01788 541076

BRITISH WOOD PRESERVING AND DAMP-PROOFING ASSOCIATION
1 Gleneagles House, Vernon Gate, South Street, Derby DE1 1UP
Tel 020 8519 2588 Fax 020 8519 3444

BROOKING COLLECTION, THE
University of Greenwich, Oakfield Lane, Dartford, Kent DA1 2SZ
Tel 020 8331 9897

BUILDING CENTRE LTD, THE
26 Store Street, London WC1E 7BT
Tel 020 7692 4000 Fax 020 7580 9641

BUILDING CRAFTS & CONSERVATION TRUST, THE
27 Orchard Street, Canterbury, Kent CT2 8AP
Tel 01227 451795 Fax 01233 861265

BUILDING LIMES FORUM, THE
c/o Roger Dowley Court, Russia Lane, London E2 9NJ

BUILDING OF BATH MUSEUM, THE
Countess of Huntingdon's Chapel, The Vineyards, The Paragon, Bath BA1 5NA
Tel 01225 333895 Fax 01225 445473

BUILDING RESEARCH ESTABLISHMENT
Bucknalls Lane, Garston, Watford, Hertfordshire WD2 7JR
Tel 01923 664000 Fax 01923 664010

CADW: WELSH HISTORIC MONUMENTS
Crown Building, Cathays Park, Cardiff CF10 3NQ
Tel 029 2050 0200 Fax 029 2050 0300

CAPEL – CHAPELS HERITAGE SOCIETY
c/o RCAHMW, Crown Building, Plas Crug, Aberystwyth, Ceredigion SY23 1NJ
Tel 01970 621210 Fax 01970 627701

CARPENTERS' AWARD, THE
The Worshipful Company of Carpenters, Throgmorton Avenue, London EC2N 2JJ
Tel 020 7727 9474 Fax 020 7727 1687

CATHEDRAL ARCHITECTS' ASSOCIATION
46 St Marys Street, Ely, Cambridgeshire CB7 4EY
Tel 01353 660660 Fax 01353 660661

CENTRE FOR EARTHEN ARCHITECTURE
University of Plymouth, Faculty of Technology, Drake Circus, Plymouth PL4 8AA
Tel 01752 233630 Fax 01752 233310

CHAPELS SOCIETY, THE
1 Newcastle Avenue, Beeston, Nottinghamshire NG9 1BT
Tel 0115 922 4930

CHARLESTOWN WORKSHOPS AT THE SCOTTISH LIME CENTRE
The Schoolhouse, 4 Rocks Road, Charlestown, Fife KY11 3EN
Tel 01383 872722 Fax 01383 872744

CHARTERED INSTITUTE OF BUILDING HERITAGE GROUP, THE
Englemere, Kings Ride, Ascot, Berkshire SL5 7TB
Tel 01344 630700 Fax 01344 630777

CHILTERN OPEN AIR MUSEUM
Newland Park, Gorelands Lane, Chalfont St Giles, Buckinghamshire HP8 4AB
Tel 01494 871117 Fax 01494 872163

CHURCH MONUMENTS SOCIETY, THE
34 Bridge Street, Shepshed, Leicestershire LE12 9AD
Tel 01509 650637

CHURCHES CONSERVATION TRUST, THE
89 Fleet Street, London EC4Y 1DH
Tel 020 7936 2285 Fax 020 7936 2284

CINEMA THEATRE ASSOCIATION
56 Charrington Street, London NW1 1RD
Tel 020 7387 0528 Fax 020 7387 0528

CIVIC TRUST
17 Carlton House Terrace, London SW1Y 5AW
Tel 020 7930 0914 Fax 020 7321 0180

CIVIC TRUST AWARDS, THE
The Civic Trust, 17 Carlton House Terrace, London SW1Y 5AW
Tel 020 7930 0914 Fax 020 7321 0180

CIVIC TRUST FOR WALES
2nd Floor, Empire House, Mount Stuart Square, The Docks, Cardiff CF10 5FN
Tel 029 2048 4606 Fax 029 2048 2086

COMMISSION FOR ARCHITECTURE AND THE BUILT ENVIRONMENT (CABE)
7 St James's Square, London SW1Y 4JU
Tel 020 7839 6537 Fax 020 7839 8475

CONCRETE REPAIR ASSOCIATION, THE
Association House, 235 Ash Road, Aldershot, Hampshire GU12 4DD
Tel 01252 321322 Fax 01252 333901

CONFERENCE ON TRAINING IN ARCHITECTURAL CONSERVATION (COTAC)
Room 97a, Platform 7, St Pancras Station, Euston Road, London NW1 2QP
Tel 020 7713 0135 Fax 020 7713 0359

CONSERVATION AWARDS
Resource (Awards), 16 Queen Anne's Gate, London SW1H 9AA
Tel 020 7273 1441

CONSERVATION REGISTER, THE
UKIC, 109 The Chandlery, 50 Westminster Bridge Road, London SE1 7QD
Tel 020 7721 8246

CONSTRUCTION CONFEDERATION
Construction House, 56-64 Leonard Street, London EC2A 4JX
Tel 020 7608 5000 Fax 020 7608 5101

CONSTRUCTION HISTORY SOCIETY
c/o Chartered Institute of Building, Englemere, Kings Ride, Ascot, Berkshire SL5 7TB
Tel 01844 346270 Fax 01844 346270

COPPER DEVELOPMENT ASSOCIATION
Verulam Industrial Estate, 224 London Road, St Albans, Hertfordshire AL1 1AQ
Tel 01727 731200 Fax 01727 731216

CORPUS VITREARUM MEDII AEVI ARCHIVE
National Monuments Record Centre, Kemble Drive, Swindon, Wiltshire SN2 2GZ
Tel 01793 414600 Fax 01793 414606

CORROSION PREVENTION ASSOCIATION, THE
Association House, 235 Ash Road, Aldershot, Hampshire GU12 4DD
Tel 01252 321302 Fax 01252 333901

COUNCIL FOR BRITISH ARCHAEOLOGY
Bowes Morrell House, 111 Walmgate, York YO1 2UA
Tel 01904 671417 Fax 01904 671384

COUNCIL FOR BRITISH ARCHAEOLOGY – WALES, THE
c/o CPAT, 20 High Street, Welshpool, Powys SY21 7JP
Tel 01938 552035 Fax 01938 552179

COUNCIL FOR SCOTTISH ARCHAEOLOGY, THE
c/o National Museums of Scotland, Chambers Street, Edinburgh EH1 1JF
Tel 0131 247 4119 Fax 0131 247 4126

USEFUL CONTACTS Further information about the organisations in this section, including links to website and e-mail addresses, can be found at www.buildingconservation.com

COUNCIL FOR THE CARE OF CHURCHES
Church House, Great Smith Street, London SW1P 3NZ
Tel 020 7898 1866 Fax 020 7898 1881

COUNCIL FOR THE PREVENTION OF ART THEFT
The Estate Office, Stourhead Park, Stourton, Warminster, Wiltshire BA12 6QD
Tel 01747 841540

COUNCIL FOR THE PROTECTION OF RURAL ENGLAND
Warwick House, 25 Buckingham Palace Road, London SW1W 0PP
Tel 020 7976 6433 Fax 020 7976 6373

CUSTOMS & EXCISE, LANDFILL TAX COPE
Dobson House, Regent Centre, Gosforth, Newcastle upon Tyne NE3 3PF
Tel 0845 010 9000

DEPARTMENT FOR CULTURE, MEDIA AND SPORT (DCMS)
2-4 Cockspur Street, London SW1Y 5DH
Tel 020 7211 6200 Fax 020 7211 6210

DEPARTMENT OF THE ENVIRONMENT FOR NORTHERN IRELAND
Environment and Heritage Service, 5-33 Hill Street, Belfast BT1 2LA
Tel 028 9023 5000 Fax 028 9054 3111

THE DEPARTMENT FOR TRANSPORT, LOCAL GOVERNMENT AND THE REGIONS
Eland House, Bressenden Place, London SW1E 3EB
Tel 020 7944 3000

DOCOMOMO – UK
70 Cowcross Street, London EC1M 6EJ
Tel 020 7490 7243 Fax 01223 311166

DRY STONE WALLING ASSOCIATION OF GREAT BRITAIN
PO Box 8615, Sutton Coldfield B75 7HQ
Tel 0121 378 0493 Fax 0121 378 0493

DÚCHAS – THE HERITAGE SERVICE (IRELAND)
Department of Arts, Heritage Gaeltacht and Islands, 6 Ely Place Upper, Dublin 2
Tel +353 (0)1 647 3000 Fax +353 (0)1 661 6764

EAST ANGLIAN EARTH BUILDINGS GROUP (EARTHA)
Ivy Green, London Rd, Wymondham, Norfolk NR18 9JD
Tel 01953 601701 Fax 01953 601701

ECCLESIASTICAL ARCHITECTS AND SURVEYORS ASSOCIATION
Property Department, Diocese of Salisbury, Church House, Crane Street, Salisbury, Wiltshire SP1 2QB
Tel 01722 411933 Fax 01722 329833

EDINBURGH WORLD HERITAGE TRUST
5 Charlotte Street, Edinburgh EH2 4DR
Tel 0131 220 7720 Fax 0131 220 7730

ENGLISH HERITAGE
23 Savile Row, London W1X 1AB
Tel 020 7973 3000 Fax 020 7973 3001

East Midlands Region
Hazelrigg House, 33 Marefair, Northampton NN1 1SR
Tel 01604 735422 Fax 01604 730321

East of England Region
62-74 Burleigh Street, Cambridge CB1 1DJ
Tel 01223 582700 Fax 01223 582701

London Region
23 Savile Row, London W1S 2ET
Tel 020 7973 3000 Fax 020 7973 3534

National Monuments Record
National Monuments Record Centre, Kemble Drive, Swindon, Wiltshire SN2 2GZ
Tel 01793 414600 Fax 01793 414804

North East Region
Bessie Surtees House, 41 Sandhill, Newcastle upon Tyne NE1 8JF
Tel 0191 261 1585 Fax 0191 261 1130

North West Region
Suites 3.3 and 3.4, Canada House, 3 Chepstow Street, Manchester M1 5FW
Tel 0161 242 1400 Fax 0161 242 1401

South East Region
2nd and 3rd Floor, Eastgate Court, 195-205 High Street, Guildford GU1 3EH
Tel 01483 252000 Fax 01483 252001

South West Region
29-30 Queen Square, Bristol BS1 4ND
Tel 0117 975 0700 Fax 0117 975 0701

West Midlands Region
112 Colmore Row, Birmingham B3 3AG
Tel 0121 625 6820 Fax 0121 625 6821

Yorkshire Region
37 Tanner Row, York YO1 6WP
Tel 01904 601901 Fax 01904 601999

Please note:
English Heritage now includes the former Royal Commission on the Historical Monuments of England (RCHME).

ENGLISH HISTORIC TOWNS FORUM
PO Box 22, Bristol BS16 1RZ
Tel 0117 975 0459 Fax 0117 975 0460

ENTRUST
5th Floor, Suite 2, Acre House, 2 Town Square, Sale, Cheshire M33 7WZ
Tel 0161 972 0055

EUROPA NOSTRA
Lange Voorhout 35, 2514 EC The Hague, The Netherlands
Tel (31)70 3024050 Fax (31)70 3617865

EUROPEAN CONFEDERATION OF CONSERVATOR-RESTORERS' ORGANISATIONS (ECCO)
Secretariat, Diepestraat 18, B-3061 Leefdaal, Belgium
Tel 00 (32)2 767 9780

EUROPEAN FOUNDATION FOR HERITAGE SKILLS
c/o Palais de l'Europe, F-67075 Strasbourg Cedex, France
Tel 00 33 (0) 3 90 21 45 37 Fax 00 33 (0) 3 88 41 2755

FEDERATION OF MASTER BUILDERS
Gordon Fisher House, 14-15 Great James Street, London WC1N 3DP
Tel 020 7242 7583 Fax 020 7404 0296

FIRE PROTECTION ASSOCIATION, THE
Bastille Court, 2 Paris Garden, London SE1 8ND
Tel 020 7902 5300 Fax 020 7902 5301

FOLLY FELLOWSHIP, THE
7 Inch's Yard, Market Street, Newbury, Berkshire RG14 5DP
Tel 01635 42864 Fax 01635 552366

FOUNTAIN SOCIETY, THE
Flat 10, 81 Onslow Square, London SW7 3LT
Tel 020 7584 9004 Fax 020 7584 2917

FRIENDS OF FRIENDLESS CHURCHES
St Ann's Vestry Hall, 2 Church Entry, London EC4V 5HB
Tel 020 7236 3934 Fax 020 7329 3677

FURNITURE HISTORY SOCIETY, THE
1 Mercedes Cottages, St John's Road, Haywards Heath, West Sussex RH16 4EH
Tel 01444 413845 Fax 01444 413845

GARDEN HISTORY SOCIETY IN SCOTLAND, THE
The Glasite Meeting House, 33 Barony Street, Edinburgh EH3 6NX
Tel 0131 557 5717 Fax 0131 557 5717

GARDEN HISTORY SOCIETY, THE
70 Cowcross Street, London EC1M 6BP
Tel 020 7608 2409 Fax 020 7490 2974

GEORGIAN GROUP, THE
6 Fitzroy Square, London W1T 5DX
Tel 020 7387 1720 Fax 020 7387 1721

HERITAGE BUILDING CONTRACTORS GROUP (UK)
c/o Linford Group Ltd, Quonians, Lichfield, Staffordshire WS13 7LB
Tel 01543 414234 Fax 01543 410065

HERITAGE CONSERVATION TRUST
2 Chester Street, London SW1X 7BB
Tel 020 7259 5688 Fax 020 7259 5590

HERITAGE COUNCIL, THE (IRELAND)
Rothe House, Kilkenny, Ireland
Tel +353 (0)56 70777 Fax +353 (0)56 70788

HERITAGE EDUCATION TRUST
Boughton House, Kettering, Northants NN14 1BJ
Tel 01536 515731 Fax 01536 417255

HERITAGE LOTTERY FUND
7 Holbein Place, London SW1W 8NR
Tel 020 7591 6000 Fax 020 7591 6255

HISTORIC BUILDINGS COUNCIL FOR WALES
National Asembly for Wales, Crown Building, Cathays Park, Cardiff CF10 3NQ
Tel 029 2082 6311 Fax 029 2082 6375

HISTORIC BURGHS ASSOCIATION OF SCOTLAND
PO Box 1124, Stirling FK9 4ZW
Tel 01786 833318 Fax 01786 833318

HISTORIC CHAPELS TRUST
29 Thurloe Street, Kensington, London SW7 2LQ
Tel 020 7584 6072 Fax 020 7225 0607

HISTORIC CHURCHES PRESERVATION TRUST
Fulham Palace, London SW6 6EA
Tel 020 7736 3054 Fax 020 7736 3880

HISTORIC FARM BUILDINGS GROUP
c/o Museum of English Rural Life, University of Reading, Whiteknights, PO Box 229, Reading, Berkshire RG6 6AG
Tel 0118 931 8663 Fax 0118 975 1264

HISTORIC GARDENS FOUNDATION
34 River Court, Upper Ground, London SE1 9PE
Tel 020 7633 9165 Fax 020 7401 7072

HISTORIC HOUSES ASSOCIATION
2 Chester Street, London SW1X 7BB
Tel 020 7259 5688 Fax 020 7259 5590

HISTORIC ROYAL PALACES
Surveyor of the Fabric Department, Hampton Court Palace, Surrey KT8 9AU
Tel 020 8781 9829 Fax 020 8781 9809

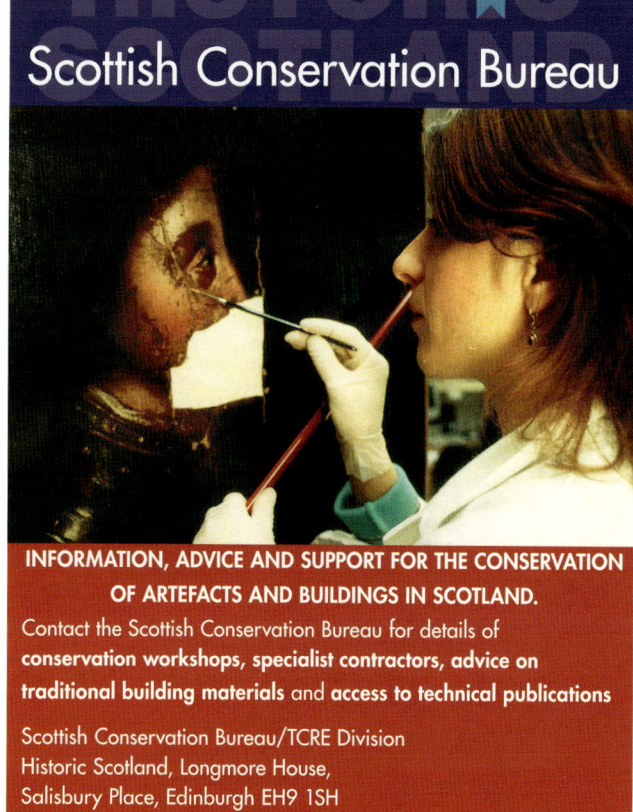

HISTORIC SCOTLAND
Longmore House, Salisbury Place, Edinburgh EH9 1SH
Tel 0131 668 8600 Fax 0131 668 8788

INSTITUTE OF FIELD ARCHAEOLOGISTS
The University of Reading, 2 Earley Gate, PO Box 239, Reading RG6 6AU
Tel 0118 931 6446 Fax 0118 931 6448

INSTITUTE OF HISTORIC BUILDING CONSERVATION
3 Stafford Road, Tunbridge Wells, Kent TN2 4QZ
new address and phone number from 1 January 2002
Jubilee House, High Street, Tisbury, Wiltshire SP3 6HA
Tel 01747 873133

INSTITUTE OF HISTORIC BUILDING CONSERVATION (SCOTLAND)
The Glasite Meeting House, 33 Barony Street, Edinburgh EH3 6NX
Tel 0131 529 3913 Fax 0131 529 7478

INSTITUTE OF MAINTENANCE AND BUILDING MANAGEMENT
Leigh Lodge, Leigh, Worcester WR6 5LB
Tel 01886 832323

INSTITUTE OF PAPER CONSERVATION
Keets House, 30 East Street, Farnham, Surrey GU9 7SW
Tel 01252 734062 Fax 01252 737741

INSTITUTION OF CIVIL ENGINEERS, PANEL FOR HISTORICAL ENGINEERING WORKS
1 Great George Street, London SW1P 3AA
Tel 020 7665 2250 Fax 020 7976 7610

INSTITUTION OF STRUCTURAL ENGINEERS
11 Upper Belgrave Street, London SW1X 8BH
Tel 020 7235 4535 Fax 020 7235 4294

INTERIOR DECORATORS AND DESIGNERS ASSOCIATION
1/4 Chelsea Harbour Design Centre, Lots Road, London SW10 0XE
Tel 020 7349 0800 Fax 020 7349 0500

INTERNATIONAL CENTRE FOR THE STUDY OF THE PRESERVATION AND RESTORATION OF CULTURAL PROPERTY (ICCROM)
13 Via de San Michele, I-00153 Rome, Italy
Tel 00 39 06 58553 1 Fax 00 39 06 58553 349

INTERNATIONAL COUNCIL ON MONUMENTS & SITES (ICOMOS)
49-51, rue de la Fédération, 75015 Paris, France
Tel (33 1) 45 67 67 70 Fax (33 1) 45 66 06 22

INTERNATIONAL COUNCIL ON MONUMENTS & SITES UK (ICOMOS UK)
10 Barley Mow Passage, Chiswick, London W4 4PH
Tel 020 8994 6477 Fax 020 8747 8464

INTERNATIONAL INSTITUTE FOR CONSERVATION OF HISTORIC AND ARTISTIC WORKS, THE
6 Buckingham Street, London WC2N 6BA
Tel 020 7839 5975 Fax 020 7976 1564

IRISH GEORGIAN SOCIETY
74 Merrion Square, Dublin 2
Tel +353 (0)1 676 7053 Fax +353 (0)1 662 0290

IRONBRIDGE GORGE MUSEUM TRUST
Ironbridge, Telford, Shropshire TF8 7AW
Tel 01952 433522 Fax 01952 432204

IRONBRIDGE INSTITUTE
Ironbridge Gorge Museum, Ironbridge, Telford, Shropshire TF8 7AW
Tel 01952 432751 Fax 01952 432237

JOHN BETJEMAN MEMORIAL AWARD, THE
37 Spital Square, London E1 6DY
Tel 020 7377 1644 Fax 020 7247 5296

KING OF PRUSSIA AWARDS, THE
Fulham Palace, London SW6 6EA
Tel 020 7736 3054 Fax 020 7736 3880

LANDFILL TAX CREDIT SCHEME – *see Entrust*

LANDMARK TRUST, THE
Shottesbrooke, Maidenhead, Berkshire SL6 3SW
Tel 01628 825920 Fax 01628 825417

LANDSCAPE INSTITUTE, THE
6-8 Barnard Mews, London SW11 1QU
Tel 020 7350 5200 Fax 020 7350 5201

LEAD SHEET ASSOCIATION, THE
Hawkwell Business Centre, Maidstone Road, Pembury, Tunbridge Wells, Kent TN2 4AH
Tel 01892 822773 Fax 01892 823003

LEAGUE OF PROFESSIONAL CRAFTSMEN
York House, Empire Way, Wembley, Middlesex HA9 0PA
Tel 020 8782 7494 Fax 020 8902 7557

LONDON HISTORIC PARKS AND GARDENS TRUST, THE
Duck Island Cottage, c/o The Store Yard, St James's Park, London SW1A 2BJ
Tel 020 7839 3969 Fax 020 7839 3969

LONDON STAINED GLASS REPOSITORY, THE
Glaziers' Hall, 9 Montague Close, London Bridge, London SE1 9DD
Tel 020 7403 3300 Fax 020 7407 6036

USEFUL CONTACTS Further information about the organisations in this section, including links to website and e-mail addresses, can be found at www.buildingconservation.com

MAINTAIN OUR HERITAGE
Weymouth House, Beechen Cliff Road, Bath BA2 4QS
Tel 01225 482074 Fax 0870 137 3805

MASTER CARVERS ASSOCIATION
Unit 20, 21 Wren Street, London WC1X 0HF
Tel 020 7278 8759 Fax 020 7278 8759

METAL ROOFING CONTRACTORS ASSOCIATION
c/o Alexio Metal Roofing, 3 Blondin Street, London E3 2TR
Tel 020 8981 6080 Fax 020 8981 4614

MILLENNIUM COMMISSION, THE
26th Floor, Portland House, Stag Place, London SW1E 5EZ
Tel 020 7880 2001 Fax 020 7880 2000

MONUMENTAL BRASS SOCIETY
c/o Society of Antiquaries of London, Burlington House, Piccadilly, London W1V 0HS
Tel 020 8520 5249 Fax 0208 521 8387

MUSEUM OF GARDEN HISTORY
Lambeth Palace Road, London SE1 7LB
Tel 020 7401 8865 Fax 020 7401 8869

NATIONAL ARCHIVES OF SCOTLAND
HM General Register House, 2 Princes Street, Edinburgh EH1 3YY
Tel 0131 535 1330 Fax 0131 535 1360

NATIONAL ASSEMBLY FOR WALES, THE
Crown Building, Cathays Park, Cardiff CF10 3NQ
Tel 029 2050 0200 Fax 029 2082 6375

NATIONAL ASSOCIATION OF DECORATIVE AND FINE ARTS SOCIETIES
NADFAS House, 8 Guilford Street, London WC1N 1DA
Tel 020 7430 0730 Fax 020 7242 0686

NATIONAL CONSERVATION CONFERENCE, THE
9-10 Old Stone Link, Ship Street, East Grinstead, West Sussex RH19 4EF
Tel 01342 410242 Fax 01342 313493

NATIONAL COUNCIL FOR THE CONSERVATION OF PLANTS AND GARDENS
The Stable Courtyard, RHS Gardens, Wisley, Woking, Surrey GU23 6QP
Tel 01483 211465 Fax 01483 212404

NATIONAL COUNCIL OF CONSERVATION-RESTORATION
c/o Institute of Paper Conservation, Leigh Lodge, Leigh, Worcester WR6 5LB
Tel 01886 832332 Fax 01886 833688

NATIONAL COUNCIL OF MASTER THATCHERS ASSOCIATIONS
17 Oldbutt Road, Shipston on Stour, Warwickshire CV36 4EG
Tel 07000 781909

NATIONAL FEDERATION OF BUILDERS
Construction House, 56-64 New Cavendish Street, London EC2A 4JX
Tel 020 7608 5150 Fax 020 7608 5151

NATIONAL FEDERATION OF MASTER STEEPLEJACKS AND LIGHTNING CONDUCTOR ENGINEERS
4d St Mary Place, The Lace Market, Nottingham NG1 1PH
Tel 0115 955 8818 Fax 0115 941 2238

NATIONAL FEDERATION OF ROOFING CONTRACTORS LTD
24 Weymouth Street, London W1N 4LX
Tel 020 7436 0387 Fax 020 7637 5215

NATIONAL HERITAGE MEMORIAL FUND
7 Holbein Place, London SW1W 8NR
Tel 020 7591 6000 Fax 020 7591 6001

NATIONAL MONUMENTS RECORD
– see English Heritage

NATIONAL MONUMENTS RECORD OF SCOTLAND
John Sinclair House, 16 Bernard Terrace, Edinburgh EH8 9NX
Tel 0131 662 1456 Fax 0131 662 1477

NATIONAL MONUMENTS RECORD OF WALES
Crown Building, Plas Crug, Aberystwyth, Ceredigion SY23 1NJ
Tel 01970 621233 Fax 01970 627701

NATIONAL PHYSICAL LABORATORY
Queens Road, Teddington, Middlesex TW11 0LW
Tel 020 8977 3222 Fax 020 8977 6458

NATIONAL PIERS SOCIETY
4 Tyrrell Road, South Benfleet, Essex SS7 5DH
Tel 01268 757291 Fax 020 7483 1902

NATIONAL PRESERVATION OFFICE
The British Library, 96 Euston Road, London NW1 2DB
Tel 020 7412 7612 Fax 020 7412 7796

NATIONAL SOCIETY OF MASTER THATCHERS
73 Hughenden Avenue, High Wycombe, Buckinghamshire HP13 5SL
Tel 01494 443198

NATIONAL TRUST FOR SCOTLAND, THE
28 Charlotte Square, Edinburgh EH2 4ET
Tel 0131 243 9300 Fax 0131 243 9301

NATIONAL TRUST, THE
33 Sheep Street, Cirencester, Gloucestershire GL7 1RQ
Tel 01285 651818 Fax 01285 657935

NATIONAL TRUST, THE
36 Queen Anne's Gate, Westminster, London SW1H 9AS
Tel 020 7222 9251 Fax 020 7222 5097

NATURAL STONE AWARDS, THE
The Stone Federation of Great Britain, 56-64 Leonard Street, London EC2

NEW STUDY CENTRE, THE
21 Palace Gardens Terrace, London W8 4SA
Tel 020 7229 3393 Fax 020 7229 4220

NEWCOMEN SOCIETY, THE
The Science Museum, London SW7 2DD
Tel 020 7371 4445 Fax 020 7371 4445

ORTON TRUST, THE
20 Copelands Road, Desborough, Kettering, Northants NN14 2QF
Tel 01536 761303

PASSIVE FIRE PROTECTION FEDERATION
Association House, 235 Ash Road, Aldershot, Hampshire GU12 4DD
Tel 01252 321322 Fax 01252 333901

PHILIP WEBB AWARD, THE
37 Spital Squre, London E1 6DY
Tel 020 7377 1644 Fax 020 7247 5296

PRISM FUND
Science Museum/Museums & Galleries Commission, South Kensington, London SW7 2DD
Tel 020 7942 4104 Fax 020 7942 4102

PROFESSIONAL ACCREDITATION OF CONSERVATOR-RESTORERS
c/o Institute of Paper Conservation, Leigh Lodge, Leigh, Worcester WR6 5LB
Tel 01886 832332 Fax 01886 833688

PUBLIC MONUMENTS AND SCULPTURE ASSOCIATION
72 Lissenden Mansions, Lissenden Gardens, London NW5 1PR
Tel 020 7485 0566 Fax 020 7267 1742

PUBLIC RECORD OFFICE
Kew, Richmond, Surrey TW9 4DU
Tel 020 8392 5200 Fax 020 8878 8905

RAILWAY HERITAGE TRUST
PO Box 686, Melton House, 65-67 Clarendon Road, Watford, Hertfordshire WD17 1XZ
Tel 01923 240250 Fax 01923 207079

REGENERATION THROUGH HERITAGE
The Prince's Foundation, 19-22 Charlotte Road, London EC2A 3SG
Tel 020 7613 8518 Fax 020 7613 8599

REGIONAL DEVELOPMENT AGENCIES
South-west Regional Development Agency, Sterling House, Dix's Field, Exeter, Devon EX1 1QA
Tel 01329 214747 Fax 01329 214848

East Midlands
Apex Court, City Link, Nottingham NG2 4LA
Tel 0115 988 8300 Fax 0115 853 3666

East of England
Compass House, Chivers Way, Histon, Cambridge CB4 9ZR
Tel 01223 713900 Fax 01223 713940

North East
Great North House, Sandyford Road, Newcastle Upon Tyne NE1 8ND
Tel 0191 261 2000 Fax 0191 232 9069

North West
New Town House, Buttermarket Street, Warrington WA1 2LF
Tel 01925 400100 Fax 01925 400400

South East
Crosslanes, Guildford GU1 1YA
Tel 01483 484226 Fax 01483 484247

South West
Sterling House, Dix's Field, Exeter, Devon EX1 1QA
Tel 01392 214747 Fax 01392 214848

West Midlands
3 Priestley Wharf, Holt Street, Aston Science Park, Birmingham B7 4BN
Tel 0121 380 3500 Fax 0121 380 3501

Yorkshire & The Humber
Victoria House, 2 Victoria Place, Leeds LS11 5AE
Tel 0113 243 9222 Fax 0113 243 1088

RESOURCE: THE COUNCIL FOR MUSEUMS, ARCHIVES AND LIBRARIES
16 Queen Anne's Gate, Westminster, London SW1H 9AA
Tel 020 7273 1444 Fax 020 7273 1404

ROYAL ARCHAEOLOGICAL INSTITUTE
c/o Society of Antiquaries of London, Burlington House, Piccadilly, London W1V 0HS
Tel 020 7479 7092

ROYAL COMMISSION ON THE HISTORICAL MONUMENTS OF ENGLAND *– now part of English Heritage*

ROYAL COMMISSION ON THE ANCIENT AND HISTORICAL MONUMENTS OF SCOTLAND
John Sinclair House, 16 Bernard Terrace, Edinburgh EH8 9NX
Tel 0131 662 1456 Fax 0131 662 1477

ROYAL COMMISSION ON THE ANCIENT AND HISTORICAL MONUMENTS OF WALES
Crown Building, Plas Crug, Aberystwyth, Ceredigion SY23 1NJ
Tel 01970 621200 Fax 01970 627701

ROYAL FINE ART COMMISSION FOR SCOTLAND
Bakehouse Close, 146 Canongate, Edinburgh EH8 8DD
Tel 0131 556 6699 Fax 0131 556 6633

ROYAL HORTICULTURAL SOCIETY
80 Vincent Square, London SW1P 2PE
Tel 020 7834 4333 Fax 020 7630 6060

ROYAL INCORPORATION OF ARCHITECTS IN SCOTLAND
15 Rutland Square, Edinburgh EH1 2BE
Tel 0131 229 7545 Fax 0131 228 2188

ROYAL INSTITUTE OF BRITISH ARCHITECTS CONSERVATION GROUP
66 Portland Place, London W1N 4AD
Tel 020 7580 5533 Fax 020 7255 1541

ROYAL INSTITUTION OF CHARTERED SURVEYORS AWARDS, THE
The Awards Officer, Communications Department, The Royal Institution of Chartered Surveyors, 12 Great George Street, Parliament Square, London SW1P 3AD
Tel 020 7222 7000 Fax 020 7222 9430

ROYAL INSTITUTION OF CHARTERED SURVEYORS, BUILDING CONSERVATION GROUP
12 Great George Street, Parliament Square, London SW1P 3AD
Tel 020 7222 7000 Fax 020 7222 9430

ROYAL SOCIETY OF ARCHITECTS IN WALES
Bute Buildings, King Edward VII Avenue, Cathays Park, Cardiff CF10 3NB
Tel 029 2087 4753 Fax 029 2087 4926

ROYAL SOCIETY OF ULSTER ARCHITECTS
2 Mount Charles, Belfast BT7 1NZ
Tel 028 9032 3760 Fax 028 9023 7313

ROYAL TOWN PLANNING INSTITUTE
26 Portland Place, London W1N 4BE
Tel 020 7636 9107 Fax 020 7323 1582

RUG RESTORERS ASSOCIATION, THE
Lower Clatcombe House, Sherborne, Dorset DT9 4RH
Tel 01935 813274

SALVO
PO Box 333, Cornhill on Tweed, Northumberland TD12 4YJ
Tel 01890 820333 Fax 01890 820499

SAVE BRITAIN'S HERITAGE
70 Cowcross Street, London EC1M 6EJ
Tel 020 7253 3500 Fax 020 7253 3400

SCOTTISH CIVIC TRUST, THE
The Tobacco Merchant's House, 42 Miller Street, Glasgow G1 1DT
Tel 0141 221 1466 Fax 0141 248 6952

SCOTTISH CONSERVATION BUREAU
Historic Scotland, Longmore House, Salisbury Place, Edinburgh EH9 1SH
Tel 0131 668 8668 Fax 0131 668 8669

SCOTTISH EXECUTIVE DEVELOPMENT DEPARTMENT
Victoria Quay, Edinburgh EH6 6QQ
Tel 0131 556 8400

SCOTTISH HISTORIC BUILDINGS TRUST, THE
33 High Street, Cockenzie, East Lothian EH32 0HP
Tel 01875 813608 Fax 01875 813608

SCOTTISH RECORD OFFICE –
see National Archives of Scotland

SCOTTISH REDUNDANT CHURCHES TRUST, THE
14 Long Row, New Lanark ML11 9DD
Tel 01555 666023 Fax 01555 665738

SCOTTISH SOCIETY FOR CONSERVATION AND RESTORATION
The Glasite Meeting House, 33 Barony Street, Edinburgh EH3 6NX
Tel 01506 811777 Fax 01506 8118887

SIR JOHN SOANE'S MUSEUM
13 Lincoln's Inn Fields, London WC2A 3BP
Tel 020 7405 2107 Fax 020 7831 3957

SOCIETY FOR THE PROTECTION OF ANCIENT BUILDINGS, THE (SPAB)
37 Spital Square, Spitalfields, London E1 6DY
Tel 020 7377 1644 Fax 020 7247 5296

SOCIETY FOR THE PROTECTION OF ANCIENT BUILDINGS IN SCOTLAND, THE
The Glasite Meeting House, 33 Barony Street, Edinburgh EH3 6NX
Tel 0131 557 1551 Fax 0131 557 1551

SOCIETY OF ARCHITECTURAL HISTORIANS OF GREAT BRITAIN
115 Henderson Row, Edinburgh EH3 5BB

STAINED GLASS MUSEUM
The Chapter House, Ely Cathedral, Ely CB7 4DN
Tel 01353 660347 Fax 01353 665025

STATIONERY OFFICE, THE
PO Box 29, Norwich NR3 1GN
Tel 0870 6005522

STEEL WINDOW ASSOCIATION
The Building Centre, 26 Store Street, London WC1E 7BT
Tel 020 7637 3571 Fax 020 7637 3572

STONE FEDERATION GREAT BRITAIN
Construction House, 56-64 Leonard Street, London EC2A 4JX
Tel 020 7608 5094 Fax 020 7608 5081

STONE ROOFING ASSOCIATION
Ceunant, Caernarfon, Gwynedd LL55 4SA
Tel 01286 650402 Fax 01286 650402

THEATRES TRUST, THE
22 Charing Cross Road, London WC2H 0QL
Tel 020 7836 8591 Fax 020 7836 3302

TILES AND ARCHITECTURAL CERAMICS SOCIETY
Decorative Art Department, Liverpool Museum, Liverpool L3 8EN
Tel 0151 207 0001

TIMBER RESEARCH AND DEVELOPMENT ASSOCIATION
Stocking Lane, Hughenden Valley, High Wycombe, Buckinghamshire HP14 4ND
Tel 01494 563091 Fax 01494 565487

TIMBER TRADE FEDERATION, THE
4th Floor Clareville House, 26-27 Oxendon Street, London SW1Y 4EL
Tel 020 7839 1891 Fax 020 7930 0094

TOOL AND TRADES HISTORY SOCIETY, THE
Swanton Lodge, Swanton Street, Bredgar, Sittingbourne, Kent ME9 8AS
Tel 01622 884434 Fax 01622 884434

TOWN & COUNTRY PLANNING ASSOCIATION
17 Carlton House Terrace, London SW1Y 5AS
Tel 020 7930 8903 Fax 020 7930 3280

TRADITIONAL PAINT FORUM, THE
c/o The National Trust for Scotland, 28 Charlotte Square, Edinburgh EH2 4ET
Tel 0131 243 9449 Fax 0131 243 9599

TWENTIETH CENTURY SOCIETY, THE
70 Cowcross Street, London EC1M 6EJ
Tel 020 7250 3857 Fax 020 7251 8985

ULSTER ARCHITECTURAL HERITAGE SOCIETY
66 Donegall Pass, Belfast BT7 1BU
Tel 028 9055 0213 Fax 028 9055 0214

UNESCO (UK)
The British Council, 10 Spring Gardens, London SW1A 2BN
Tel 020 7389 4683 Fax 020 7389 4497

UNITED KINGDOM INSTITUTE FOR CONSERVATION OF HISTORIC AND ARTISTIC WORKS (UKIC)
109 The Chandlery, 50 Westminster Bridge Road, London SE1 7QY
Tel 020 7721 8721 Fax 020 7721 8722

VERNACULAR ARCHITECTURE GROUP
c/o Mrs Brenda Watkin, Ashley, Willows Green, Great Leighs, Chelmsford, Essex CM3 1QD
Tel 01245 361408

VICTORIA AND ALBERT MUSEUM, THE
Cromwell Road, South Kensington, London SW7 2RL
Tel 020 7942 2000 Fax 020 7942 2092

VICTORIAN SOCIETY, THE
1 Priory Gardens, Bedford Park, London W4 1TT
Tel 020 8994 1019 Fax 020 8995 4895

VIVAT TRUST, THE
61 Pall Mall, London SW1Y 5HZ
Tel 020 7930 8030 Fax 020 7930 2295

WALLPAPER HISTORY SOCIETY
c/o 49 Glenpark Drive, Southport, Merseyside PR9 9FA
Tel 01704 225429

WEALD & DOWNLAND OPEN AIR MUSEUM
Singleton, Chichester, West Sussex PO18 0EU
Tel 01243 811363 Fax 01243 811475

WORLD MONUMENTS FUND IN BRITAIN
2 Grosvenor Gardens, London SW1W 0DH
Tel 020 7730 5344 Fax 020 7730 5355

WORSHIPFUL COMPANY OF GLAZIERS AND PAINTERS OF GLASS, THE
Glaziers Hall, 9 Montague Close, London Bridge, London SE1 9DD
Tel 020 7403 3300 Fax 020 7407 6036

SPECIALISTS INDEX

USEFUL INFORMATION

SPECIALISTS INDEX

SPECIALISTS INDEX

When contacting companies listed here, please let them know that you found them through *The Building Conservation Directory*

USEFUL INFORMATION

PRODUCTS & SERVICES INDEX